Optical Radiation and Matter

Optical Radiation and Matter

Robert J Brecha and J Michael O'Hare

Physics Department and Electro-optics Program, University of Dayton, Dayton, OH 45469, USA

IOP Publishing, Bristol, UK

ISBN   978-0-7503-2624-7 (ebook)
ISBN   978-0-7503-2622-3 (print)
ISBN   978-0-7503-2625-4 (myPrint)
ISBN   978-0-7503-2623-0 (mobi)

DOI   10.1088/978-0-7503-2624-7

Version: 20210501

IOP ebooks

British Library Cataloguing-in-Publication Data: A catalogue record for this book is available from the British Library.

Published by IOP Publishing, wholly owned by The Institute of Physics, London

IOP Publishing, Temple Circus, Temple Way, Bristol, BS1 6HG, UK

US Office: IOP Publishing, Inc., 190 North Independence Mall West, Suite 601, Philadelphia, PA 19106, USA

This book is dedicated to our late colleagues Peter Powers and Joe Haus.

Contents

6 Optical properties of simple systems

Preface

The Electro-Optics Program Master's Degree at the University of Dayton originated as a joint project between the Physics Department and the Electrical and Computer Engineering Department in 1983. Since 2016, the Electro-Optics and Photonics Department in the School of Engineering has been the home of both MS and PhD degrees. From the beginning, it was recognized that students with differing backgrounds would be the enrolling in the program, and therefore some preparatory classwork would be beneficial for all students to arrive at a common basis of knowledge before proceeding to more advanced coursework. This textbook, begun as a set of notes generated by one of us (JMO) has been used, in various and expanding forms since the beginning of the program and has consistently been cited by students as a valuable starting point for their studies.

We begin with the basics in chapter 1 with Maxwell's Equations and a derivation of the wave equation and examine the propagation of electromagnetic waves in free space. In chapter 2 the polarization of light waves is covered, a detailed description of the vector nature of the electromagnetic field that will become important for later chapters. Whereas the first two chapters essentially assume the existence of fields as solutions to Maxwell's equations, chapter 3 investigates the origin of electromagnetic waves within the classical framework of an oscillating electric dipole. Along the way, the scattering of light waves by dipoles is covered, followed in chapter 4 by a slight detour into optical spectroscopy. In this chapter we begin the real heart of the material, looking at the interaction of optical radiation with dilute matter, such as a molecular gas. In chapter 5, the propagation of electromagnetic waves is revisited, now in dense media. These considerations lead naturally to the derivation of laws of reflection and refraction, both the familiar version of Snell's Law, but also more complex propagation relations at the interface between dielectrics and metals. Chapter 6 takes us into the realm of condensed matter physics and the properties of materials, but also the interactions between light and solid materials of different types. Chapter 7 looks at more complex interactions of polarized light waves with materials that are often not isotropic and can thus change the state of polarization. With chapter 8, a primary goal is reached with the discussion of the electro-optic effect—the namesake of the program. The material in the text comes together in this examination of the interaction of applied, low-frequency electric fields with anisotropic crystals through which light waves propagate. Finally, one last subject, the acousto-optic effect is presented in chapter 9—the interaction of sound waves with crystals and light.

Acknowledgements

The authors would like to acknowledge our many colleagues who created, built, taught courses in, advised students for and carried out research in the Electro-Optics Program and (now) Department of Electro-Optics and Photonics. We both profited greatly from our interactions with students and colleagues, JMO since the beginning of the program in 1983 and RJB after joining the university a decade later.

Author biographies

Bob Brecha

Robert (Bob) Brecha graduated from Wright State University (BS in Physics, 1983) and from the University of Texas at Austin (PhD in Physics, 1990). Since 1993 he has been at the University of Dayton where he is Professor of Physics and was for many years a member of the Electro-Optics Program (now Department). His research was mainly in the areas of experimental and theoretical quantum optics as well as magnetic rotation spectroscopy of molecular oxygen. He is currently affiliated with the Renewable and Clean Energy Program, the Hanley Sustainability Institute, and was a founding coordinator of the Sustainability, Energy, and the Environment (SEE) initiative from 2007 to 2015. From 2006 to 2017 he was a regular visiting scientist at the Potsdam Institute for Climate Impact Research (PIK) in Germany, including one year as a Fulbright Fellow (2010–2011). More recently he has been affiliated with the Berlin think-tank Climate Analytics as a visiting scientist including a two-year period as a European Union Marie Curie Fellow. His current research publications focus on energy efficiency in buildings, climate change mitigation strategies, and energy needs for sustainable development.

Mike O'Hare

Dr J Michael O'Hare is a Distinguished Service Professor and Professor Emeritus of Physics and Electro-Optics at the University of Dayton. A native of Cedar Rapids, Iowa, he received his BS degree from Loras College in 1960, his MS from Purdue University in 1962, and his PhD in theoretical physics from The State University of New York at Buffalo in 1966. He joined the faculty of The University of Dayton in 1966 and since then has taught physics courses at all levels. He served as Chair of the Department of Physics from 1983 to 2007. He has research experience in atomic, molecular, solid state, and optical physics. While at the University of Dayton he has carried out theoretical and experimental work on the optical properties of materials, and was the principal investigator on various research contracts with the Air Force Materials Laboratory at the Wright-Patterson Air Force Base. Dr O'Hare was the principal investigator on a NASA sponsored NOVA grant for the design of curricula for developing scientific literacy in preservice teachers.

IOP Publishing

Optical Radiation and Matter

Robert J Brecha and J Michael O'Hare

Chapter 1

Review of electromagnetic radiation

The main part of this text deals with the propagation of electromagnetic waves in matter. In this first chapter, however, after a brief historical introduction, we begin by considering the properties of electromagnetic waves propagating in a vacuum, ignoring for the moment the possible presence of any material medium. Equivalently, we can imagine that our observation point for the waves under consideration is very far from any possible sources of radiation, such as oscillating charges (more on this in the next chapter) or currents. First we will take a look at how Maxwell's equations relate the electric and magnetic fields of radiation. We will see how Maxwell's equations, in turn, can lead to the free-space wave equation, which connects the spatial and time dependence of the electric field. This equation may be used to find solutions describing several different types of waves. We will learn to describe the phase and group velocities of waves, which tell us about the information transmission speed possible by a wave. Finally, we discuss the energy flux and the time-averaged power carried by a wave. The last section of the chapter provides an example of how to tie some of these bits of information together, through an example studying the electric field of an optical cavity or resonator.

1.1 Historical introduction

Although it is impossible to pick out any one given point in time to begin the study of the development of electromagnetic theory, a fairly good argument can be made that the beginning of the nineteenth century was a pivotal epoch. We can start with the Danish physicist Hans Christian Oersted (1777–1851), who hypothesized that there might be a relation between electric and magnetic phenomena [1]. In fact, Oersted performed a series of experiments in 1820 in which the effects of current flow on the deflection of a compass needle were observed. A few months later, in France, André-Marie Ampère (1775–1836) was able to show that the current in an electrical circuit was equivalent to a magnet oriented in a plane perpendicular to that of the circuit [2]. He even came up with a first attempt at a molecular theory of magnetism

doi:10.1088/978-0-7503-2624-7ch1

based on these early observations. Another Frenchman, Dominique François Jean Arago (1786–1853), and independently, Humphry Davy (1778–1829) in the UK, soon showed that a helically wound wire would be able to magnetize a piece of iron placed inside the helix [3, 4]. The result of all these investigations was the basis of the first qualitative understanding of the connection between what had previously been seen as two distinct physical phenomena.

The concept of 'fields' was still relatively new, but Michael Faraday (1791–1867) was able to use this idea to construct an experiment to demonstrate the nature of the lines of magnetic force surrounding a straight wire [5]. It should be noted that Faraday's experiments were performed in late 1821; all of the above-mentioned progress had taken place in the remarkably short period of about one year! Faraday was also key in making one of the next important discoveries linking electricity and magnetism. He found (in 1831) that a changing magnetic flux can create an electric current in a circuit. From start to finish Faraday needed about ten days to perform his experiments on this phenomenon and to arrive at what are essentially the correct conclusions, results which in turn form the basis for all electric motors and generators, for example.

James Clerk Maxwell (1831–1879) wished to find a mathematical framework capable of containing all of the aforementioned experimental results, and in particular the concept of electric and magnetic fields [6]. Maxwell postulated an 'electric displacement', the time rate of change of which makes a contribution to the total current and thus to the magnetic field. As with all of the previous workers, one fundamental conviction of Maxwell's was that all of space is pervaded by the 'ether', undetected but crucial conceptually for all theoretical models. The experimental and theoretical accomplishments of this period still stand unchanged, although based on the false premise of the ether.

One of the key results of Maxwell's work was the realization that changes in magnetic and electric fields propagate through the surrounding medium at a finite speed. His theory showed further that this speed of propagation was exactly the same as the known speed of light propagation. The predictions arising from Maxwell's theory led to a great deal of experimental activity attempting to verify the speed of wave propagation in various materials.

1.2 Maxwell's equations in free space

We begin our study of the interaction between optical radiation and matter with a review of the basic elements of electromagnetic theory, starting with the Maxwell equations for fields in the absence of currents and charges. These equations, formulated in terms of $\vec{E}$, the electric field vector, and $\vec{B}$, the magnetic induction vector, are all we need to understand some of the fundamental properties of electromagnetic radiation. In MKS units (the units which will be used throughout this text) the free-space Maxwell equations are

$$\nabla \cdot \vec{E} = 0 \tag{1.1}$$

$$\nabla \cdot \vec{B} = 0 \tag{1.2}$$

$$\nabla \times \vec{E} = -\frac{\partial \vec{B}}{\partial t} \tag{1.3}$$

$$\nabla \times \vec{B} = \frac{1}{c^2}\frac{\partial \vec{E}}{\partial t}. \tag{1.4}$$

In writing the above we have assumed that neither charges nor currents are present. The constant c is the speed of light in a vacuum, considered one of the fundamental physical constants and *defined* to be exactly 299 792 458 m s^{-1}. As a side note, the meter itself, which was once defined by the length of a standard bar, is now the distance light travels in vacuum in 1/299792458 s. To complete this somewhat circular set of definitions, the second is defined in terms of a specific transition in atomic cesium (an atomic oscillator, if you like) as follows: 'The second is the duration of 9 192 631 770 periods of the radiation corresponding to the transition between the two hyperfine levels of the ground state of the cesium 133 atom' [7].

1.3 The free-space wave equation

The free-space electromagnetic wave equation is readily obtained from Maxwell's equations by taking the curl of equation (1.3) and substituting equation (1.4):

$$\nabla \times \left(\nabla \times \vec{E}\right) = -\frac{\partial}{\partial t}[\nabla \times \vec{B}]$$

$$= -\frac{\partial}{\partial t}\left[\frac{1}{c^2}\frac{\partial \vec{E}}{\partial t}\right] \tag{1.5}$$

$$= -\frac{1}{c^2}\frac{\partial^2 \vec{E}}{\partial t^2}.$$

Using a standard vector identity (see appendix A, 'Vector theorems and identities') on the left-hand side of the above equation, we obtain

$$\nabla(\nabla \cdot \vec{E}) - \nabla^2 \vec{E} = -\frac{1}{c^2}\frac{\partial^2 \vec{E}}{\partial t^2}. \tag{1.6}$$

Next, using equation (1.1) we obtain the following form for the free-space electromagnetic wave equation:

$$\nabla^2 \vec{E} = \frac{1}{c^2}\frac{\partial^2 \vec{E}}{\partial t^2}. \tag{1.7}$$

This result is one of the key starting points for what follows in the rest of this chapter. To summarize the assumptions we have thus far made in arriving at equation (1.7), we recall that the fields in which we are interested are far from any sources of radiation and thus the Maxwell equation source terms were taken to be zero. In the following sections we will investigate three different solutions to the wave equation, starting with the one that will be used most frequently throughout this text.

1.3.1 Plane wave solution to the wave equation

The most straight-forward and algebraically simple solutions to equation (1.7) are the so-called plane wave solutions. A monochromatic plane wave in free space can be written as

$$\vec{E}(\vec{r}, t) = \frac{1}{2}\vec{\mathcal{E}}(\vec{k}, \omega) \, e^{i(\vec{k}_0 \cdot \vec{r} - \omega t)} + \text{c.c.} \tag{1.8}$$

$$= \Re\left[\vec{\mathcal{E}}(\vec{k}_0, \omega) \, e^{i(\vec{k}_0 \cdot \vec{r} - \omega t)}\right], \tag{1.9}$$

where c.c. stands for complex conjugate and $\Re$ indicates the real part of a complex number. $\vec{\mathcal{E}}(\vec{k}_0, \omega)$ is a complex vector amplitude, $\vec{k}_0$ is the free-space wavevector, with ($|\vec{k}_0| = 2\pi/\lambda_0$) being the spatial frequency of the wave, and ω is the angular frequency of the wave ($\omega = f/2\pi$, $k_0 = \omega/c$). The complex vector amplitude is a notational device used to designate both the amplitude and the polarization of the wave; this will be discussed in much greater detail in chapter 2. The wave in equation (1.9) is a harmonic wave and could just as easily have been represented by a sine or cosine rather than the complex expression that we have chosen. Let us write this out once for the sake of completeness and to get a better feeling for the notation. We consider a monochromatic (single frequency) wave traveling in the z-direction, which means that $\vec{k}_0 \cdot \vec{r}$ becomes $k_z z$ and that we can drop the arguments for $\vec{\mathcal{E}}$. From Maxwell's equations we find that the electric field vector must lie in a plane perpendicular to the propagation direction, i.e. in the $x-y$ plane (the proof is left as an exercise). Thus we write the vector amplitude as

$$\vec{\mathcal{E}} = A_x \, e^{i\delta_x} \, \hat{x} + A_y \, e^{i\delta_y} \, \hat{y}.$$

In the above expression, δ_x and δ_y represent a relative phase of the wave with respect to a chosen starting point. Substituting this into equation (1.9) gives

$$\vec{E}(\vec{r}, t) = \Re\left\{\left(A_x \, e^{i\delta_x} \, \hat{x} + A_y \, e^{i\delta_y} \, \hat{y}\right)e^{i(\vec{k}_0 \cdot \vec{r} - \omega t)}\right\}$$

$$= A_x \cos\left(k_z z - \omega t + \delta_x\right)\hat{x} + A_y \cos\left(k_z z - \omega t + \delta_y\right)\hat{y},$$

which is the final result we are after.

Alternatively we could have used equation (1.8) to find

$$\vec{E}(\vec{r},\,t) = \frac{1}{2}\left(A_x\, e^{i\delta_x}\, \hat{x} + A_y\, e^{i\delta_y}\, \hat{y}\right)e^{i(\vec{k}_0\cdot\vec{r}-\omega t)}$$

$$+ \frac{1}{2}\left(A_x\, e^{-i\delta_x}\, \hat{x} + A_y\, e^{-i\delta_y}\, \hat{y}\right)e^{-i(\vec{k}_0\cdot\vec{r}-\omega t)}$$

$$= \frac{1}{2}A_x\left\{\exp\left[i\left(k_z\, z - \omega t + \delta_x\right)\right] + \exp\left[-i\left(k_z\, z - \omega t + \delta_x\right)\right]\right\}\hat{x}$$

$$+ \frac{1}{2}A_y\left\{\exp\left[i\left(k_z\, z - \omega t + \delta_y\right)\right] + \exp\left[-i\left(k_z\, z - \omega t + \delta_y\right)\right]\right\}\hat{y}$$

$$= A_x \cos\left(k_z\, z - \omega t + \delta_x\right)\hat{x} + A_y \cos\left(k_z\, z - \omega t + \delta_y\right)\hat{y}.$$

The reason for using the complex notation is that it offers algebraic simplicity for use with Maxwell's equations. From now on we will write equation (1.9) without the term $\Re$ in the expression but with the convention *that taking the real part is implied*.

Equation (1.9) is referred to as a plane wave solution because the surfaces of constant phase, i.e. for which $\vec{k}\cdot\vec{r}$ is a constant and used to characterize the wave geometrically, are planes. This is illustrated in figure 1.1 where we have drawn the spatial variables of the phase of the wave of equation (1.9). As shown in the figure, for any radius vector $\vec{r}$ drawn from the origin to a point on a plane which is perpendicular to the wavevector, the phase of the wave, $\vec{k}\cdot\vec{r}$, will have the same value.

We will return often to the plane wave solution to the wave equation. Although idealized, in the sense that no real electromagnetic wave is truly a plane wave, the mathematical simplicity of starting out by treating plane waves is a great bonus. For most problems we will treat in this text, the results obtained by using plane waves is sufficient for our purposes. However, it is also useful to look at two other wave equation solutions commonly encountered in optical and laser physics.

Figure 1.1. Surfaces of constant phase for the plane wave solution to the wave equation. To be a truly accurate representation, the planes should extend to infinity.

1.3.2 Spherical wave solution to the wave equation

In this section we will briefly present the so-called spherical wave solution to the wave equation. An application of spherical waves will be encountered in chapter 3.

In complex notation the spherical wave is represented by

$$E(r,\ t) = \frac{1}{r} \exp i(kr - \omega t), \tag{1.10}$$

where r is the spherical coordinate ($r = \sqrt{x^2 + y^2 + z^2}$). This solution represents a wave emanating from a point and propagating uniformly in all directions in space. The amplitude of the wave falls off as $1/r$ and, as we shall see shortly, this leads to a total propagating energy which is constant when integrated over a sphere of radius R, independent of the magnitude of R. We can show that the spherical wave is a solution to the wave equation by using the Laplacian in spherical coordinates (see appendix B):

$$\nabla^2 E = \frac{1}{r^2} \frac{\partial}{\partial r} \left(r^2 \frac{\partial E}{\partial r} \right)$$
$$+ [\text{derivatives with respect to } \Theta \text{ and } \phi]$$
$$= \frac{1}{c^2} \frac{\partial^2 E}{\partial t^2}.$$

We ignore the derivatives with respect to the angular variables θ and ϕ since our 'guessed' solution has no dependence on those variables. Substituting the spherical wave solution into the above,

$$\frac{1}{r^2} \frac{\partial}{\partial r} r^2 \left(\frac{ik}{r} - \frac{1}{r^2} \right) e^{i(kr - \omega t)} = \frac{1}{c^2} \frac{\partial^2 E}{\partial t^2}$$
$$-\frac{k^2}{r} e^{i(kr - \omega t)} = \frac{1}{r} \left(\frac{-\omega^2}{c^2} \right) e^{i(kr - \omega t)},$$

which confirms that equation (1.10) satisfies the wave equation, since $k^2 = \omega^2/c^2$.

1.3.3 Gaussian beam solution to the wave equation

There is another set of solutions to the wave equation that is very important, and that must be treated extensively when considering optical resonators and lasers. These fields are known as Gaussian beams, the derivation of which will be the subject of this section. Further information on Gaussian beams and more generalized solutions to the free-space wave equations can be found in many textbooks, for example, [8–10]. An early and valuable review of this material is found in a paper by Kogelnik and Li [11]. In chapter 2 we will look at a way to think of a Gaussian beam as being a superposition of plane waves; that result will serve to connect some of the concepts covered in these first two chapters.

We begin by showing in figure 1.2 some features of Gaussian beams. When seen from the side, a Gaussian beam with waist w_0 propagates to the right, increasing in

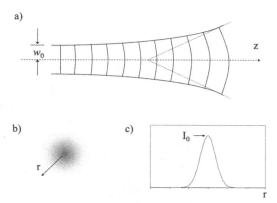

Figure 1.2. A qualitative illustration of different Gaussian beam characteristics. (a) Viewed from the side, with spot size as a function of position z. The curved lines represent surfaces of constant phase, each with radius of curvature $R(z)$. (b) Viewed head-on, the beam intensity is highest in the center and decreases radially. For the only profile considered here, there is no dependence of the intensity on the azimuthal angle ϕ. (c) A sketch of the intensity as a function of radial distance for the beam shown in part (b), showing the Gaussian functional dependence.

radius as a function of position z. The surfaces of constant phase are shown as curved lines. The intensity of the beam has a maximum at the center, and decreases radially with a Gaussian functional dependence, as shown in figure 1.2(b), which might represent a burn pattern formed by an intense laser incident on an absorbing medium. Figure 1.2(c) represents a plot of the intensity of a Gaussian beam that would be measured by scanning a pinhole or slit across the beam profile and recording the intensity transmitted through the aperture.

With the preceding qualitative discussion, we can now start the mathematical formalism needed to arrive at the specific result we are seeking. We begin by assuming a harmonic wave proportional to $e^{-i\omega t}$. The wave equation for the electric field, equation (1.7), may be written as

$$\nabla^2 \vec{E} + k^2 \vec{E} = 0,$$

where $k^2 = \left(\frac{\omega^2}{c^2}\right)$. Before proceeding toward a solution to the wave equation, we consider a concrete physical situation to which the mathematical formalism will commonly apply. A laser, such as a common red helium–neon laser, has a necessary component of two round mirrors, through one of which the laser light is emitted. We will make use of the fact that for such an optical cavity, or resonator, a natural coordinate system to choose might be one with cylindrical symmetry. Thus we will rewrite the wave equation in cylindrical coordinates yielding (see appendix C for the coordinate transformation)

$$\nabla^2 \vec{E}(r, \phi, z) + k^2 \vec{E} = \frac{1}{r}\frac{\partial}{\partial r}\left(r\frac{\partial \vec{E}}{\partial r}\right) + \frac{1}{r^2}\frac{\partial^2 \vec{E}}{\partial \phi^2} + \frac{\partial^2 \vec{E}}{\partial z^2} + k^2 \vec{E} = 0. \qquad (1.11)$$

We postulate that $\vec{E}$ depends only on r in the transverse direction, i.e. that there is cylindrical symmetry. Thus

$$\frac{\partial^2 \vec{E}}{\partial \phi^2} = 0.$$

This assumption is not a general one and will restrict us in the type of solutions we will find to the very simplest solutions. We can now write

$$\frac{\partial^2 \vec{E}}{\partial r^2} + \frac{1}{r}\frac{\partial \vec{E}}{\partial r} + \frac{\partial^2 \vec{E}}{\partial z^2} + k^2\vec{E} = 0. \tag{1.12}$$

If we were to consider, for example, a laser beam propagating in free space after leaving a laser cavity, the energy flow will be predominantly in one direction, the z-direction as we have the problem set up here. Thus we can use the intuition we have built up in the previous sections and assume a solution to the wave equation which is a 'modified' plane wave:

$$E(r, z) = \Psi(r, z)e^{ikz}. \tag{1.13}$$

Here we have dropped the vector notation for simplicity; we can assume that the field is polarized in, for example, the $\hat{y}$-direction, but this will play no further role in what follows. The amplitude $\Psi(r, z)$ varies slowly in space, i.e. it represents an 'envelope' in the z-direction along with variations in the direction transverse to the propagation direction with a spatial scale large compared to the wavelength, while the exponential factor takes into account rapid spatial variations in the z-direction. As we shall see later, the possibility of absorption of energy from the beam may be included in this term as well. In the context of laser physics, the amplification, or gain, of beam energy propagating through the medium may also be part of this latter term.

Now we can substitute this ansatz into the wave equation, yielding

$$\frac{\partial^2 \Psi}{\partial r^2} + \frac{1}{r}\frac{\partial \Psi}{\partial r} + 2ik\frac{\partial \Psi}{\partial z} + \frac{\partial^2 \Psi}{\partial z^2} = 0. \tag{1.14}$$

Our assumption of the slowly varying field allows us to make further simplifications in the above. If Ψ is a slowly varying function of z, we can drop the second derivative term (with respect to z) as being small compared to the first derivative or Ψ itself. This leaves us with

$$\frac{\partial^2 \Psi}{\partial r^2} + \frac{1}{r}\frac{\partial \Psi}{\partial r} + 2ik\frac{\partial \Psi}{\partial z} = 0. \tag{1.15}$$

To solve the above we try a solution of the form

$$\Psi(r, z) = \exp\left[-i\left(P(z) - \frac{k}{2q(z)}r^2\right)\right], \tag{1.16}$$

where $P(z)$ and $q(z)$ may both be complex numbers. Although not necessarily the most obvious choice for a trial solution, we are here most interested in arriving at a result, and not in the mathematical details behind solving a partial differential equation. Substituting this into the above form of the wave equation gives

$$\left[\frac{-k^2 r^2}{q^2(z)} + \frac{2ik}{q(z)} + 2k\frac{\partial P(z)}{\partial z} + \frac{k^2 r^2}{q^2(z)}\frac{\partial q(z)}{\partial z} \right]\Psi(r, z) = 0. \tag{1.17}$$

For this relation to be satisfied the expression in brackets must vanish and for that to happen the individual powers of r must vanish separately. This leads to two conditions,

$$\frac{-k^2}{q^2(z)} + \frac{k^2}{q^2(z)}\frac{\partial q(z)}{\partial z} = 0$$

$$\frac{2ik}{q(z)} + 2k\frac{\partial P(z)}{\partial z} = 0.$$

Simplifying these gives two partial differential equations which can be solved easily. From the first of the above equations we have

$$\frac{\partial q(z)}{\partial z} = 1 \implies q(z) = z + q_0. \tag{1.18}$$

This solution can in turn be used in the second equation:

$$\frac{\partial P(z)}{\partial z} = \frac{-i}{q(z)} = \frac{-i}{z + q_0} \tag{1.19}$$

or

$$P(z) = -i \ln\left(1 + \frac{z}{q_0}\right). \tag{1.20}$$

Now we need to start working backward through all of the various substitutions we have made, so that we determine the form of the electric field for this solution. Recalling how we defined $\Psi(r, z)$ initially, we can substitute the expressions for $q(z)$ (equation (1.18)) and $P(z)$ (equation (1.20)) into equation (1.16) which gives, to within an overall amplitude and phase factor,

$$\Psi(r, z) = \exp\left[-\ln\left(1 + \frac{z}{q_0}\right) + \frac{ikr^2}{2q(z)} \right]. \tag{1.21}$$

As $r \to \infty$ we require that $\Psi(r, z) \to 0$ since physically the fields must vanish at very large distances from the axis. Examining the expression above, we see that this requires that q_0 be imaginary. Setting $q_0 = iz_0$, where z_0 is a real variable, we can write equation (1.21) as

$$\Psi(r, z) = \frac{1}{\sqrt{1 + \left(\frac{z}{z_0}\right)^2}} e^{i \tan^{-1}(z/z_0)} e^{+\frac{ikr^2}{2q(z)}}.$$

(1.22)

To arrive at this result, we have used the relation

$$\ln\left(1 + \frac{z}{iz_0}\right) = \ln\left(1 - \frac{iz}{z_0}\right) = \ln\left(\sqrt{1 + \left(\frac{z}{z_0}\right)^2}\right) - i \tan^{-1}\left(\frac{z}{z_0}\right).$$

Finally, we may define $q(z)$ in terms of two real parameters $R(z)$ and $w(z)$ such that

$$\frac{1}{q(z)} = \frac{1}{R(z)} - \frac{i\lambda}{\pi w^2(z)},$$

(1.23)

we can rewrite equation (1.22) as

$$\Psi(r, z) = \frac{1}{\sqrt{1 + \left(\frac{z}{z_0}\right)^2}} e^{i \tan^{-1}(z/z_0)} e^{+ikr^2/2R(z)} e^{-r^2/w^2(z)}.$$

(1.24)

It will be left to the problems to show that

$$R(z) = z\left[1 + \left(\frac{\pi w_0^2}{\lambda z}\right)^2\right] = z\left[1 + \left(\frac{z_0}{z}\right)^2\right]$$

(1.25)

$$w^2(z) = w_0^2\left[1 + \left(\frac{\lambda z}{\pi w_0^2}\right)^2\right] = w_0^2\left[1 + \left(\frac{z}{z_0}\right)^2\right].$$

(1.26)

In the above we have set $z_0 = \pi w_0^2/\lambda$. The real quantities $R(z)$ and $w(z)$ represent, respectively, the radius of curvature of the wavefront of the beam and the spot size of the beam. Of particular interest are the values of these parameters at the position $z = 0$, namely $R(z) = \infty$, thus representing the plane wave, and $w(z) = w_0$, the minimum value of the beam spot, called the beam waist. Also from the above we can gain a feeling for the parameter z_0, called the Rayleigh parameter or Rayleigh range. For $z = z_0$ we have a beam radius equal to $\sqrt{2}$ of the waist size, or an area of twice the minimum spot size. Thus, the Rayleigh range gives an indication of how quickly the beam is diverging.

Returning to the definition of $E(r, z)$ in terms of $\Psi(r, z)$, we can write finally

$$E(r, z) = E_0\frac{1}{\sqrt{1 + \left(\frac{z}{z_0}\right)^2}} e^{i \tan^{-1}(z/z_0)} e^{+ik\left(z + \frac{r^2}{2R(z)}\right)} e^{-r^2/\omega^2(z)}.$$

(1.27)

Here the factor E_0 represents the maximum amplitude of the field, i.e. the amplitude at $z = 0$ and $r = 0$. This result represents a wave which varies slowly in amplitude along the propagation direction, z. In addition, the field amplitude decreases as a function of the distance away in a radial direction from the z-axis. The Gaussian functional dependence gives the name to this solution to the wave equation.

The phase of the electric field, given by the imaginary arguments in the exponential of equation (1.27), is now a function of not only the position along the z-axis, but also of the distance from the axis, as well as of the beam radius, $R(z)$. Two limiting cases are of some interest to us. We have already seen that the radius of curvature of the wavefront is infinite at the beam waist position $z = 0$. However, the phase of the wave, due to the factor $\exp(i \tan^{-1}(z/z_0))$, shifts as the beam passes through the focus. In any case, we see that the waist of a focused beam provides one example of a good approximation to a plane wave. Equivalently, for a beam that is well-collimated, we have the same result.

In the opposite extreme, for $z \gg z_0$ the Gaussian solution reduces on axis ($r = 0$) to

$$E' \propto \frac{1}{z} e^{+ikz}, \tag{1.28}$$

which is just the result for a spherical wave. In figure 1.2(a) the tangent lines drawn to the profile for large values of z appear to converge at a point more or less near to the beam waist, depending on the Rayleigh parameter. Looking at the beam at a distance $z \gg z_0$ one could approximate the wavefronts by spherical waves with an effective origin shown schematically by this convergence point. Off-axis there is an additional r-dependent phase shift, but the main characteristic of the $1/z$ amplitude dependence remains.

If we had not thrown away the ϕ-dependence early on in our derivation, the situation would have been more complicated. The interested reader may wish to consult a textbook on laser theory to see examples of some other modes which appear when the total solution including ϕ is considered [8].

1.4 Phase and group velocity

We return now to consider the relationship between the frequency of an electro-magnetic wave and its wavevector. The goal of this section is to investigate the speed of travel of a plane wave, or of a group of plane waves. Two key concepts are needed at the outset. First, the fact that the electromagnetic wave equation is a linear function of the electric field implies mathematically that, for any two solutions to the wave equation $\vec{E}_1$ and $\vec{E}_2$, a linear superposition of the two, $\vec{E} = a\vec{E}_1 + b\vec{E}_2$ is also a valid solution. This 'principle of superposition' is a powerful tool to which we will refer repeatedly. For the purposes of the present chapter we wish to consider the superposition of two fields that can be taken as a description of a pulse of electromagnetic radiation. The second concept needed for this section is that of the index of refraction. We will discuss the index of refraction in far more detail later; for now it suffices to note that the index of refraction describes the ratio of the speed of propagation of light in vacuum to the speed of propagation in a medium.

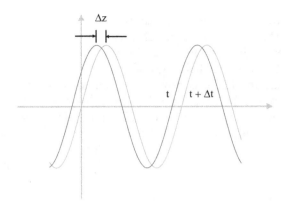

Figure 1.3. Two snapshots of the motion of a plane wave along the z-axis, taken at times separated by an interval Δt.

The index of refraction is usually a (weak) function of the frequency of radiation propagating in a medium, a phenomenon known as dispersion.

For simplicity we will only look at a wave propagating in the z-direction, $\vec{E} = \vec{E}_0 \cos(kz - \omega t)$. In figure 1.3 a sinusoidal wave is sketched at two different times, t and $t + \Delta t$, a time interval during which the wave moves to the right ($+z$-direction) a distance Δz. If we consider any given point on the wave and ask how fast it is moving, we will thereby define the 'phase velocity' of the wave. Since we are looking at a point on the wave at which the phase is constant, we must have, for the two different times,

$$[kz - \omega t] = [k(z + \Delta z) - \omega(t + \Delta t)].$$

Canceling common terms, we find for the displacement Δz

$$\Delta z = \frac{\omega}{k}\Delta t = v_p \Delta t,$$

which gives the definition of the phase velocity we seek, $v_p = \omega/k$. We have already seen that in free space, $\omega/k_0 = c$, and will find later that for waves in matter, $v_p = c/n$, where n is the index of refraction of the medium.

Since a purely monochromatic wave is an idealization, and because such an unchanging, unmodulated wave does not carry useful information in any case, we can extend our considerations to a more complicated situation in which two waves of slightly differing frequencies are considered. In the process we will define the group, or information velocity of the wave. For two waves, of wavenumbers $k \pm \Delta k$ and frequencies $\omega \pm \Delta \omega$, we can write (assuming here unit amplitude)

$$\cos\left[(k + \Delta k)z - (\omega + \Delta \omega)t\right] + \cos\left[(k - \Delta k)z - (\omega - \Delta \omega)t\right], \qquad (1.29)$$

which can be rewritten using trigonometric relations in the form

$$2\cos(\Delta kz - \Delta \omega t)\cos(kz - \omega t). \qquad (1.30)$$

To arrive at the above result, we assumed that we could write our wave as the sum of two different waves with differing frequencies and wavenumbers. Mathematically speaking, we took advantage of the linear nature of the wave equation which gives rise to the principle of superposition, as mentioned in the introduction to this section. Again, we can state this principle as follows: if A is a solution to the wave equation and B is a solution to the wave equation, than any linear superposition $aA + bB$ is also a valid solution to the wave equation.

The result given in equation (1.30) shows a wave propagating with frequency ω and wavenumber k, but with a modulated amplitude. Looking at the result as shown in figure 1.4, we see that two different time scales are present. For the purposes of communication of information, it is most useful to determine the rate of arrival of pulses, or in the case shown here, of peaks in the modulated amplitude. In analogy with the approach taken to determining the phase velocity, we can define the 'group velocity' of the wave as being the rate of advance of the pulse peaks,

$$v_g \equiv \frac{\Delta\omega}{\Delta k}, \tag{1.31}$$

which becomes, in the limit $\Delta k \to 0$, $v_g = d\omega/dk$.

There are various forms in which we can express the group velocity, and more specifically, to relate the phase and group velocities. For example, using the fact that $\omega = v_p k$ we can write immediately

$$v_g = \frac{d\omega}{dk} = v_p + k\frac{dv_p}{dk} = v_p\left(1 - \frac{k}{n}\frac{dn}{dk}\right), \tag{1.32}$$

which shows us that the group and phase velocities are in general different. For example, in the presence of material 'dispersion', or dependence of the index of refraction on wavelength, the phase velocity will also depend on wavelength and the two velocities will not be equal. Expressing the result of equation (1.32) directly in terms of the index of refraction and wavelength, we find

$$v_g = v_p\left(1 + \frac{\lambda}{n}\frac{dn}{d\lambda}\right). \tag{1.33}$$

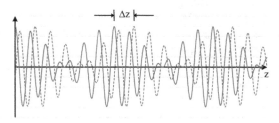

Figure 1.4. Two snapshots of the motion of a wavepacket along the z-axis, taken at times separated by an interval Δt. It is the motion of these packets that is equivalent to the transmission of information.

Likewise, we may express the group velocity in terms of the frequency,

$$v_g = \frac{c}{n + \omega\frac{dn}{d\omega}}.$$ (1.34)

We will leave to a later chapter the detailed discussion of index of refraction, stating here only that usually (so-called 'normal' dispersion) $dn/d\omega$ is positive and that both $dn/d\lambda$ and dv_p/dk are negative, implying that the group velocity is less than the phase velocity.

To close out this section we note one additional important point. The principle of superposition may be extended to any number of solutions; one example is that of a pulse of light (we referred to something like a pulse in the group velocity development), which may be considered as a sum of many different frequencies. The mathematical machinery used to rigorously describe functions of many frequencies is that of Fourier analysis, the details of which we leave for an appendix.

Example. A specific example may help with the visualization of the idea of group velocity and phase velocity. We will have to introduce the concept of index of refraction, which will be discussed in more detail in later chapters and will be taken as a characterization of a medium in which a wave propagates. Furthermore, we will see in later chapters that the index of refraction may vary depending on the frequency of the wave traveling through a given medium, a concept known as dispersion. A wave propagating in the x-direction has a two-frequency electric field given by

$$\vec{E} = \left[\frac{\sin(k_1x - \omega_1t)}{k_1x - \omega_1t} + \frac{\sin(k_2x - \omega_2t)}{k_2x - \omega_2t}\right]\hat{z}.$$ (1.35)

Note that the expression we are using as a description of the field is not simply a plane wave. However, substitution into the wave equation shows that this field is a valid solution to the equation; instead of a plane wave of infinite extent, it represents a localized electric field pulse. The frequencies, chosen simply to correspond to two visible wavelengths, are given by $\omega_1 = 2\pi c/600$ nm and $\omega_2 = 2\pi c/550$ nm. The wavevector magnitude is $k = n\omega/c$, where n is the index of refraction and is equal to one for propagation in a vacuum. First we can plot the electric field as a function of the distance propagated along the x-axis at two different times. Assume a propagation time of 20 fs (1 fs $= 10^{-15}$ s). The plot of $E(t)$ is shown in figure 1.5. We can calculate the velocity of propagation of the 'pulse' from the change in position and the elapsed time and find that the peak is traveling at $v_g = 3 \times 10^8$ m s^{-1}. Thus, in a vacuum the group velocity is identical to the phase velocity.

We may repeat this calculation making the additional assumption that $n \neq 1$ and that n depends on frequency, for example as $n = 1 + \omega^2/\omega_r^2$, where $\omega_r = 2\pi c/500$ nm. This frequency dependence of the index of refraction, and thus of the phase velocity of a wave, corresponds approximately to what one finds for many materials such as glasses in the visible part of the spectrum. Now the result is

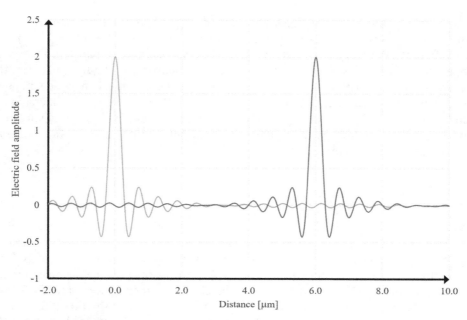

Figure 1.5. Two snapshots of the electric field $E(t)$ as a function of position along the x-axis, taken at times separated by a time interval $\Delta t = 20$ fs. The field is propagating in a vacuum ($n = 1$).

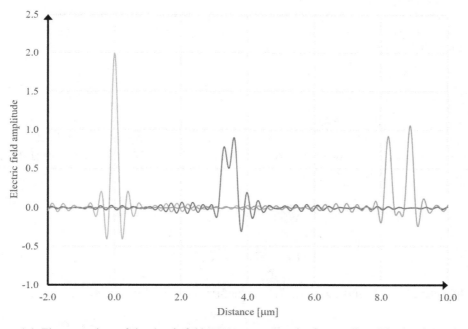

Figure 1.6. Three snapshots of the electric field $E(t)$, now propagating in a medium. The time intervals are $\Delta t_1 = 20$ fs and $\Delta t_2 = 40$ fs.

as shown in figure 1.6. The pulse has been deformed, but we can still make an estimate of the group velocity: $v_g = 3.4 \times 10^{-6}$ m/20×10^{-15} s $= 1.7 \times 10^8$ m s^{-1}. The different frequency components of the pulse begin to separate, since each propagates at a slightly different speed. The third snapshot in figure 1.6 shows the two frequencies nearly separated. In a more realistic situation in which a pulse consists of many frequencies that are closely spaced, the net result of dispersion, or index of refraction dependence on frequency, is that any pulse will begin to spread as it propagates through a medium. This is of particular importance in the world of fiber optics and telecommunications.

1.5 Energy flux

Up to this point we have considered only the electric field as an entity by itself. Now that we have expressions for the E-field vector as a function of time and spatial variables, we would like to know how energy is actually carried by these fields. This means that we will need an expression for energy transport by the electromagnetic wave or an expression for power flux of the electromagnetic wave. This is readily dealt with through the Poynting theorem and a quantity called the Poynting vector. Our interest at this point is with the Poynting vector, which will give us the appropriate expression for the transport of electromagnetic radiation. In what follows we will consider only plane waves.

From elementary electricity and magnetism texts [12] we can find an expression for the energy density (J m^{-3}) stored in electric and magnetic fields, which for vacuum fields we express as

$$u = \frac{1}{2}\left(\varepsilon_0 E^2 + \frac{1}{\mu_0}B^2\right). \tag{1.36}$$

In the end we will see that this result holds not only for static fields, but for time-varying fields as well. To connect the expression for the energy density in the above form to Maxwell's equations we form the dot products of $\vec{E}$ with equation (1.4) and of $\vec{B}$ with (1.3), yielding

$$\vec{E} \cdot \left(\nabla \times \vec{B}\right) = \frac{1}{c^2}\vec{E} \cdot \frac{\partial \vec{E}}{\partial t} \tag{1.37}$$

$$\vec{B} \cdot \left(\nabla \times \vec{E}\right) = -\vec{B} \cdot \frac{\partial \vec{B}}{\partial t}. \tag{1.38}$$

Subtracting the second equation from the first and dividing by μ_0 we arrive at

$$\varepsilon_0\vec{E} \cdot \frac{\partial \vec{E}}{\partial t} + \frac{1}{\mu_0}\vec{B} \cdot \frac{\partial \vec{B}}{\partial t} = -\frac{1}{\mu_0}\nabla \cdot \left(\vec{E} \times \vec{B}\right) \tag{1.39}$$

$$= -\nabla \cdot \vec{S},$$

where we have used a vector identity (see appendix A) to arrive at the right-hand side, and then defined the Poynting vector to be

$$\vec{S} = \frac{1}{\mu_0}\left(\vec{E} \times \vec{B}\right). \tag{1.40}$$

The left-hand side of equation (1.39) can be seen to be the time derivative of the energy density defined by equation (1.36). Integrating both sides over the volume and using the divergence theorem we can write

$$\frac{1}{2}\frac{\partial}{\partial t}\int_V dV\left(\varepsilon_0 E^2 + \frac{1}{\mu_0}B^2\right) = \int_V dV\left(-\nabla \cdot \vec{S}\right) \tag{1.41}$$

$$= -\oint_S \vec{S} \cdot \hat{n}\, dA.$$

Equation (1.41) is referred to as Poynting's theorem and essentially describes conservation of energy. The left-hand side represents the time rate of-change of energy stored in a given volume. If that energy changes it must be due to a flux of energy through the surface bounding the same volume, as given by the expression on the right-hand side of equation (1.41).

1.5.1 The free-space Poynting vector

The Poynting vector defined in equation (1.40) is defined as the power flux of an electromagnetic field; that is, it is the energy carried by the field per unit area per unit time (energy/time/area) [12, 13]. Since the Poynting vector involves the vector product of two waves, the use of complex notation involves a little pitfall of which we need to be aware. We must keep in mind the fact that $\vec{S}$ is the product of the real part of each function rather that the real part of the product. An example of how the complex fields notation can lead one into an error is given below. Consider a harmonic function and the use of complex notation to represent it:

$$E(t) = |\mathcal{E}|\cos(\omega t + \delta)$$
$$= \Re[\mathcal{E}e^{-i\omega t}],$$

where $\mathcal{E} = |\mathcal{E}|e^{-i\delta}$ so that the quantity $E(t)$ can be written

$$E(t) = \mathcal{E}e^{-i\omega t} \quad \text{(where Re is implied)}. \tag{1.42}$$

Examine the product of two functions (such as those encountered with the Poynting vector) of this form using the sin/cos notation rather than complex notation:

$$E(t) = |\mathcal{E}|\cos(\omega t + \delta_1)$$
$$B(t) = |\mathcal{B}|\cos(\omega t + \delta_2)$$
$$E(t)B(t) = |\mathcal{E}||\mathcal{B}|\cos(\omega t + \delta_1)\cos(\omega t + \delta_2).$$

Using the identity

$$\cos\alpha\cos\beta = \frac{1}{2}(\cos(\alpha - \beta) + \cos(\alpha + \beta))$$

we obtain

$$E(t)B(t) = \frac{|\mathcal{E}||\mathcal{B}|}{2}[\cos(\delta_1 - \delta_2) + \cos(2\omega t + \delta_1 + \delta_2)]. \qquad (1.43)$$

Now using complex notation,

$$E(t)B(t) = \mathcal{E}\,e^{-i\omega t}\mathcal{B}e^{-i\omega t} = \mathcal{E}\,\mathcal{B}e^{-2i\omega t}$$
$$= |\mathcal{E}||\mathcal{B}|e^{i(-2\omega t + \delta_1 + \delta_2)}.$$

Taking the real part of this last expression yields

$$\mathfrak{R}\,[E(t)B(t)] = |\mathcal{E}||\mathcal{B}|\cos(2\omega t + \delta_1 + \delta_2). \qquad (1.44)$$

Thus when comparing equations (1.43) and (1.44) we see that the constant term $\frac{1}{2}|\mathcal{E}||\mathcal{B}|\cos(\delta_1 - \delta_2)$ is missing in equation (1.44). Again, the reason that we obtained this result is because $\vec{S}$ is the product of the real part of *each* harmonic function, *not* the real part of the product. In obtaining equation (1.44) we incorrectly calculated the result by taking the real part of the product of the complex functions. In figure 1.7 the relationship between $\vec{E}$, $\vec{B}$, and $\vec{S}$ is shown for a given point in time.

Example. Given the electromagnetic wave,

$$\vec{E} = \hat{x}E_0\cos(kz - \omega t) + \hat{y}E_0\cos(kz - \omega t),$$

where E_0 is a constant, find the corresponding magnetic induction and the free-space Poynting vector.

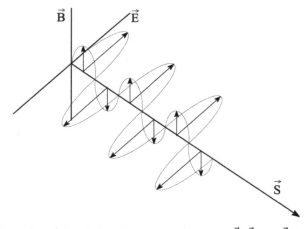

Figure 1.7. Illustration of the relative orientation of the vectors $\vec{E}$, $\vec{B}$, and $\vec{S}$ for a plane wave.

Solution. We can write $\vec{B} = (1/\omega)\vec{k} \times \vec{E}$ (the proof is left for the homework problems) using Maxwell's equations (equation (1.3)). For this example, $\vec{k}$ has only one component, in the z-direction. Thus

$$\vec{B} = \frac{1}{\omega}k\,\hat{z} \times [\hat{x}\cos(kz - \omega t) + \hat{y}\cos(kz - \omega t)]\,E_0$$

$$= \frac{E_0 k}{\omega}[\hat{y}\cos(kz - \omega t) - \hat{x}\cos(kz - \omega t)]$$

$$= \frac{E_0}{c}[\hat{y}\cos(kz - \omega t) - \hat{x}\cos(kz - \omega t)].$$

Note first of all that the magnitude of the B-field is smaller than the magnitude of the E-field by a factor of c. The fields $\vec{E}$ and $\vec{B}$ are seen from the above to be mutually orthogonal and thus $\vec{k}$, $\vec{E}$, and $\vec{B}$ form a triad of mutually orthogonal vectors, a general result for plane waves in free space.

For the second part of the problem,

$$\vec{S} = \frac{1}{\mu_0}\vec{E} \times \vec{B}$$

$$= \frac{1}{\mu_0}[\hat{x}E_0\cos(kz - \omega t) + \hat{y}E_0\cos(kz - \omega t)]$$

$$\times \frac{E_0}{c}[\hat{y}\cos(kz - \omega t) - \hat{x}\cos(kz - \omega t)]$$

$$= \frac{E_0^2}{\mu_0 c}[2\cos^2(kz - \omega t)]\,\hat{z}.$$

The key result here is that the direction of energy flow ($\vec{S}$) corresponds to the direction of propagation since in this case $\vec{k} = k\,\hat{z}$.

1.5.2 Time averaging of sinusoidal products: irradiance

For electromagnetic waves in optics we are most interested in the time average of the Poynting vector rather than the instantaneous value. This can be understood by considering the relevant frequencies involved. Optical waves have frequencies on the order of $\omega/2\pi \sim 10^{14}$ Hz; a fairly fast detector has a bandwidth on the order of 1 GHz, several orders of magnitude slower than the field oscillations. Thus, the signal actually observed by a detector is an averaged, or filtered version of the actual field oscillation.

For the product of $\vec{E}$ and $\vec{B}$ over one period $T = 2\pi/\omega$ this is given by

$$
\begin{aligned}
\langle E(t)B(t) \rangle &\equiv \frac{1}{T} \int_0^T |\mathcal{E}| \cos(\omega t + \delta_1)|\mathcal{B}| \cos(\omega t + \delta_2)dt \\
&= \frac{1}{T} \frac{|\mathcal{E}||\mathcal{B}|}{2} \int_0^T (\cos(2\omega t + \delta_1 + \delta_2) + \cos(\delta_1 - \delta_2))dt \\
&= \frac{1}{T} \frac{|\mathcal{E}||\mathcal{B}|}{2} \left(\frac{1}{2\omega} \sin(2\omega t + \delta_1 + \delta_2)|_0^{2\pi/\omega} \right. \\
&\qquad \left. + T \cos(\delta_1 - \delta_2) \right) \\
&= \frac{|\mathcal{E}||\mathcal{B}|}{2} \cos(\delta_1 - \delta_2).
\end{aligned}
\tag{1.45}
$$

If we use complex notation it is easy to show that this result is obtained from the following:

$$
\langle E(t)B(t) \rangle = \frac{1}{2}\Re[EB^*] = \frac{1}{2}\Re[E^*B].
\tag{1.46}
$$

Using the previous argument it is possible to show that the time average of equation (1.40) is

$$
\langle \vec{S} \rangle = \frac{1}{2\mu_0}\Re(\vec{E} \times \vec{B}^*).
\tag{1.47}
$$

The irradiance or intensity is the magnitude of $\langle \vec{S} \rangle$. Using the results from above

$$
I = |\langle \vec{S} \rangle| = \frac{|\vec{E}|^2}{2\mu_0 c}.
\tag{1.48}
$$

Example. Using the fields found in the previous example, calculate $\langle E(t)B(t) \rangle$ and $\langle \vec{S} \rangle$.
  Solution. For

$$
\vec{E} = \hat{x}E_0 \cos(kz - \omega t) + \hat{y}E_0 \cos(kz - \omega t)
$$

and

$$
\vec{B} = \frac{E_0}{c}[\hat{y} \cos(kz - \omega t) - \hat{x} \cos(kz - \omega t)]
$$

we can write the complex fields, respectively, as

$$
\vec{E} = \hat{x}E_0 e^{i(kz-wt)} + \hat{y}E_0 e^{i(kz-wt)}
$$

and

$$\vec{B} = -\hat{x}\frac{E_0}{c}e^{i(kz-wt)} + \hat{y}\frac{E_0}{c}e^{i(kz-wt)}.$$

The magnitude of the electric field is given by $|\vec{E}|^2 = \vec{E}^* \cdot \vec{E} = E_0^2 + E_0^2 = 2E_0^2$. Likewise, for the magnitude of the magnetic field we have $|\vec{B}|^2 = E_0^2/c^2$. Thus, from equation (1.46) we can write $\langle E(t)B(t) \rangle = 2E_0^2/2c$ and from equation (1.48) $\langle \vec{S} \rangle = 2E_0^2/2\mu_0 c$.

Example. A diode laser has a beam radius of 2.0 mm and an output power of 5.0 mW. What is its irradiance (in W m^{-2})? Calculate the amplitudes of the electric and magnetic fields.

Solution. The irradiance is the power per unit area. Thus

$$I \equiv |\langle \vec{S} \rangle| = \frac{\text{power}}{\text{area}} = \frac{5 \times 10^{-3}\ \text{W}}{\pi \times (2 \times 10^{-3}\ \text{m})^2} = 398\ \text{W m}^{-2}.$$

To find the amplitudes of the $\vec{E}$ and $\vec{B}$ fields, we use results from this section, along with the fact that the field amplitudes are related by

$$|\vec{E}| = c\ |\vec{B}|,$$

where c is the speed of light. Using equation (1.47) we find

$$|\langle \vec{S} \rangle| = \frac{1}{2\mu_0}|\vec{E}||\vec{B}| = \frac{c}{2\mu_0}|\vec{B}|^2$$

and therefore

$$|\vec{B}|^2 = \frac{2 \times (4\pi \times 10^{-7}\ \text{Wb}\ (\text{A} \cdot \text{m})^{-1}) \times (398\ \text{W m}^{-2})}{3 \times 10^8\ \text{m s}^{-1}}$$

$$|\vec{B}| = 1.83 \times 10^{-6}\ \text{T}.$$

Finally, the magnitude of the electric field is given by $|\vec{E}| = 3 \times 10^8\ \text{m s}^{-1} \times 1.83 \times 10^{-6}\ \text{T} = 549\ \text{V m}^{-1}$.

We can compare the results calculated above to some common values so as to gain a physical feeling for these magnitudes. For example, the solar constant, the amount of sun reaching the top of the Earth's atmosphere (an average value, of course) is about 1370 W m^{-2}, several times the irradiance of the diode laser in the example. As a comparison for the magnetic field magnitude, we note that the Earth's magnetic field is on the order of 10^{-5} T.

One last point to notice about this example is that we have presented the diode laser on the one hand as a plane wave, which by definition has an infinite extent, and on the other hand we have given the beam a definite radius. Referring back to our discussion of Gaussian beams, we saw that near the focus the Gaussian beam electric field is very nearly equivalent to a plane wave, and at very large distances from the beam waist the approximate spherical wave character of the beam for which the radius of curvature is small can also be treated to first approximation as a plane wave. Thus we take this as our justification for the approximations used in this example.

1.6 Resonator electric field

In our earlier discussion about phase and group velocities it was pointed out that the description of the electric field of a temporal pulse of light is best thought of in terms of a superposition of many electric fields of differing frequencies. An example at the end of chapter 2 will demonstrate how we can describe a Gaussian beam in terms of a spatial superposition of plane waves. In this section we present an example of superposition in the time domain of electric fields to describe the transient behavior of an optical resonator, such as a laser cavity. The type of resonator we will be investigating in this example is often known as a Fabry–Perot cavity or resonator; if one considers a solid piece of material such as quartz, the treatment below remains much the same, but one refers in that case to a Fabry–Perot 'etalon'.

First, we need to give a small amount of background. In its simplest form a laser consists of a collection of atoms residing in a resonant optical cavity. The purpose of the resonator is to provide feedback for the laser atomic gain medium. Here we will concentrate on the properties of the resonator in the absence of any atomic medium. The resonator consists of two mirrors facing each other, each of which reflects a fraction of the incident field, given by r_1 and r_2, and transmits a fraction t_1 or t_2. As we will see in chapter 5, the intensity *reflectance* is given by $R = r^2$ for each mirror. Throughout, we assume identical mirrors. Our aim is to evaluate the time dependence of the output intensity of the resonator, with a given input intensity.

We proceed as follows. The characteristics of a resonator can be found in somewhat simplified form if one assumes a *confocal* configuration for the resonator, with the mirrors' radii of curvature equal to the cavity length d as shown in figure 1.8. The resulting expressions may easily be generalized to a cavity of arbitrary configuration. For an incident field of amplitude $E_{in}(t)$ the transmitted

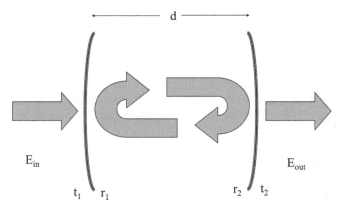

Figure 1.8. A Fabry–Perot cavity of resonator is made up of two mirrors of field reflectivity r_1 and r_2 and field transmissivities t_1 and t_2, separated by a distance d.

field amplitude $E_{out}(t)$ is simply the superposition of fields transmitted by the second mirror after different numbers of round trips within the resonator [14]:

$$E_{out}(t) = t_1 t_2 \sum_{n=0}^{\infty} E_{in}(t - \tau_0 - n\tau)(r_1 r_2)^n e^{i(\delta_0 + n\delta)}. \tag{1.49}$$

In equation (1.49) $E_{in}(t - \tau_0 - n\tau)$ is the input field evaluated earlier in time by an amount corresponding to one single pass and n round trips. τ_0 is the time taken for light to make a single pass from the entrance mirror to the exit mirror and τ is the round-trip time, δ_0 is the single-pass phase shift for the electric field and δ is the round-trip phase shift (relative to that of a particular cavity resonance), t_1 and t_2 are the mirror field transmissivities, and r_1 and r_2 are the mirror field reflectivities.

In the steady state, when E_{in} and E_{out} are independent of time, the output intensity can be found analytically by realizing that the sum given in equation (1.49) is a geometric series. This is the key step for all that follows. We realize that the output field (again, considering that the input has been on for an infinitely long time) is the superposition of many fields, with contributions due to (i) the input field multiplied by the transmission of each mirror, (ii) the field that did not escape after traversing the resonator once, but after being reflected once by each mirror, is transmitted on the second pass, (iii) two full round trips, etc. The result is given by (the derivation is left for the problems)

$$T(\delta) = \left| \frac{E_{out}}{E_{in}} \right|^2 = \frac{t_1^2 t_2^2}{(1 - (r_1 r_2)^2)^2 + 4(r_1 r_2)^2 \sin^2(\delta/2)}, \tag{1.50}$$

where δ, introduced as a round-trip phase shift of the field, can also be interpreted as a detuning of the resonator from its transmission maximum, which occurs for $\delta = 2m\pi$. The cavity intensity transmission function $T(\delta)$ is not to be confused with the transmittance of a single mirror, e.g. $T_1 = t_1^2$ or $T_2 = t_2^2$. The resonator transmission as a function of the phase argument δ is shown in figure 1.9.

The phase dependence is equivalent to a dependence of the transmission of the resonator on pathlength, since the accumulated phase delay of the wave is proportional to the distance traveled, i.e. $\delta = k(2d) = (2\frac{\pi}{\lambda_0})(2d)$. The transmission reaches a maximum for $\sin(\delta/2) = 0$, which implies $d = m\lambda_0/2$, meaning that the 'resonance' condition is one for which there is a standing wave in the cavity. Converting to frequencies using $\nu = c/\lambda$, we can write the condition for the discrete frequencies at which the cavity is resonant are given by $\nu_m = m\left(\frac{c}{2d}\right)$. The separation between successive transmission maxima is given by the 'free spectral range' of the resonator,

$$\Delta\nu_{FSR} = \nu_{m+1} - \nu_m = \frac{c}{2d}. \tag{1.51}$$

We will now assume that the two mirrors are identical, $r_1 = r_2 \equiv r$ and $t_1 = t_2 \equiv t$, and use the fact that the intensity reflectance and transmittance are given by $R \equiv r^2$ and $T \equiv t^2$, respectively. In the limit of high reflectivity and small detuning $(1 - R \ll 1, \delta \ll 1)$ each of the cavity transmission peaks shown in figure 1.9

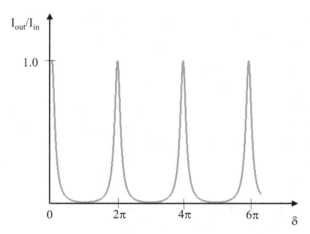

Figure 1.9. The steady-state transmission of an ideal Fabry–Perot resonator as a function of detuning (δ) from the cavity resonance frequency.

show approximately what is known as a 'Lorentzian' lineshape as a function of phase:

$$T(\delta) = \left| \frac{E_{\text{out}}}{E_{\text{in}}} \right|^2 \simeq \frac{T^2}{(1 - R)^2 + R\delta^2}. \tag{1.52}$$

Using the above relation we may define several quantities. The full-width at half-maximum (FWHM) of the transmission function is defined from equation (1.52) as (twice) the value of δ for which the function $T(\delta) = 1/2$, and is found to be $\delta_{1/2} = 2\frac{1 - R}{\sqrt{R}}$. The 'finesse' F is the ratio of the phase shift between transmission peaks (2π) (the free spectral range), to the FWHM of a single peak,

$$F = \frac{2\pi}{2\frac{1 - R}{\sqrt{R}}} = \frac{\pi\sqrt{R}}{1 - R}. \tag{1.53}$$

The cavity loss rate is defined to be $\kappa = (1 - R)/\tau$, with τ being the round-trip time for light circulating in the cavity.

The peak transmission is given by

$$T_0 = \frac{T^2}{(1 - R)^2}. \tag{1.54}$$

Note that the peak transmission T_0, an easily measured parameter, allows a determination of the value of the cavity loss rate κ. This description of the ideal resonator does not take into account any loss due to absorption and scatter in the mirror substrates and their optical coatings and, in fact, since $R + T = 1$ for an ideal cavity, we see that the peak transmission will always be unity. For a real resonator losses can be considered by writing $R + T + A = 1$ from which the peak transmission becomes

$$T_0 = \frac{T^2}{(T + A)^2}.$$

To finish off this chapter and the introduction to electromagnetic fields, we look at the time-dependent response of the resonator. If a previously constant input field is cut off suddenly at time $t = 0$, we will show that the transmitted intensity will decay in a stepwise fashion which, for short round-trip time and high finesse, will appear to be exponential [14]:

$$\frac{I_{\text{out}}(t = \tau_0 + n\tau)}{I_{\text{out}}(\tau_0)} = |\, R^n e^{in\delta}\,|^2 \simeq \exp\left[-2n(1 - R)\right]$$

$$= \exp\left[-2\kappa(t - \tau_0)\right],$$

$\qquad(1.55)$

where, for $1 - R \ll 1$, the approximation $R \simeq \exp[-(1 - R)]$ has been used along with the definition of the cavity decay rate given previously. Since 2κ is seen to be the decay rate of the transmitted intensity (or of the stored energy), κ itself must represent the decay rate for the field inside the cavity. Note also that the result in equation (1.55) is independent of cavity detuning δ. The light output from the cavity as it decays will be at the cavity's resonant frequency, also independent of δ. This is analogous to the decay of a weakly damped oscillator after a driving force is turned off; regardless of the driving frequency, oscillation during the decay takes place at the oscillator's natural frequency.

If the cavity's input field is suddenly turned on at $t = 0$, the transmission will turn on in a stepwise manner, and the field will begin to build up in the cavity with a characteristic time we refer to as the 'filling' time. We may approximate the filling behavior for short round-trip time and high finesse by an expression that depends on δ [14]:

$$\frac{I_{\text{out}}(t = \tau + n\tau)}{I_{\text{out}}(\infty)} = |1 - R^n e^{in\delta}|^2 \simeq |1 - \exp\left[-\kappa(t - n\tau_0)\right]e^{in\delta}|^2.$$

$\qquad(1.56)$

On resonance ($\delta = 0$) this becomes

$$\frac{I_{\text{out}}(t)}{I_{\text{out}}(\infty)} = (1 - \exp[-\kappa(t - \tau_0)])^2.$$

$\qquad(1.57)$

Thus κ can also be found from the cavity's transient behavior, according to equations (1.55) (decay) and (1.57) (filling).

Taking a concrete example of the concepts discussed above, if the resonator's round-trip time τ is not short compared to the time over which the input field switches on or off, the cavity transmission will show distinct steps, as is illustrated in figure 1.10 for the case of a cavity excited on resonance. Here we have chosen parameters for the cavity with mirrors of 88% reflectivity and a round-trip time of 12 ns. The input intensity pulse might by produced by an electro-optic modulator, to be discussed later in this text and has a length of 300 ns as shown in figure 1.10(a). In figure 1.10(b) is shown the result of a numerical calculation of equation (1.49) for the

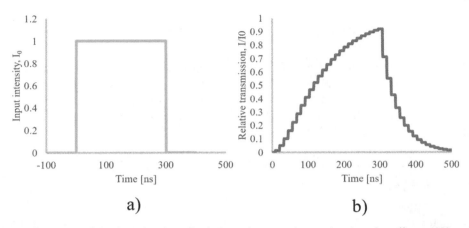

Figure 1.10. (a) Input intensity pulse of amplitude I_0, turning on at time $t = 0$ and turning off at $t = 300$ ns. (b) The calculated time-dependent transmission of an ideal Fabry–Perot resonator for mirror reflectance $R = 0.88$ and resonator round-trip time $\tau = 12$ ns.

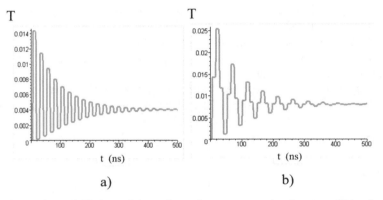

Figure 1.11. For an input field detuned from the cavity resonance, the character of the time-dependent transmission of an ideal Fabry–Perot resonator is more complicated than for zero detuning. For mirror reflectance $R = 0.88$ and resonator round-trip time $\tau = 12$ ns and detunings (a) $\delta = \pi/2$ and (b) $\delta = \pi/4$ the transmitted intensity shows oscillations occurring at a period of twice and four times the cavity round-trip time, respectively.

given cavity parameter values. Since we assume lossless mirrors, the output intensity slowly increases to asymptotically reach that of the input pulse, here equal to unity. The filling time and decay times, $1/2\kappa$, are equal to 50 ns for this example; the intensity falls to $1/e = 0.37$ of its initial value after this time as can be checked from figure 1.10, with the stepwise decay here approximating an exponential function.

In the non-resonant ($\delta \neq 0$) filling of a resonator after switch-on of the driving radiation, we see the presence of beating, or interference, between the input and cavity fields quite dramatically, as shown theoretically in figure 1.11(a). From equation (1.56) we see that deep modulation at the frequency difference between the incident light and the cavity resonance may be present while the cavity fills if that offset frequency is a submultiple of the round-trip frequency, that is, if the detuning δ

is equal to π divided by a small integer. This modulation is illustrated in the calculations of figure 1.11(a), for a detuning of π, where the cavity output oscillates with a period of twice the round-trip time ('period-2' oscillations), and of figure 1.11(b), for a detuning of $\pi/2$, where the oscillations have a period of four times the round-trip time ('period-4' oscillations). These oscillations are nothing more than a manifestation of interference between fields, either constructive or destructive, depending on the observation time.

1.7 Problems

1. The current density and charge density at a point in space are related by the continuity equation

$$\frac{\partial \rho}{\partial t} + \nabla \cdot \vec{j} = 0.$$

   This equation is just a statement of conservation of charge. Derive this equation from Maxwell's equations.

2. A sphere of radius a is uniformly charged with a density ρ. If the sphere rotates with a constant angular velocity ω, show that the magnetic induction, B, at the center of the sphere is $(\mu_0/3)\rho a^2 \omega$. Hint: Use the Biot–Savart law for the magnetic induction.

3. A particle with mass m and charge q moves in a constant magnetic field B. Show that, if the initial velocity is perpendicular to B, the path is circular and the angular velocity is

$$\omega = \frac{q}{mv}B.$$

4. Show that the relation $j = \sigma E$ is equivalent to the usual statement of Ohm's law, $V = IR$, or (voltage) = (current) × (resistance).

5. Starting with the Lorentz force and Maxwell's equations for vacuum, derive Coulomb's law.

6. Prove the following relationships for plane wave fields:

$$\nabla \cdot \vec{E}(\vec{r}, t) = \Re\left[i\vec{k} \cdot \vec{\mathcal{E}}(\vec{k}, \omega)e^{i(\vec{k}\cdot\vec{r}-\omega t)} \right]$$

$$\nabla \times \vec{E}(\vec{r}, t) = \Re\left[i\vec{k} \times \vec{\mathcal{E}}(\vec{k}, \omega)e^{i(\vec{k}\cdot\vec{r}-\omega t)} \right]$$

$$\nabla^2 \vec{E}(\vec{r}, t) = \Re\left[(ik)^2 \vec{\mathcal{E}}(\vec{k}, \omega)e^{i(\vec{k}\cdot\vec{r}-\omega t)} \right].$$

7. Using Maxwell's equations show that

$$\vec{E} = -\frac{c^2}{\omega}\vec{k} \times \vec{B}$$

   for a free-space plane wave solution to the wave equation.

8. Using Maxwell's equations show that

$$\vec{B} = \frac{1}{\omega}\vec{k} \times \vec{E}$$

for a free-space plane wave solution to the wave equation.

9. Change the expression below for a transverse sinusoidal traveling wave moving in the positive x-direction into an expression involving k, x, ω, and t in the complex notation,

$$E = E_0 \sin\left[\frac{2\pi}{\lambda}(x - vt)\right],$$

where v is the wave velocity.

10. Use Maxwell's equations to show that free-space plane waves are transverse, i.e. that $\vec{k}$, $\vec{E}$, and $\vec{B}$ are all mutually perpendicular.

11. Show that for a Gaussian beam

$$R(z) = z\left[1 + \left(\frac{\pi w_0^2}{\lambda z}\right)^2\right]$$

and

$$w^2(z) = w_0^2\left[1 + \left(\frac{\lambda z}{\pi w_0^2}\right)^2\right].$$

Hint: try taking the inverse of $q(z) = q_0 + z$ and relate the real and imaginary parts to $R(z)$ and $w(z)$.

12. Make plots of $R(z)$ and $w(z)$ for a Gaussian beam, for $0 < z < 5$ cm, assuming a helium–neon laser with beam waist $w_0 = 50$ μm and $\lambda = 632.8$ nm. Be especially careful when plotting $R(z)$. Your plot will look different depending on the number of points you use. Explain physically what is happening for $z \to 0$. If you extend the plot for $R(z)$ to $z = 50$ cm, what happens? Explain physically.

13. Show that for the electric field given by equation (1.27) there is non-transverse component.

14. The peak power of a certain CO_2 laser ($\lambda = 10.6$ μm) is 100 W. If the beam is focused to a spot 10 μm in diameter, find the irradiance and the amplitude of the electric field of the light wave at the focal point.

15. The radiant energy from the Sun is about 8 J cm^{-2} per minute. Assuming this to be in the form of a plane wave traveling in the z-direction in a vacuum, plane polarized with its electric field in the x–y plane, and having a wavelength of 6000 Å, find expressions for the electric and magnetic induction fields as functions of z and t and with all constants evaluated.

16. (a) Find the magnitude of the magnetic induction, B, of a free-space electromagnetic wave in terms of the amplitude of the electric field, E, of

that wave. (b) Find an expression for the free-space irradiance in terms of the amplitude of the electric field.

17.

(a) Given $\vec{E} = \hat{x}25\cos(kz - \omega t)$ V m^{-1} in free space, find the average power passing through a circular area of diameter 5 m in the plane where $z = 5$ m.

(b) Given that in free space

$$\vec{E} = \left[200\,\frac{\sin\Theta}{r}\cos(kr - \omega t)\right]\hat{\Theta}.$$

Determine the average power passing through a hemispherical shell of radius $r = 10^2$ m and $0 \leqslant \Theta \leqslant \pi/2$.

18. Show that when complex notation is used, the time-averaged Poynting vector can be written,

$$\langle \vec{S} \rangle = \Re\,(\vec{E} \times \vec{B}^*/2\mu_0).$$

19. The energy flow associated with sunlight, striking the surface of the Earth in a normal direction is 1.33 kW m^{-2}.

(a) If the corresponding electromagnetic wave is taken to be a plane polarized monochromatic wave, determine the maximum values of E and B.

(b) Taking the distance from the Earth to the Sun as 1.5×10^{11} m, find the total power radiated by the Sun.

20.

(a) Consider a long straight conductor, of conductivity σ and radius r, carrying a current density j. Find the magnitude and direction of the Poynting vector in the conductor in terms of r, σ, and j.

(b) Suppose that the very long coaxial line is used as a transmission line between a battery and resistor. Designate the EMF of the battery as V. The battery is connected to one end of the coaxial line while the other end of the coaxial line is connected to a resistor of resistance R. Find the magnitude and direction of the Poynting vector in the region between the conductors of the coaxial cable. Find the total power passing through a cross section of the line. Will the direction of energy flow change if the connections to the battery are interchanged?

21. Compute the irradiance of a plane wave with electric field vector given by

$$\vec{E} = (\hat{x}E_x e^{i\alpha} + \hat{y}E_y e^{i\beta})e^{i(kz - \omega t)}.$$

22. Consider a superposition of two independent orthogonal plane waves:

$$\vec{E} = \hat{x}\,E_1 e^{i(kz - \omega t + \phi_1)} + \hat{y}\,E_2 e^{i(kz - \omega t + \phi_2)}.$$

Find $\langle \vec{S} \rangle$ and show that it is equal to the sum of the average Poynting vectors for each component.

23. Consider the superposition of parallel independent plane waves

$$\vec{E} = \hat{x}\, E_1 e^{i(kz - \omega t + \phi_1)} + \hat{x}\, E_2 e^{i(kz - \omega t + \phi_2)}.$$

Find $\langle \vec{S} \rangle$ and show that it is *not* equal to the sum of the average Poynting vectors for each component. What is the difference physically between this result and that of the previous problem?

24. Consider an electromagnetic wave,

$$\vec{E} = \hat{x} E_0 e^{i(kz - \omega t)} + \hat{x} E_0 e^{i(-kz - \omega t)}.$$

Find $\vec{S}$ and $\langle \vec{S} \rangle$.

25. Derive the result given by equation (1.50)

26. Show that the transmission function for a Fabry–Perot resonator can be described approximately by 'Lorentzian' lineshape as a function of phase as in equation (1.52):

$$T(\delta) = \left| \frac{E_{\text{out}}}{E_{\text{in}}} \right|^2 \simeq \frac{T^2}{(1 - R)^2 + R\delta^2}. \tag{1.58}$$

27. Plot the time dependence of the output intensity of a resonator with identical mirrors of field reflectivity $r = 0.9$ and with a separation of $d = 100$ cm between the mirrors.

References

[1] Wikipedia: Hans Christian Ørsted https://en.wikipedia.org/wiki/Hans_Christian_%C3%98rsted (Accessed: 11 Jan. 2021)
[2] Wikipedia: André-Marie Ampère https://en.wikipedia.org/wiki/Andr%C3%A9-Marie_Amp%C3%A8re (Accessed: 11 Jan. 2021)
[3] Wikipedia: François Arago https://en.wikipedia.org/wiki/Fran%C3%A7ois_Arago (Accessed: 11 Jan. 2021)
[4] Wikipedia: Humphry Davy https://en.wikipedia.org/wiki/Humphry_Davy (Accessed: 11 Jan. 2021)
[5] Wikipedia: Michael Faraday https://en.wikipedia.org/wiki/Michael_Faraday (Accessed: 11 Jan. 2021)
[6] Wikipedia: James Clerk Maxwell https://en.wikipedia.org/wiki/James_Clerk_Maxwell (Accessed: 11 Jan. 2021)
[7] NIST Reference on Constants, Units and Uncertainty International System of Unit (SI) http://physics.nist.gov/cuu/Units/second.html
[8] Yariv A 1975 *Quantum Electronics* 2nd edn (New York: Wiley)
[9] Davis C 2014 *Lasers and Electro-optics* 2nd edn (Cambridge: Cambridge University Press)
[10] Siegman A E 1986 *Lasers* (Sausalito, CA: University Science Books)
[11] Kogelnik H and Li T 1966 Laser beams and resonators *Appl. Opt.* **5** 1550–67

[12] Griffiths D 2017 *Introduction to Electrodynamics* 4th edn (Cambridge: Cambridge University Press)

[13] Lorrain P and Corson D 1988 *Electromagnetic Fields and Waves* 3rd edn (San Francisco, CA: W H Freeman)

[14] Hernandez G 1986 *Fabry–Perot Interferometers* (Cambridge: Cambridge University Press)

IOP Publishing

Optical Radiation and Matter

Robert J Brecha and J Michael O'Hare

Chapter 2

Polarization of light

In chapter 1 we reviewed and derived some of the basic properties of commonly encountered solutions to the electromagnetic wave equation. Our emphasis in chapter 1 was mainly on free-space propagation properties of these waves and on the spatial profile of the wavefronts. In this second chapter we will concentrate on the vector nature of the (usually transverse) electromagnetic waves, as manifested in the phenomena of polarization.

We first introduce the definitions and algebraic formalism used to treat the polarization of electromagnetic waves. Later in the chapter we present a convenient matrix formalism that can be used for treating problems involving polarized light. Specific applications of this formalism will be encountered in chapters 7 and 8. To close out chapter 2 we will look in some detail at the polarization of a Gaussian beam, with the specific aim of showing that, in contrast to plane waves, there must be a component of polarization in the direction of wave propagation. Although this is a minor point, it does serve to remind us of the fundamental properties of waves and of the approximations we will be using for most of our work.

2.1 Historical introduction

Etienne Louis Malus (1775–1812) first gave the name 'polarization' to properties of light that had been discovered earlier, for example by Christiaan Huygens (1629–1695) in the Netherlands [1, 2]. Malus was experimenting with light transmitted by crystals known to split an incident beam into two parts. (This property is known as birefringence; we will return to this in chapter 7.) Interestingly enough, although polarization is now seen as one of the clearest indicators of the transverse wave nature of light, Malus chose a name reflecting the still-dominant belief that light consists of a stream of particles or corpuscles with a given polarity, much like charged particles.

doi:10.1088/978-0-7503-2624-7ch2

Malus noticed that if polarized light is incident on a surface, the reflected intensity depends on the square of the cosine of an angle relating the polarization direction and the surface. We would express this a bit differently today, writing

$$I = I_0 \cos^2 \theta,$$

where θ is the angle between the polarization direction and the axis of a polarizing element.

As the wave theory of light gained in acceptance at the beginning of the nineteenth century, due in great measure to the work of Thomas Young (1773–1829) on interference, Augustin Jean Fresnel (1788–1827) and François Arago (1786–1853) were able to conclude from a series of experiments that light from sources with mutually orthogonal polarizations shows no interference effects, whereas parallel polarized light beams do interfere [3–5]. Through these and other related experiments, they were also able to hypothesize that light may be described by a vector quantity and that physically the direction of vibration or polarization is transverse to the propagation direction. These conclusions were drawn decades before Maxwell's full theory of the electromagnetic field had been developed, and formed one of the starting points for Maxwell. In what follows we will concentrate on the vector amplitude of plane waves.

2.2 Polarization of light waves

The polarization of a light wave is determined by the behavior of the amplitude of the E-vector. If we simplify for the moment to propagation in the z-direction, the transverse nature of the plane wave requires that the E-field be confined to the $x-y$ plane. We can write the components of the field as

$$E_x = A_x \cos (kz - \omega t + \delta_x)$$
$$E_y = A_y \cos (kz - \omega t + \delta_y).$$

The above represents a wave traveling in the $+z$-direction; compared to our previous notation we have defined $A_x \equiv |\mathcal{E}_x|$ and $A_y \equiv |\mathcal{E}_y|$ and the overall amplitude of the field is given by $|\vec{\mathcal{E}}| = \sqrt{A_x^2 + A_y^2}$. We can eliminate the explicit space and time dependence in the above equations (by rewriting the cos terms using trigonometric identities and then squaring and adding the resulting equations) to find an expression for a curve traced out by the tip of the $\vec{E}$-vector, arriving at, after some algebraic steps left for the problems,

$$\left(\frac{E_x}{A_x}\right)^2 + \left(\frac{E_y}{A_y}\right)^2 - 2\frac{E_x E_y}{A_x A_y} \cos \delta = \sin^2 \delta \qquad (2.1)$$

in which $\delta = \delta_y - \delta_x$ $(-\pi \leqslant \delta \leqslant \pi)$. Geometrically, equation (2.1) represents the general equation of an ellipse in the $E_x - E_y$ plane.

Equation (2.1) defines the general state of an electromagnetic wave, which we refer to as being *elliptically polarized*. In the following sections we examine electric fields that represent differing degrees of ellipticity, as described by (2.1). For example, if $\delta = \pm\frac{\pi}{2}$ and $A_x = A_y$, equation (2.1) represents *circular polarization*, and if $\delta = m\pi$ ($m = 0, 1, \ldots$) and $E_y/E_x = (-1)^m(A_y/A_x)$ the state of the electromagnetic field is termed *linear polarization*.

In this text we will consider only states of the electromagnetic field for which the polarization is well-defined, i.e. we will work with various types of polarized light. In general, one must be concerned as well about unpolarized light, since the output of most light sources other than some lasers has a rapidly varying, and thus on average, poorly defined polarization state.

2.2.1 Elliptical polarization

Given the above general result for the polarization state of the field, it is always possible to find a new coordinate system x', y' such that the new coordinate axes correspond to the axes of the ellipse. In terms of these new coordinates the ellipse equation becomes

$$\left(\frac{E_{x'}}{a}\right)^2 + \left(\frac{E_{y'}}{b}\right)^2 = 1,$$

where $E_{x'}$ and $E_{y'}$ are the components of the electric field along the new coordinate axes x' and y'. The lengths of the principal axes of the ellipse are given by $2a$ and $2b$ and, as illustrated in figure 2.1, the new coordinate system is rotated with respect to the original system by an angle ϕ, which is given by (the derivation is again left for a homework problem)

$$\tan 2\phi = \frac{2A_x A_y}{A_x^2 - A_y^2} \cos \delta. \tag{2.2}$$

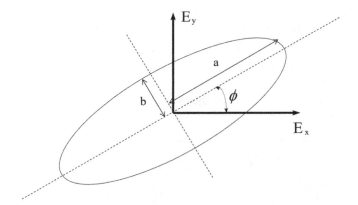

Figure 2.1. Illustration of the polarization ellipse described by equation (2.2).

The ellipticity is defined in terms of the lengths of the major and minor axes

$$\xi = \pm \frac{b}{a}$$

assuming that we choose ϕ such that $b \leqslant a$. This convention implies that the ellipticity satisfies

$$0 \leqslant |\xi| \leqslant 1.$$

The complete description of the general polarization state of an electric field requires, in addition to the ellipticity and the orientation of the ellipse, one more parameter to be specified. We must decide on a convention defining the sense of rotation or handedness of the light. In the definition of ξ there is an arbitrary sign assignment that can be made. Here we choose to use the convention that the '+' direction corresponds to right-elliptically polarized (REP) light, that is, for an observer facing the source of the wave, 'right-handed' means that the electric field vector rotates in a counterclockwise direction. Likewise, for an observer facing the source left-handed polarization means that the electric field vector appears to rotate in a clockwise direction. Put in another way, for REP light, an observer riding on the light wave would observe a sense of electric (or magnetic) field rotation equivalent to that of a 'normal' screw. This convention is chosen because of a correspondence to the quantum-mechanical definition of the angular momentum of a photon for which the z-component, $L_z = \hbar$ for right-circularly polarized (RCP) light, and $L_z = -\hbar$ for left circularly polarized (LCP) light.

The handedness is determined by the sign of $\sin \delta$. For $\sin \delta > 0$ the sense of rotation is counterclockwise ($0 < \delta < \pi$) and we refer to the light as REP; conversely for $\sin \delta < 0$ the rotation sense is clockwise and one refers to left elliptically polarized (LEP) light. Several examples of phase angles and the corresponding polarization states are illustrated in figures 2.2 and 2.3. For each of these figures, the propagation direction of the field is to be taken as out of the page, i.e. we are looking into the approaching beam of light.

The distinction between right and left circular (elliptical) polarization is found by noting that if the vertical component lags in phase, the E-field vector rotates in a counterclockwise direction, which we have defined to be right elliptical polarization. If the vertical component leads in phase the E-field vector rotates in a clockwise direction, giving left elliptical polarization. Clockwise and counterclockwise refer to the rotation direction as seen by an observer looking into the approaching light wave, i.e. looking along the $-\hat{z}$ direction. All of this, of course, is within the framework of the conventions adapted for this course. Note that many optics texts use a convention which is exactly opposite to ours, while engineering texts tend to use a convention based on the sense of rotation as seen by an observer riding with the wave.

A method to determine with certainty the polarization of a given field is as follows. For RCP light we can begin with equation (1.9), dropping the subscript '0':

$$\vec{E} = \mathfrak{R}\left[\vec{\mathcal{E}}(\vec{k}, \omega)\, e^{i(\vec{k}\cdot\vec{r}-\omega t)}\right] = \mathfrak{R}\left\{\frac{1}{\sqrt{2}}(\hat{x} + i\hat{y})\,|\vec{\mathcal{E}}|\, e^{i(\vec{k}\cdot\vec{r}-\omega t)}\right\},$$

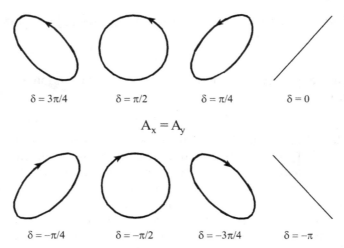

Figure 2.2. Polarization ellipses for different phase angles δ, with field amplitudes $A_x = A_y$. The orientation and sense of rotation correspond to our conventions, as described in the text. We imagine ourselves to be looking along the $-z$-axis, i.e. into an oncoming elliptically polarized plane wave.

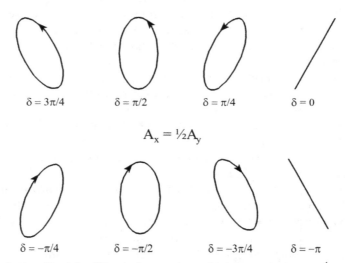

Figure 2.3. Polarization ellipses for different phase angles δ, with field amplitudes $A_x = \frac{1}{2}A_y$. Note that in this case it is not possible to achieve circular polarization.

in which we have defined explicitly the vector nature of the electric field amplitude $\mathcal{E}$ by writing the amplitude in terms of the magnitude $|\vec{\mathcal{E}}|$ and the unit vectors $\hat{x}$ and $\hat{y}$. The factor of $\sqrt{2}$ arises from the normalization of the unit vector terms. Consider an observer standing at $z = 0$ ($\vec{r} = 0$ and therefore $\vec{k} \cdot \vec{r} = 0$) and looking back into the beam of light. Starting at $t = 0$, the electric field points in the $+x$-direction, along $\hat{x}$. A quarter period later, at $t = (2\pi/\omega)/4 = \pi/2\omega$, we have

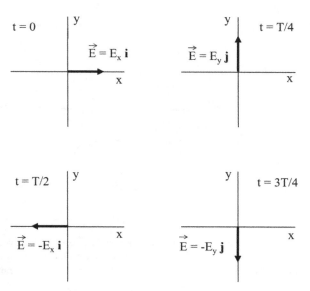

Figure 2.4. The E-vector for RCP shown in the $z = 0$ plane for one full time cycle, in increments of one-quarter of a period. At $t = 0$, A_x is a maximum and A_y is zero. At $t = T/4$, A_x is zero and A_y is a maximum. As time progresses, the E-vector rotates counterclockwise. This is a consequence of the $\pi/2$ phase difference between the x- and y-components of $\vec{E}$.

$$\vec{E} = \Re\left\{\frac{1}{\sqrt{2}}(\hat{x} + i\hat{y})\,|\vec{\mathcal{E}}|\,e^{-i(\omega\cdot\pi/2\omega)}\right\}$$

$$= \Re\left\{\frac{1}{\sqrt{2}}(-i\hat{x} + \hat{y})\,|\vec{\mathcal{E}}|\right\} = \frac{|\vec{\mathcal{E}}|}{\sqrt{2}}\hat{y}.$$

Thus the electric field points along the $+y$-direction. The observer therefore sees an electric field rotating counterclockwise as a function of time; again, this is our definition of RCP light. Resorting to this type of analysis is the key to resolving any confusion which crops up because of notational differences between texts. In figure 2.4 we show an example of the time evolution of the electric field vector for right-circularly polarized light.

2.2.2 Linear or plane polarization

One limiting case of elliptically polarized light occurs for an ellipticity parameter $|\xi| = 0$. Equivalently, we can see that in terms of the phase shift, $\delta = \pm m\pi$. We refer to linear polarized light in this case. In figure 2.5 the contrast between a field of well-defined linear polarization and an unpolarized field is illustrated.

2.2.3 Circular polarization

In equation (2.2), if $A_x = A_y$, then we must have $\phi = \pi/4$, unless $\delta = \pi/2$. In the latter case ϕ is indeterminate, but if we plot the time-variation of the field vector for these

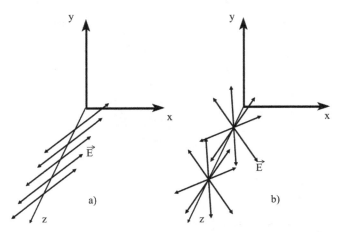

Figure 2.5. (a) Plane polarization; the wave is propagating along the z-axis and the electric field vector has a well-defined direction. (b) Unpolarized wave; the E-vector for the wave propagating in the z-direction lies in the x–y plane but varies rapidly as a function of time. The E-vector appears to have random orientations.

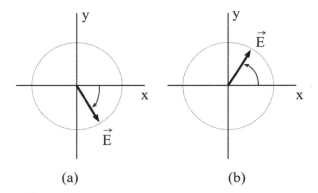

Figure 2.6. (a) Left circular polarization (LCP). As time evolves, the E-vector rotates clockwise on a circle in the ($z = 0$) plane. (b) Right circular polarization (RCP); the E-vector rotates counterclockwise on a circle. In both cases we imagine looking into the source of the light, i.e. the light is propagating along the $+z$-direction.

parameters, we see that $\delta = \pi/2$ is simply the condition that the field traces out a circle in the x–y plane, and therefore the coordinate-axis rotation is meaningless. We refer to this polarization state of the field as *circularly polarized*. In terms of the ellipticity parameter, circular polarization corresponds to $|\xi| = 1$. Pictorially, the time evolution of the electric field vector is shown in figure 2.6 for right-circularly polarized light.

2.2.4 Polarization in the complex plane

Again, the field amplitude $\vec{\mathcal{E}}(\vec{k}, \omega)$ is a complex vector quantity and contains all the polarization information for the light field. Concentrating on light propagating in the z-direction (polarization in the x–y plane), we can relate the complex amplitude to the components A_x and A_y. If we define a complex number χ as

$$\chi \equiv \frac{A_y}{A_x}e^{i\delta} = \tan\Psi\, e^{i\delta}$$

with $\delta = \delta_y - \delta_x$ and $\tan\Psi \equiv \frac{A_y}{A_x}$ for $0 \leqslant \Psi \leqslant \frac{\pi}{2}$, the parameters δ and Ψ will completely define the state of polarization. Although it may seem an additional complication to introduce yet another notation for the polarization, the representation in terms of the 'azimuthal' angle Ψ will be useful in discussions of ellipsometry occurring in chapter 5, and thus we include it at this point for the sake of completeness. The representation in the complex plane is shown in figure 2.7.

Relating these parameters back to the ellipticity ξ and the inclination angle ϕ, we note that

$$|\chi| = \frac{A_y}{A_x}$$

and thus, from equation (2.2),

$$\tan 2\phi = \frac{2\Re\{\chi\}}{1 - |\chi|^2}$$
$$\sin 2\theta = \frac{2\Im\{\chi\}}{1 + |\chi|^2}$$

with θ being the ellipticity angle and $\tan\theta \equiv \xi$.

In this complex-plane representation, linear polarization is represented in the imaginary plane for $\delta = 0$, i.e. along the $\pm x$-axis, as shown in figure 2.7.

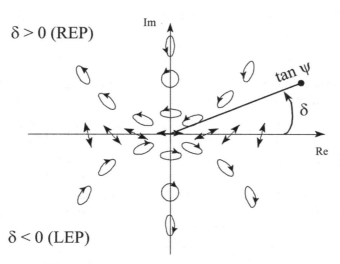

Figure 2.7. Representation of the polarization state in the complex plane. Each point in the plane represents one state of light polarization.

As mentioned above, this corresponds to the field components A_x and A_y being either in phase (+x-axis, $\delta = 0$) or 180° out-of-phase ($-x$-axis, $\delta = \pi$). Note as well that as $\tan \Psi$ becomes very large, corresponding to points far away from the origin in figure 2.7, the field approaches more and more closely a state of linear polarization. Be careful with this notation: $\tan \Psi$ represents the length of the vector in the complex plane. Relating back to the field amplitudes, a large value of $\tan \Psi$ implies that $A_x \to 0$ (or at least that $A_x \ll A_y$) and thus a field polarized in the y-direction.

For circular polarization, as we have seen, the E-vector has orthogonal components, equal in magnitude, $A_x = A_y$, with a phase difference $\delta = \pm \pi/2$. Circularly polarized light can only be represented by a point along the imaginary axis in the complex plane, as shown in figure 2.7. Furthermore, only the point with $\tan \psi = 1$ (again, $A_x = A_y$) is possible as a representation of circularly polarized light.

Now that the general definitions for various polarization states have been specified, we present in the next section a particularly easy method for carrying out calculations involving polarized light.

2.3 Jones vector representation of polarization states

The polarization of a light wave can also be conveniently represented by a 2×1 matrix, called a Jones vector, which expresses the relative amplitude and phase of the two orthogonal components of the E-vector. For example, if the complex amplitude is written as

$$\vec{\mathcal{E}} = \hat{x}\mathcal{E}_x + \hat{y}\mathcal{E}_y \tag{2.3}$$

with

$$\mathcal{E}_x = A_x e^{i\delta_x} \quad \text{and} \quad \mathcal{E}_y = A_y e^{i\delta_y}$$

a corresponding 'Jones vector' can be expressed as

$$\bar{J} \equiv \begin{pmatrix} A_x e^{i\delta_x} \\ A_y e^{i\delta_y} \end{pmatrix}.$$

Returning briefly to the full form of the plane wave electric field, so that we do not lose track of that, we would write

$$\vec{E}\,(\vec{r},\,t) = \mathfrak{R}\left\{ \bar{J}\; e^{i\left(\vec{k}\cdot\vec{r}-\omega t\right)} \right\} = \mathfrak{R}\left\{ \left[\hat{x}\; A_x e^{i\delta_x} + \hat{y}\; A_y e^{i\delta_y} \right] e^{i\left(\vec{k}\cdot\vec{r}-\omega t\right)} \right\}.$$

The important thing to realize about the Jones vectors is that the matrix elements express the relative amplitude and phase of the orthogonal components of the E-vector. By convention, the first element in the vector corresponds to the x-component of the E-field, and the second element gives the y-component, assuming propagation in the z-direction.

The Jones vectors are usually *normalized*, that is,

$$\bar{J}^{\,\dagger} \cdot \bar{J} = \left(\bar{J}^{\,T}\right)^* \cdot \bar{J} = 1,$$

where the † stands for the Hermitian conjugate of the vector, T means the transpose of the vector, and * is the complex conjugation operation.

Example. We can represent a linearly polarized electric field oriented at an angle ϕ by the vector

$$\bar{J}_1 = \begin{pmatrix} \cos\phi \\ \sin\phi \end{pmatrix}.$$

Given another linearly polarized field

$$\bar{J}_2 = \begin{pmatrix} -\sin\phi \\ \cos\phi \end{pmatrix}$$

we see first of all that both represent normalized states, e.g.

$$\left(\bar{J}_1{}^T\right)^* \cdot \bar{J}_1 = (\cos\phi \quad \sin\phi) \cdot \begin{pmatrix} \cos\phi \\ \sin\phi \end{pmatrix}$$
$$= \cos^2\phi + \sin^2\phi = 1.$$

Furthermore, these two states are *orthogonal*, i.e.

$$\left(\bar{J}_1{}^T\right)^* \cdot \bar{J}_2 = (\cos\phi \quad \sin\phi) \cdot \begin{pmatrix} -\sin\phi \\ \cos\phi \end{pmatrix}$$
$$= -\sin\phi\cos\phi + \sin\phi\cos\phi$$
$$= 0.$$

Whenever a pair of states satisfies the orthogonality condition $\left(\bar{J}_1{}^T\right)^* \cdot \bar{J}_2 = 0$, that pair may be used as a *basis set*: any polarization lying in that same plane may be described as a linear combination of the two basis vectors.

Example. For an angle $\phi = \pi/4$, the Jones vector for linearly polarized light is

$$\bar{J}_{\pi/4} = \begin{pmatrix} \cos\phi \\ \sin\phi \end{pmatrix} = \begin{pmatrix} 1/\sqrt{2} \\ 1/\sqrt{2} \end{pmatrix} = \frac{1}{\sqrt{2}} \begin{pmatrix} 1 \\ 1 \end{pmatrix}.$$

This can be described alternatively as a combination of x- and y-polarized light, for which the Jones vectors are given by

$$\hat{e}_x = \begin{pmatrix} 1 \\ 0 \end{pmatrix} \quad \text{and} \quad \hat{e}_y = \begin{pmatrix} 0 \\ 1 \end{pmatrix},$$

respectively. Thus we can write

$$\bar{J}_{\pi/4} = \frac{1}{\sqrt{2}}\hat{e}_x + \frac{1}{\sqrt{2}}\hat{e}_y.$$

A few examples of Jones vectors which will illustrate this point are given below:

$$\begin{pmatrix} 1 \\ 0 \end{pmatrix} \qquad \Longrightarrow \qquad \text{Wave linearly polarized in the } x\text{-direction.}$$

$$\begin{pmatrix} 0 \\ 1 \end{pmatrix} \qquad \Longrightarrow \qquad \text{Wave linearly polarized in the } y\text{-direction.}$$

$$\frac{1}{\sqrt{2}}\begin{pmatrix} 1 \\ 1 \end{pmatrix} \qquad \Longrightarrow \qquad \text{Wave linearly polarized at } 45° \text{ to the } x\text{-axis.}$$

$$\frac{1}{\sqrt{2}}\begin{pmatrix} 1 \\ \pm i \end{pmatrix} \qquad \Longrightarrow \qquad \begin{array}{l} (-) \text{ left-circularly polarized (LCP);} \\ (+) \text{ right-circularly polarized (RCP).} \end{array}$$

We can summarize by giving the Jones vector in slightly more general terms by calling the y-axis the vertical component of the E-vector and the x-axis the horizontal component. Thus the vector can be written schematically as

$$\begin{pmatrix} \text{Horizontal} \\ \text{Vertical} \end{pmatrix}.$$

The most general form for the Jones vector is that for elliptically polarized light. As we have already seen, elliptically polarized light allows for arbitrary relative amplitude of the two field components as well as arbitrary relative phase; the Jones vector can be written in several different forms:

$$\bar{J} = \begin{pmatrix} \cos\phi \\ \sin\phi \, e^{i\delta} \end{pmatrix} = \begin{pmatrix} A_x \\ A_y e^{\pm i\delta} \end{pmatrix} = \begin{pmatrix} A_x \\ b \pm ic \end{pmatrix}, \tag{2.4}$$

where we have written the polar-form complex number $A_y e^{\pm i\delta}$ in the alternative Cartesian form with $A_y = \sqrt{b^2 + c^2}$, and with the relative phase angle between the A_x and A_y components given by

$$\delta = \delta_y - \delta_x = \tan^{-1}\left(\frac{c}{b}\right).$$

The ellipse is oriented with respect to the x-axis such that the angle between the major axis and the x-axis is ϕ as shown earlier.

Example. Given the Jones vector $\begin{pmatrix} 3 \\ 2 + i \end{pmatrix}$ which is in the standard form $\begin{pmatrix} A_x \\ b + ic \end{pmatrix}$, find the relative phase angle between the two field components, along with the ellipse orientation.

Solution. First we can identify the given field as being REP using our sign convention. The relative phase angle is given by

$$\delta = \tan^{-1}\left(\frac{c}{b}\right) = \tan^{-1}\left(\frac{1}{2}\right) = 26.6°.$$

The amplitude A_y is given by $A_y = \sqrt{b^2 + c^2}$. The ellipse orientation is found from

$$\phi = \frac{1}{2}\tan^{-1}\left[\frac{2A_xA_y\cos\delta}{A_x^2 - A_y^2}\right] = \frac{1}{2}\tan^{-1}\left[\frac{2(3)\left(\sqrt{5}\right)\cos\left(26.6°\right)}{9 - 5}\right] = 35.8°.$$

This result is similar to that illustrated in figure 2.1.

2.3.1 Superposition of waves using Jones vectors

The superposition of two waves with a definite state of polarization can be easily represented using Jones vectors. For example, consider the superposition of an LCP wave and an RCP wave:

$$\frac{1}{\sqrt{2}}\begin{pmatrix}1\\-i\end{pmatrix} + \frac{1}{\sqrt{2}}\begin{pmatrix}1\\i\end{pmatrix} = \begin{pmatrix}\sqrt{2}\\0\end{pmatrix} = \sqrt{2}\begin{pmatrix}1\\0\end{pmatrix}.$$

This represents linearly polarized light with an amplitude $\sqrt{2}$ larger than that of the components of the circularly polarized wave, and thus an intensity of twice the initial individual intensities.

2.4 Optical elements and Jones matrices

When a light wave has a definite polarization state so that it can be represented by a Jones vector, the effect on the electric field of various optical elements can be determined by using a 2×2 matrix. In general we can write the matrix operation as

$$\bar{E}_{\text{out}} = \bar{M}\,\bar{E}_{\text{in}}$$
$$\begin{pmatrix}E_x^{\text{out}}\\E_y^{\text{out}}\end{pmatrix} = \begin{bmatrix}m_{xx} & m_{xy}\\m_{yx} & m_{yy}\end{bmatrix}\begin{pmatrix}E_x^{\text{in}}\\E_y^{\text{in}}\end{pmatrix},$$

with, for example, $E_x^{\text{in}} = A_x^{\text{in}}\,e^{i\delta_x^{\text{in}}}$, etc.

Linear polarizers
For example, consider a linear polarizer with transmission axis oriented along the x-axis, which can be represented by

$$P_x = \begin{bmatrix}1 & 0\\0 & 0\end{bmatrix}$$

and a linear polarizer with its transmission axis oriented along the y-axis given by

$$P_y = \begin{bmatrix}0 & 0\\0 & 1\end{bmatrix}.$$

Example. Consider the result of passing light which is linearly polarized along the x-axis through a polarizer which has its transmission axis oriented along the y-axis:

$$\underbrace{\begin{bmatrix} 0 & 0 \\ 0 & 1 \end{bmatrix}}_{\text{Polarizer}} \ \underbrace{\begin{bmatrix} 1 \\ 0 \end{bmatrix}}_{\text{Initial state}} = \underbrace{\begin{bmatrix} 0 \\ 0 \end{bmatrix}}_{\text{Output}}.$$

The physical meaning of the above matrix operation is that the polarization state represented by the vector $\begin{bmatrix} 1 \\ 0 \end{bmatrix}$ is operated on by the optical element, in this case a linear polarizer represented by the matrix $\begin{bmatrix} 0 & 0 \\ 0 & 1 \end{bmatrix}$. The ordering of the matrix multiplication is from right to left, with the state of the input field right-most, followed by matrices representing each successive optical element. For this example, this means that the matrix for the polarizer multiplies (operates on) the vector for the polarization state. The result in this case is, obviously, no transmission, $\begin{bmatrix} 0 \\ 0 \end{bmatrix}$.

For a perfect linear polarizer with transmission axis oriented at some angle Θ with respect to the polarization direction of the incident field, we have

$$P_\Theta = \begin{bmatrix} \cos^2 \Theta & \sin \Theta \cos \Theta \\ \sin \Theta \cos \Theta & \sin^2 \Theta \end{bmatrix}.$$

This reduces to the special cases given above for $\Theta = 0°$ and $\Theta = 90°$. The above polarizers select one polarization and eliminate the orthogonal one.

We will present here without proof the Jones matrices for several other optical elements, using the conventions that were mentioned earlier in the chapter. All of these results can be confirmed by using the matrices to operate on an arbitrary state of the field, and proving that the expected result is obtained.

Partial or imperfect polarizer
A partial polarizer can be represented by

$$P_p = \begin{bmatrix} \alpha & 0 \\ 0 & \beta \end{bmatrix},$$

where $0 \leqslant \alpha \leqslant 1$ and $0 \leqslant \beta \leqslant 1$. A common application for such a Jones matrix would be to analyze a system containing a polarizer which is not quite ideal. For example, if we have a polarizer which nominally passes only x-polarized light, but allows as well a small amount of y-polarized light through, it could be represented by the above matrix, with $\beta \ll \alpha$ and $\alpha = 1$. Another example might be a polarizer that does not pass any of the 'wrong' (e.g. x-) component, but also does not perfectly transmit the 'right' (e.g. y-) component—here we might have $\alpha = 0$ and β close to but less than unity.

Polarization rotator
A *polarization rotator* takes a given linear polarization and rotates it by the angle β according to

$$R(\beta) = \begin{bmatrix} \cos \beta & -\sin \beta \\ \sin \beta & \cos \beta \end{bmatrix}.$$

The difference between these two cases is in the amount of light which is transmitted through the optical element. The polarization rotator $R(\beta)$ transmits all of the incident light, whereas the matrix for P_θ only transmits a portion of the incident field. The following example illustrates this point.

Example. In this two-part example, we consider a linearly polarized field incident first on a polarizer oriented at 30° with respect to the incident polarization and calculate the field and the intensity transmitted through the polarizer. In the second part of the example, we repeat the calculation for light incident on a polarization rotator.

We take the incident field to be polarized in the x-direction, and the polarizer angle to be 30°. To find the transmitted field we write, using the definition of P_Θ

$$P_\Theta J_{in} = \begin{bmatrix} \cos^2 30° & \sin 30° \cos 30° \\ \sin 30° \cos 30° & \sin^2 30° \end{bmatrix} \begin{pmatrix} 1 \\ 0 \end{pmatrix} = \begin{pmatrix} \dfrac{3}{4} \\ \dfrac{\sqrt{3}}{4} \end{pmatrix}. \tag{2.5}$$

The transmitted intensity is $\sqrt{\left(\frac{3}{4}\right)^2 + \left(\frac{\sqrt{3}}{4}\right)^2} = \frac{3}{4}$ in the dimensionless units used here, as compared to the incident intensity of unity.

For the polarization rotator of 30° we can likewise write

$$R(30°) J_{in} = \begin{bmatrix} \cos 30° & -\sin 30° \\ \sin 30° & \cos 30° \end{bmatrix} \begin{pmatrix} 1 \\ 0 \end{pmatrix} = \begin{pmatrix} \dfrac{\sqrt{3}}{2} \\ \dfrac{1}{2} \end{pmatrix} \tag{2.6}$$

and the transmitted intensity is $\sqrt{\left(\frac{\sqrt{3}}{2}\right)^2 + \left(\frac{1}{2}\right)^2} = 1$, and thus the transmitted and incident intensities are identical. Again, the rotator simply changes the orientation of the electric field vector, without any attenuation of the components, whereas the polarizer will transmit only that component of the incident field oriented along its transmission axis, while blocking the orthogonal component.

Phase retarders
We can also have optical elements which introduce only a relative phase shift between different polarization components. In chapter 7 the origin of these phase shifts will be covered in detail. The Jones matrix for the phase retarder is given by

$$W = \begin{bmatrix} e^{i\delta_x} & 0 \\ 0 & e^{i\delta_y} \end{bmatrix}.$$

In figure 2.8 we illustrate the relevant parameters for light propagating along the z-axis through a phase retarder of thickness L. What we mean by 'slow axis' and 'fast axis' will become more clear in chapter 7; suffice it to say now that relative

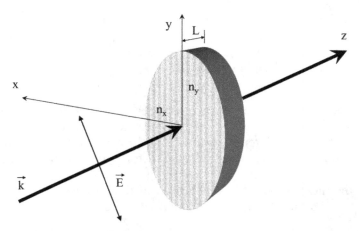

Figure 2.8. A phase retarder of thickness L introduces a relative phase shift between the x- and y-components of an incident electric field due to differing indices of refraction, and thus speeds of light propagation, along those two axes.

retardation occurs because of 'birefringence', or differing indices of refraction for different polarizations. Since the speed of light in a medium depends inversely on the index of refraction ($v = c/n$), different indices imply different phase velocities, one of which is necessarily slower than the other. Thus the 'slow' axis refers to the axis corresponding to a larger index of refraction and 'fast' axis refers to a smaller index.

Quarter-wave plate
Special cases of the above occur for $\delta = \pi/2$, where $\delta = \delta_y - \delta_x$. This corresponds physically to a 'quarter-wave plate' (QWP). (Measured in radians, a full cycle corresponds to 2π; thus a relative phase difference of $\pi/2$ is a quarter of a cycle, or 'wave'.) If $\delta_y - \delta_x = \pi/2$, we refer to this as 'slow-axis vertical' and the Jones matrix is given by

$$W_{\text{QWP}} = \begin{bmatrix} e^{-i\pi/4} & 0 \\ 0 & e^{i\pi/4} \end{bmatrix} = e^{-i\pi/4} \begin{bmatrix} 1 & 0 \\ 0 & i \end{bmatrix}.$$

For a QWP with the slow-axis horizontal the Jones matrix is

$$W_{\text{QWP}} = \begin{bmatrix} e^{i\pi/4} & 0 \\ 0 & e^{-i\pi/4} \end{bmatrix} = e^{i\pi/4} \begin{bmatrix} 1 & 0 \\ 0 & -i \end{bmatrix}.$$

There are several important properties of QWPs:
- Linearly polarized light incident with the E-field direction along one of the principal axes of the QWP undergoes no change in polarization state. There will be an overall phase shift of the field which depends on which axis is involved.
- If the QWP axes are aligned with the principal axes of elliptically polarized light, the transmitted light will be linearly polarized. The orientation of the resulting linearly polarized light depends on the initial ellipticity and ellipse orientation of the input field.

- Linearly polarized light oriented at 45° with respect to the axes of a QWP will emerge circularly polarized. Whether RCP or LCP will depend on the absolute orientation of the QWP.
- Circularly polarized light incident on a QWP will always emerge linearly polarized, with the direction of polarization depending on the orientation of the QWP and on the initial helicity or sense of rotation of the initially circularly polarized light.

Half-wave plate

When $\delta = \pi$ the optical element represented is the 'half-wave plate' and can be represented by

$$W_{\text{HWP}} = \begin{bmatrix} e^{-i\pi/2} & 0 \\ 0 & e^{i\pi/2} \end{bmatrix} = \begin{bmatrix} -i & 0 \\ 0 & i \end{bmatrix} = e^{-i\pi/2} \begin{bmatrix} 1 & 0 \\ 0 & -1 \end{bmatrix}.$$

Again, there are several properties of half-wave plates with which one should become familiar:

- If the HWP is aligned with linearly polarized light, the resultant is linearly polarized light with a net absolute phase shift dependent on whether the input polarization is along the fast or slow axis of the wave plate.
- If the HWP is aligned with the principal axes of incident elliptically polarized light, the emerging light will be elliptically polarized with the opposite helicity (REP→LEP and LEP→REP). The ellipticity and orientation are unchanged.
- If circularly polarized light is incident on a HWP the result is circularly polarized light with the opposite helicity. This is just a special case of the previous property.
- Linearly polarized light incident on a HWP at some azimuthal angle θ will undergo a rotation through an angle 2θ. The direction of rotation depends on the absolute orientation of the slow and fast axes of the HWP.

In figure 2.9 the rotation of input linear polarization by an angle 2θ is illustrated. This is a very common application of a half-wave plate. Consider in particular the case $\theta = 45°$, for which the rotation angle is then 90°. The HWP thus provides a simple way to rotate, e.g. linearly polarized light oscillating horizontally to vertically polarized.

Example. Let linearly polarized light oriented at 45° pass through a $\lambda/4$-plate, which has its 'slow-axis vertical'. What is the resultant output light polarization state?

Solution. We can perform the following matrix multiplication:

$$\frac{1}{\sqrt{2}} e^{-i\pi/4} \begin{pmatrix} 1 & 0 \\ 0 & i \end{pmatrix} \begin{pmatrix} 1 \\ 1 \end{pmatrix} = \frac{1}{\sqrt{2}} e^{-i\pi/4} \begin{pmatrix} 1 \\ i \end{pmatrix}.$$

The output light is thus RCP, with an overall phase shift. Note as well that the amplitude of the input and output fields are the same in this case; i.e. that the QWP has not attenuated the field in any way, but only introduced a relative phase shift between two mutually orthogonal field components.

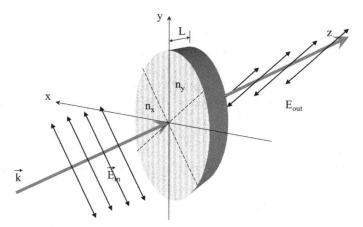

Figure 2.9. Illustration of the action of a half-wave plate on incident linearly polarized light. By convention one refers the rotation to the slow axis of the HWP. Recall that the polarization vector is 'double-headed', i.e. the electric field is oscillating back and forth in time. Thus one could refer the rotation angle to the fast axis and arrive at the same result.

Example. Starting with linearly polarized light oriented at an angle of 30° with respect to the *x*-axis, we send it through a QWP oriented with the slow-axis horizontal. What is the polarization state of the light emerging from the QWP?

Solution. The initial state of light is found by taking the components with respect to the *x*- and *y*-axes. This leads to

$$\begin{pmatrix} \cos 30° \\ \sin 30° \end{pmatrix}.$$

(Alternatively we could have used the polarization rotator Jones matrix, $R(\beta)$) operating on *x*-polarized light to come up with the same result.) Sending the light through the QWP as described is represented by

$$\begin{pmatrix} E_{x,\text{ out}} \\ E_{y,\text{ out}} \end{pmatrix} = e^{i\pi/4} \begin{pmatrix} 1 & 0 \\ 0 & -i \end{pmatrix} \begin{pmatrix} \cos 30° \\ \sin 30° \end{pmatrix} = e^{i\pi/4} \begin{pmatrix} 0.866 \\ -0.5\,i \end{pmatrix}$$

$$= 0.866 e^{i\pi/4} \begin{pmatrix} 1 \\ -0.577\,i \end{pmatrix}.$$

Thus the emerging light is LEP.

One point should be made clear here. The matrices corresponding to optical elements such as wave plates are defined with reference to a specific coordinate system, intrinsic to that element. Thus, 'fast' and 'slow' axes are built into the wave plates. The lab coordinate system may be another one, referenced for example to the optical table, with polarizations referred to as 'parallel' and 'perpendicular' to that reference frame. To treat the effect of a wave plate on an electric field with lab-referenced coordinates, it is convenient to think of a three step process. First, the initial electric field is rotated to a new coordinate system, that of the optical element. The coordinate systems are illustrated in figure 2.10. The components of some vector

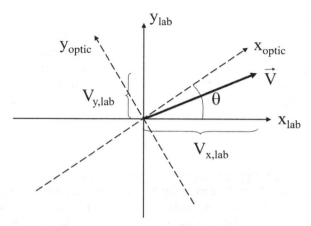

Figure 2.10. A vector defined with respect to a lab-fixed coordinate system can be transformed into the optical-element-fixed coordinates using this sketch.

$\vec{V}$ in the laboratory system (where they are given by $V_{x,\,lab}$ and $V_{y,\,lab}$) can be transformed to the new coordinate system using the rotation matrix given by

$$R(\theta) = \begin{pmatrix} \cos\theta & \sin\theta \\ -\sin\theta & \cos\theta \end{pmatrix}.$$

The second stage of the mathematical process is to operate on the resulting electric field with the Jones matrix for the optical element, in its own coordinate system. Finally, it is necessary to rotate back to the lab system using the rotation matrix $R(-\theta)$.

A thought experiment. Given the following series of 'operations' on an initial light field, you should be able to give the resulting output state of light *without writing down any matrices at all.* You can check your work by going through the calculation.
1. Initial state of light: linearly polarized in the *x*-direction
2. Passes through a linearly polarizer with its transmission axis at 45°.
3. Next comes a QWP with SA horizontal.
4. Now a linear polarizer with TA horizontal.
5. Next a HWP with FA horizontal.
6. Finally, a linear polarizer with TA vertical.

What is your result?

2.5 Longitudinal field components

As an example illustrating various concepts we have covered in this chapter, we will now consider in some detail a problem that is not usually included in laser physics and electro-optics texts. We have already mentioned the fact that any linear superposition of solutions to the wave equation is itself a solution to the wave equation. By superposing plane waves of various frequencies we were able to create a temporal pulse of light. Now we would like to use the superposition principle to 'construct' a Gaussian beam out of plane waves, and to show in the process that any Gaussian beam must have a longitudinal component of the electric field, i.e. the field is not purely transverse. References [6–8] include various pieces of the following treatment, with a great deal more additional detail that need not concern us here. We begin by writing a solution to the wave equation with the electric field oriented (polarized) in the y-direction, predominantly,

$$\vec{E} = \left(\hat{x} \times \hat{k} \right) E_0 e^{i\left(\vec{k} \cdot \vec{r} - \omega t \right)}.$$

In what follows we will suppress the time dependence, as it will be of no consequence to our spatial superposition. The propagation direction is roughly in the z-direction, however, we will consider the beam to be made up of a bundle of rays propagating at various angles α with respect to the z-axis. We may write the propagation unit vector in terms of the α and β, with β being the azimuthal angle in the x–y plane,

$$\hat{k} = \hat{x} \sin \alpha \cos \beta + \hat{y} \sin \alpha \sin \beta + \hat{z} \cos \alpha.$$

In figure 2.11 we show such a superposition in the y–z plane. The beam is assumed to be azimuthally symmetric.

Before moving on with the calculations, it is possible to look at the sketch in figure 2.11 and deduce some properties of the solution we will find. First, we can see that there will be a longitudinal component in the field that will vary radially from the propagation axis. On the axis the longitudinal component disappears, and furthermore, we see that the sign of the longitudinal component changes as one passes through $r = 0$. Finally, we can predict as well that the longitudinal component of the Gaussian field will be relatively small, except for the case of a tightly focused beam.

Our goal is to come up with a superposition, weighting the individual components of the superposition such that the end result is equivalent to the Gaussian beam solution we found in chapter 1. The summation in the superposition will becomes an

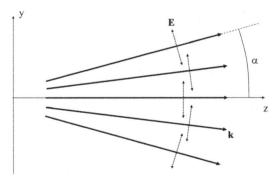

Figure 2.11. Superposing plane waves of different k-vectors, it is possible to create a representation for a Gaussian beam. Here we show a two-dimensional view of a beam propagating along the z-direction and polarized predominantly in the y-direction. We see from the sketch that there must also be a small longitudinal component of the field in a Gaussian beam.

integration over the azimuthal angle β and the spherical polar angle α, where the latter is restricted to angles between 0 and $\pi/2$ to include only rays propagating in the forward direction. We can write the 'summation' (i.e. integration) as

$$\vec{E} = \int_0^{2\pi} d\beta \int_0^{\pi/2} \sin \alpha d\alpha (\hat{x} \times \hat{k}) e^{i\vec{k}\cdot\vec{r}} f(\alpha, \beta), \qquad (2.7)$$

where $f(\alpha, \beta)$ is the weighting function for various plane wave directions. We will show that the field satisfies Maxwell's equation $\nabla \cdot \vec{E} = 0$ and, most importantly, has a longitudinal component.

The weighting function is chosen to give the correct final result, i.e. the standard Gaussian beam solution to the wave equation, and takes the form

$$f(\alpha, \beta) = C \exp\left[-\frac{1}{4}k^2w^2 \sin^2 \alpha\right], \qquad (2.8)$$

where the constant $C = E_0 \frac{\pi w^2 n^2}{\lambda^2}$. We can find the various components of the field by taking projections of equation (2.7) onto the coordinate axes,

$$\hat{x} \cdot \vec{E} \propto \hat{x} \cdot (\hat{x} \times \hat{k}) = 0$$
$$\hat{y} \cdot \vec{E} \propto \hat{y} \cdot (\hat{x} \times \hat{k}) = -\hat{z} \cdot \hat{k} = -\cos \alpha$$
$$\hat{z} \cdot \vec{E} \propto \hat{z} \cdot (\hat{x} \times \hat{k}) = \hat{y} \cdot \hat{k} = \sin \alpha \sin \beta.$$

Now we can write the transverse component of the field in the $\hat{y}$-direction as

$$E_y = - C \int_0^{2\pi} d\beta \int_0^{\pi/2} d\alpha \sin \alpha \cos \alpha \exp\left[-\frac{1}{4}k^2w^2 \sin^2 \alpha\right]$$
$$\exp\left[ik(x \sin \alpha \cos \beta + y \sin \alpha \sin \beta + z \cos \alpha)\right],$$

where we have used

$$\hat{k} \cdot \vec{r} = x \sin \alpha \cos \beta + y \sin \alpha \sin \beta + z \cos \alpha.$$

The coordinates (x, y, z) can be written in terms of circular cylindrical coordinates as $x = R \cos \phi$, $y = R \sin \phi$, with $R = \sqrt{x^2 + y^2}$. In addition, we rename variables, setting $\sin \alpha = \rho$, thereby making ρ the variable describing the divergence of the plane waves with respect to the z-axis. Making these substitutions and using the trigonometric relation $\cos \phi \cos \beta + \sin \phi \sin \beta = \cos(\beta - \phi)$, we arrive at

$$E_y = - C \int_0^{2\pi} d\beta \int_0^1 \rho \, d\rho e^{\frac{-k^2w^2}{4}\rho^2} e^{ikR\rho \cos(\beta-\phi)+ikz\sqrt{1-\rho^2}}$$
$$= - E_0 \frac{2\pi^2 w^2 n^2}{\lambda^2} \int_0^1 \rho \, d\rho e^{\frac{-k^2w^2}{4}\rho^2} J_0(k\rho R) e^{ikz\sqrt{1-\rho^2}}. \qquad (2.9)$$

To arrive at equation (2.9) we have used the relation [6, 9] for the integral representation of the nth order Bessel function

$$J_n(x) = \frac{i^{-n}}{2\pi} \int e^{ix\cos\varphi + in\varphi} d\varphi, \tag{2.10}$$

where the integral is to be taken over any interval of 2π. Recalling that the variable ρ is related to the angular divergence of the plane wave components, and making the assumption that we are dealing with a paraxial wave, the fact that therefore only the region $\rho \ll 1$ contributes significantly to the integral allows us to extend the integration limit to ∞ and use the relation

$$\int_0^\infty dx\, x^{\nu+1} e^{-ax^2} J_\nu(bx) = \frac{b^\nu}{(2a)^{\nu+1}} e^{-\frac{b^2}{4a}} \tag{2.11}$$

to find

$$E_y = -E_0 \frac{1}{1 + i\frac{z}{z_0}} \exp\left[-\frac{x^2 + y^2}{w^2\left(1 + i\frac{z}{z_0}\right)} \right] e^{ikz}. \tag{2.12}$$

This result is equivalent to the form for the Gaussian beam used in chapter 1.

It remains to take a closer look at the z-component of the field. We return to equation (2.7) and use the projection from equation (2.9) to arrive at

$$E_z = C \int_0^{2\pi} d\beta \int_0^{\pi/2} d\alpha\, \sin^2\alpha \sin\beta \exp\left[-\frac{1}{4} k^2 w^2 \sin^2\alpha \right]$$
$$\exp\left[ik(x \sin\alpha \cos\beta + y \sin\alpha \sin\beta + z \cos\alpha) \right].$$

Using the same coordinate changes and variable definitions as previously we can write

$$E_z = C \int_0^{2\pi} d\beta \sin\beta \int_0^1 d\rho \frac{\rho^2}{\sqrt{1 - \rho^2}} e^{-k^2 w^2 \rho^2} \tag{2.13}$$

$$\times\, e^{ikR\rho\cos(\beta - \phi)} e^{ikz\sqrt{1-\rho^2}}. \tag{2.14}$$

The integral over β can be evaluated using the Bessel function expression, equation (2.10) above, now with $n = 1$, along with equation (2.11). (We have omitted some steps; it is necessary to write $\sin\beta = (e^{i\beta} - e^{-i\beta})/2i$ and carry out two integrations. In addition, the integration variable will be $\beta' = \beta - \phi$, which gives rise to an extra factor of $\sin\phi$ in the final result.) The result is an integral over the variable ρ, similar to the expression for the y-component,

$$E_z = 2\pi C i \frac{y}{R} \int_0^1 d\rho \frac{\rho^2}{\sqrt{1 - \rho^2}} e^{-k^2 w^2 \rho^2} e^{ikz\sqrt{1-\rho^2}} J_1(kR\rho). \tag{2.15}$$

This integral can be evaluated using equation (2.11) with index $\nu = 1$. We do not need to carry out all the algebra for our desired result, however. Note that the only difference between the expressions for the y- and z-components ($\nu = 0$ and $\nu = 1$, respectively, in equation (2.11)) can be reduced to the following ratio,

$$\frac{E_z}{E_y} = i\frac{y}{R}\frac{b/4a^2}{1/2a} = i\frac{yb}{2aR} = \frac{iy/z_0}{1 + iz/z_0}. \tag{2.16}$$

The maximum longitudinal component occurs near the focus (actually, at $z = 0$) and for a y distance equal to $\pm\sqrt{2}\,w/2$. Using these two pieces, we find the ratio can be simplified to

$$\frac{E_z}{E_y} = -i\frac{\sqrt{2}}{kw}.$$

For a beam with waist $w = 100\ \mu$m at $\lambda = 633$ nm we find $\left|\frac{E_z}{E_y}\right| = 1.4 \times 10^{-3}$. Longer wavelengths and tighter focusing lead to a relatively larger longitudinal electric field component.

This example, while somewhat complicated mathematically, serves to illustrate several points we have made in this chapter, as well as to re-visit the Gaussian beam wave equation solution. Although we will be dealing with plane waves in this text, it is useful to keep in mind that what we become accustomed to with plane waves is a useful approximation and nearly always adequate, but that it does not represent the whole truth about waves and polarization.

2.6 Problems

1. A linearly polarized plane electromagnetic wave is traveling in the positive z-direction in free space and has its plane of polarization in the x–y plane. Its frequency is 10 MHz and its amplitude is 1.0 V m^{-1},
    (a) find the period and wavelength of the wave,
    (b) write an expression for E and B,
    (c) find the irradiance of the wave.
2. Derive the result for the polarization ellipse equation, equation (2.1),

$$\left(\frac{E_x}{A_x}\right)^2 + \left(\frac{E_y}{A_y}\right)^2 - 2\frac{E_xE_y}{A_xA_y}\cos\delta = \sin^2\delta. \tag{2.17}$$

3. Derive the result for the ellipse angle, equation (2.2),

$$\tan 2\phi = \frac{2A_xA_y}{A_x^2 - A_y^2}\cos\delta. \tag{2.18}$$

4. A linearly polarized wave with an amplitude of 1 V m^{-1} is propagating along a line in the x–y plane at an angle of 45° to the y-axis, with the x–y plane being the plane of polarization. The wavelength is $\lambda = 1.06\ \mu$m. Write an expression for the wave and find the irradiance.

5. For the electric field given by

$$\vec{\mathcal{E}} = \hat{x}(3 + i5) \text{ V cm}^{-1}.$$

Determine the magnitude of $\vec{\mathcal{E}}$, i.e. $|\vec{\mathcal{E}}|$ and the phase ϕ_E.

6. Give the general expressions for the electric fields of the following waves in free space:
    (a) A linearly polarized wave traveling in the x-direction with the electric vector making an angle of 30° with the y-axis.
    (b) A right-elliptically polarized wave traveling in the y-direction. The major axis of the ellipse is in the z-direction and is three times the minor axis.
    (c) A linear polarized wave traveling in the x–y plane in a direction making an angle of 45° with the x-axis. The direction of polarization is in the z-direction.

7. Obtain the Jones vectors for the previous problem.

8. Determine the polarization states represented by the following Jones vectors.

    (a) $\begin{pmatrix} 1 + i \\ 1 - i \end{pmatrix}$ (b) $\begin{pmatrix} 1 - i \\ 1 + i \end{pmatrix}$ (c) $\begin{pmatrix} 2 \\ -i \end{pmatrix}$ (d) $\begin{pmatrix} 1 \\ 1 - i \end{pmatrix}$.

9. The electric field vector of a certain wave is given by the real expression

$$\vec{E} = E_0[\hat{x} \cos(kz - \omega t) + \hat{y} \cos(kz - \omega t + \phi)].$$

    (a) Find the equivalent complex expression.
    (b) What is the equivalent Jones vector for this wave?

10. Explain the phase selection in LCP and RCP polarization, that is, discuss why we write the Jones vectors for these polarization states as $\begin{pmatrix} 1 \\ -i \end{pmatrix}$ and $\begin{pmatrix} 1 \\ i \end{pmatrix}$.

11. Linear polarized light $\begin{pmatrix} 1 \\ 0 \end{pmatrix}$ is sent through two polarizers. The first one is oriented with its transmission axis at 45° to the horizontal axis and the second one is oriented with its transmission axis along the vertical. Find the polarization state of the emergent light.

12. Consider a QWP with the fast axis oriented at 30° with respect to the x-axis. Linear y-polarized light is incident on the QWP. What is the polarization state of the emerging light?

13. Show that circularly polarized light emerges from a QWP linearly polarized. Treat the case of arbitrary orientation of the QWP and find the dependence of the emerging polarization on the orientation of the QWP.

14. Show that a HWP aligned with the principal axes of incident elliptically polarized incident causes a reversal in helicity.

References

[1] Wikipedia: Étienne-Louis Malus https://en.wikipedia.org/wiki/%C3%89tienne-Louis_Malus (Accessed: 11 Jan. 2021)

[2] Wikipedia: Christiaan Huygens https://en.wikipedia.org/wiki/Christiaan_Huygens (Accessed: 11 Jan. 2021)

[3] Wikipedia: Thomas Young (scientist) https://en.wikipedia.org/wiki/Thomas_Young_(scientist) (Accessed: 11 Jan. 2021)

[4] Wikipedia: Augustin-Jean Fresnel https://en.wikipedia.org/wiki/Augustin-Jean_Fresnel (Accessed: 11 Jan. 2021)

[5] Wikipedia: François Arago https://en.wikipedia.org/wiki/Fran%C3%A7ois_Arago

[6] Stratton J A 1941 *Electromagnetic Theory* 1st edn (New York: McGraw-Hill)

[7] Carter W H 1972 Electromagnetic field of a Gaussian beam with an elliptical cross section *J. Opt. Soc. Am.* **62** 1195–201

[8] Carter W H 1974 Electromagnetic beam fields *Opt. Acta* **21** 871–92

[9] Wolfram Mathworld http://mathworld.wolfram.com (Accessed: 11 Jan. 2021)

IOP Publishing

Optical Radiation and Matter

Robert J Brecha and J Michael O'Hare

Chapter 3

Radiation and scattering

In chapter 1 we examined the properties of a propagating electromagnetic field, assuming that the fields are solutions to Maxwell's equations. In effect, we asked no questions at all about the source of such fields. Now we turn our attention to that subject by examining the origin of the propagating electromagnetic fields, and take a look at some first, relatively simple, examples of the interaction of light and matter. We will first introduce the electromagnetic vector potential and use this as a calculational tool to find the electric and magnetic fields emitted by an oscillating dipole. Next we consider the result of incident radiation scattering from a dipole, including the phenomena of Rayleigh scattering. Finally, a first model of continuous matter will be presented in the form of a collection of microscopic dipoles.

3.1 Historical introduction

We have already described in the introduction to chapter 2 the connections made between the propagation of electric and magnetic 'disturbances' and the speed of light propagation, as discovered by Maxwell theoretically. In the 1880s, Heinrich Hertz (1857–1894) carried out a series of experiments in which he was able to measure the speed of propagation of waves produced by an electric spark [1]. In addition to the speed of propagation of these waves, Hertz was also able to measure their wavelengths, confirm the laws of reflection and refraction, and characterize the transverse nature of the waves. The combination of these discoveries was a clear demonstration that electric vibrations and light obey the same basic laws, a strikingly direct confirmation of Maxwell's theory.

Before Hertz had carried out his experiments, Lord Rayleigh (1842–1919) had provided an explanation for many common optical phenomena involving scattering of light from small (molecular-sized) objects [2]. Amongst the most important of the results he found was the description for the blue sky and the redness of the setting sun, which are both fundamentally related to the fact that light of shorter wavelengths is highly more likely to scatter from small objects than is long

doi:10.1088/978-0-7503-2624-7ch3

wavelength light. John Tyndall (1820–1893), who was a colleague of Faraday, was the first to verify these predictions experimentally [3]. Rayleigh won an early Nobel Prize (1904) for his work on scattering theory. We will be using some of the basic concepts from the theory of dipole radiation to arrive at the basic results found by Rayleigh.

3.2 Summary of Maxwell's equations

As a starting point we will consider the role of the charge and current densities, ρ and $\vec{j}$, respectively, in Maxwell's equations. The resulting form of Maxwell's equations, when combined with the Lorentz force equation and Newton's second law of motion, provide a complete description of interacting electromagnetic fields and microscopic charged particles, at least within a classical physics framework. The microscopic Maxwell equations when charges and currents are present are given by

$$\nabla \cdot \vec{E} = \rho/\varepsilon_0 \tag{3.1}$$

$$\nabla \cdot \vec{B} = 0 \tag{3.2}$$

$$\nabla \times \vec{E} = -\frac{\partial \vec{B}}{\partial t} \tag{3.3}$$

$$\nabla \times \vec{B} = \mu_0 \vec{j} + \frac{1}{c^2}\frac{\partial \vec{E}}{\partial t}. \tag{3.4}$$

The constants ε_0, μ_0, and c are the permittivity of free space, the permeability of free space, and the speed of light, respectively.

As mentioned above, in addition to the Maxwell equations, one additional equation is needed to make the formalism of microscopic electromagnetism complete. The equation for the Lorentz force, which describes the forces which act on microscopic charges is

$$\vec{F} = q\vec{E} + q\vec{v} \times \vec{B},$$

where $\vec{v}$ is the velocity of the particle with charge q.

These equations form our starting point; the goal is to now look for solutions to Maxwell's equations in the presence of currents and charges—matter.

3.3 Potential theory and the radiating EM field

Often the determination of electric and magnetic fields can be simplified through the use of potentials. In this section we will review the properties of the classical scalar and vector potentials, designated as Φ and $\vec{A}$, respectively. These potentials can also be used to express the Maxwell equations in a somewhat more lucid form which displays the nature of the fields.

The *vector potential* is introduced through the magnetic induction according to the following equation,

$$\vec{B} = \nabla \times \vec{A}. \tag{3.5}$$

We see that the second Maxwell equation ($\nabla \cdot \vec{B} = 0$) is satisfied identically if $\vec{B}$ has the form of equation (3.5). This follows from the fact that the divergence of the curl of any vector is identically zero. We can use this result in the third Maxwell equation:

$$\nabla \times \vec{E} = -\frac{\partial \vec{B}}{\partial t}.$$

Substitution gives

$$\nabla \times \vec{E} = -\frac{\partial}{\partial t}(\nabla \times \vec{A}) = -\nabla \times \frac{\partial \vec{A}}{\partial t}$$

or

$$\nabla \times \left(\vec{E} + \frac{\partial \vec{A}}{\partial t} \right) = 0.$$

Recalling another vector identity,

$$\nabla \times (\nabla \Phi) = 0,$$

where Φ is a scalar function, the third Maxwell equation is identically satisfied if

$$\vec{E} = -\nabla \Phi - \frac{\partial \vec{A}}{\partial t}. \tag{3.6}$$

The scalar potential Φ and the vector potential $\vec{A}$ are ambiguously defined, i.e. if we know $\vec{A}$ we can calculate $\vec{B}$ from equation (3.5); the converse is not necessarily true, however. This can be shown by the following line of reasoning. Since $\nabla \times (\nabla \Phi) = 0$ for any scalar function, we could write

$$\Phi = \Phi' + \frac{\partial \Psi}{\partial t}, \tag{3.7}$$

where Ψ is some arbitrary scalar function.
    Similarly for $\vec{A}$, we can write

$$\vec{A} = \vec{A}' - \nabla \Psi \tag{3.8}$$

with Ψ being the same scalar function as in equation (3.7).
    Using these last two relations in equation (3.6) we easily find that

$$\vec{E} = -\nabla \Phi' - \frac{\partial \vec{A}'}{\partial t}.$$

That is to say, the new 'primed' potentials satisfy the same relation as the original potentials. The question remains as to how we should choose the scalar function Ψ.

The choice made depends somewhat on what problems are to be solved, but in general there are two common choices, or 'gauge transformations'.

(a) The Coulomb gauge for which we require $\nabla \cdot \vec{A} = 0$.

(b) The Lorentz gauge for which

$$\nabla \cdot \vec{A} = -\frac{1}{c^2}\frac{\partial \Phi}{\partial t}.$$

Essentially we are able to make these choices because a vector is not fully specified by placing a requirement on the curl of the vector (here $\nabla \times \vec{A} = \vec{B}$); in addition it is necessary to specify the divergence, which is what we have done in the two above relations. In other words, specifying both the curl and the divergence of a vector fully determines that vector. We will be working exclusively in the Lorentz gauge; a parallel development in the Coulomb gauge is left for the final section of the chapter.

Taking the divergence of $\vec{A}$ in equation (3.8) and adding $\frac{1}{c^2}\frac{\partial}{\partial t}\Phi$ from equation (3.7) yields

$$\nabla \cdot A + \frac{1}{c^2}\frac{\partial \Phi}{\partial t} = \nabla \cdot \vec{A}' - \nabla^2 \Psi + \frac{1}{c^2}\frac{\partial \Phi'}{\partial t} + \frac{1}{c^2}\frac{\partial^2 \Psi}{\partial t^2}. \qquad (3.9)$$

For this relation to hold true we are required to have a function Ψ such that

$$\nabla^2 \Psi - \frac{1}{c^2}\frac{\partial^2 \Psi}{\partial t^2} = 0,$$

provided we already know that the vector potential $\vec{A}$ and scalar potential Φ are valid solutions.

What does this gauge transformation do for us? We will show now that this choice for the function Ψ leads us to separated wave equations for the two potentials. We start by using the first Maxwell equation, $\nabla \cdot \vec{E} = \rho/\varepsilon_0$. Substituting equation (3.6) we have

$$-\nabla^2 \Phi - \frac{\partial}{\partial t}(\nabla \cdot \vec{A}) = \frac{\rho}{\varepsilon_0}$$

or

$$\nabla^2 \Phi - \frac{1}{c^2}\frac{\partial^2 \Phi}{\partial t^2} = -\frac{\rho}{\varepsilon_0}. \qquad (3.10)$$

This is our first wave equation, involving Φ only. The second wave equation is given by (the proof is left to a homework problem)

$$\nabla^2 \vec{A} - \frac{1}{c^2}\frac{\partial^2 \vec{A}}{\partial t^2} = -\mu_0 \vec{j}. \qquad (3.11)$$

If there are no charges or currents, these equations reduce to

$$\nabla^2 \Phi - \frac{1}{c^2}\frac{\partial^2 \Phi}{\partial t^2} = 0$$

$$\nabla^2 \vec{A} - \frac{1}{c^2}\frac{\partial^2 \vec{A}}{\partial t^2} = 0.$$

Both of these equations have the same form as our familiar wave equation for the electric field, as seen in chapter 1.

On the other hand, for charges and currents that are not time varying we find

$$\nabla^2 \Phi = -\frac{\rho}{\varepsilon_0}$$

$$\nabla^2 \vec{A} = -\mu_0 \vec{j}.$$

The first of these equations, Poisson's equation, is very familiar from electrostatics.

The general solutions to equations (3.10) and (3.11) can be found in any electricity and magnetism textbook and will simply be stated here:

$$\Phi(\vec{r}, t) = \frac{1}{4\pi\varepsilon_0} \int_V \frac{\rho(\vec{r}', t')}{|\vec{r} - \vec{r}'|} d^3 r' \qquad (3.12)$$

$$\vec{A}(\vec{r}, t) = \frac{\mu_0}{4\pi} \int_V \frac{\vec{j}(\vec{r}', t')}{|\vec{r} - \vec{r}'|} d^3 r'. \qquad (3.13)$$

The notation used here is such that the primed coordinates refer to the source charge distribution and the unprimed coordinates refer to the observation point, as illustrated in figure 3.1.

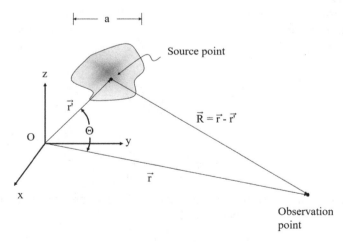

Figure 3.1. The coordinates used to describe far-field radiation from a current source of dimension a. Primed coordinates always refer to points within the source itself, while unprimed coordinates refer to the observation point.

There is one subtlety that must be considered at this point. In the above solutions for the scalar and vector potentials, the primed time variable, t', refers to the time dependence of the motion of charges at the source. However, any effect observed at the point $\vec{r}$ takes a certain amount of time to propagate from $\vec{r}'$ to $\vec{r}$; again, figure 3.1 illustrates these coordinate definitions. Since the propagation (in free space) happens at the speed of light, c, we can relate the time variables t and t' by

$$t - t' = \frac{|\vec{r} - \vec{r}'|}{c} \text{ or } t' = t - \frac{|\vec{r} - \vec{r}'|}{c}.$$

The time t' is called the *retarded* time and we will use the following simplified notation in the calculation of the fields

$$\vec{j}(\vec{r}', t') \rightarrow [\vec{j}]$$
$$\rho(\vec{r}', t') \rightarrow [\rho]$$
$$|\vec{r} - \vec{r}'| \rightarrow R.$$

3.4 Radiation from a dipole

Much of the theory of propagation of radiation through matter can be treated by using a classical physics model of radiation from an oscillating dipole. Although a quantum-mechanical description of the radiation and matter interaction is more fundamentally correct, we will demonstrate throughout this book that many qualitative properties of the interaction of radiation and matter are adequately described using the much simpler classical models, for which we can hope to gain a good intuitive picture. To begin with, we will consider the potentials, fields, and radiation from a localized system of charges and currents which have a sinusoidal time dependence according to

$$\rho(\vec{r}', t') = \rho(\vec{r}')e^{-i\omega t'} \tag{3.14}$$

$$\vec{j}(\vec{r}', t') = \vec{j}(\vec{r}')e^{-i\omega t'}. \tag{3.15}$$

Using the above relations for the current density and $t' = t - R/c$, where $R = |\vec{r} - \vec{r}'|$, the expression for the vector potential, equation (3.13), becomes

$$\vec{A}(\vec{r}, t) = \frac{\mu_0}{4\pi}e^{-i\omega t} \int d^3r' \frac{\vec{j}(\vec{r}')}{R}e^{ikR}, \tag{3.16}$$

where we have used $k = \omega/c$.

Given a current distribution $\vec{j}(\vec{r}')$, the fields can, in principle at least, be determined by calculating the integral in equation (3.16). We will examine the fields in the long wavelength limit for which the source is confined to a very small region compared to the wavelength, λ, of the light. If we let the source dimensions be designated by a, and employ the so-called far-field approximation which applies when $a \ll \lambda \ll r$ (see figure 3.1), we will see that for the far-field, in addition to being transverse to the radius vector $\vec{r}$, the fields fall off as r^{-1}. This is typical of radiation fields.

In the far-field (since $\vec{r} \gg \vec{r}'$), we can make the approximation $R = |\vec{r} - \vec{r}'| = (r^2 + r'^2 - 2rr'\cos\Theta)^{1/2} \simeq r\left(1 - 2\frac{r'}{r}\cos\Theta\right)^{1/2}$. Here we have dropped terms of order r'^2/r^2. We can now write, using a power series expansion,

$$\frac{1}{R} \simeq \frac{1}{r}\left(1 + \frac{r'}{r}\cos\Theta\right). \tag{3.17}$$

Likewise we can expand the exponential term

$$e^{ikR} \simeq e^{ikr\left(1 - 2\frac{r'}{r}\cos\Theta\right)^{1/2}} \simeq e^{ikr}e^{-ikr'\cos\Theta} \tag{3.18}$$

$$\simeq e^{ikr}[1 - ikr'\cos\Theta]. \tag{3.19}$$

We have used the 'large wavelength' approximation in the last step; since the second exponential term can be expanded only if $kr' \ll 1$, this implies $2\pi a/\lambda \ll 1$, where a is the maximum value r' can take. Combining these two results we have

$$\frac{e^{ikR}}{R} = \frac{e^{ikr}}{r}\left(1 + \frac{r'}{r}\cos\Theta\right)[1 - ikr'\cos\Theta]$$

$$\simeq \frac{e^{ikr}}{r} + \frac{e^{ikr}}{r^2}[r'\cos\Theta - ikrr'\cos\Theta] \tag{3.20}$$

$$= \frac{e^{ikr}}{r} + \frac{e^{ikr}}{r^2}[(1 - ikr)(\hat{r} \cdot \vec{r}')],$$

where in the last step $\hat{r}$ is a unit vector in the $\vec{r}$ (observation) direction. The expression for the vector potential becomes,

$$\vec{A}(\vec{r}, t) = \frac{\mu_0}{4\pi}\frac{e^{i(kr-\omega t)}}{r}\left[\int \vec{j}(\vec{r}')d^3r' + \frac{(1 - ikr)}{r}\int \vec{j}(\vec{r}')(\hat{r} \cdot \vec{r}')d^3r'\right]. \tag{3.21}$$

The first term in square brackets describes *electric dipole* radiation, as we will see shortly. The second term in brackets can be shown to give contributions to the radiation field due to *magnetic dipoles* and *electric quadrupoles*, which we will not consider in this course [4]. Remember that these last results, given in equation (3.21) are valid only in the limit in which $r \gg a$ *and* for which $\lambda \gg a$ as well. We will need to come back to look at these results more carefully when we consider the problem of radiation from an antenna, for which the extent of the source is of the same order of magnitude as the wavelength of emitted radiation. That is, for a radiating antenna with dimensions comparable to or smaller than the wavelength of the radiation we will be able to use the approximation in equation (3.18) for the exponential term, but not that of equation (3.19).

Before working a couple of example problems we will recast the above result for electric dipole radiation in another form which will be useful for some cases.

Equation (3.21) can be directly applied when the current distribution is given. Often, however, one is given the time dependent-dipole moment, $\vec{\wp}$ of the source charge distribution instead of the current density. The dipole moment is defined by

$$\vec{\wp} \equiv \sum_i q_i \vec{r}_i' \equiv \int \rho(\vec{r}') \, \vec{r}' \, d^3 r', \tag{3.22}$$

where any time dependence has been suppressed. How are these two approaches related?

Recall the continuity equation for charges:

$$\nabla \cdot \vec{j} + \frac{\partial \rho}{\partial t} = 0. \tag{3.23}$$

For our usual harmonically varying charge and current distributions, $\rho = \rho_0 e^{-i\omega t}$ and $\vec{j} = \vec{j}_0 \, e^{-i\omega t}$, we can write

$$\nabla \cdot \vec{j}_0 = i\omega\rho_0. \tag{3.24}$$

Next we will need the following vector identities, with $\vec{A}$ and $\vec{B}$ arbitrary vectors not to be confused with the vector potential and the magnetic induction:

$$\oint_S \vec{B} \left(\vec{A} \cdot d\vec{a} \right) = \int_V \left[(\vec{A} \cdot \nabla)\vec{B} + \vec{B} \left(\nabla \cdot \vec{A} \right) \right] d^3 r$$

$$(\vec{A} \cdot \nabla)\vec{B} = \hat{x} \left(A_x \frac{\partial B_x}{\partial x} + A_y \frac{\partial B_x}{\partial y} + A_z \frac{\partial B_x}{\partial z} \right)$$

$$+ \hat{y} \left(A_x \frac{\partial B_y}{\partial x} + A_y \frac{\partial B_y}{\partial y} + A_z \frac{\partial B_y}{\partial z} \right)$$

$$+ \hat{z} \left(A_x \frac{\partial B_z}{\partial x} + A_y \frac{\partial B_z}{\partial y} + A_z \frac{\partial B_z}{\partial z} \right).$$

The latter relation yields, when we identify $\vec{A}$ with $\vec{j}_0$ and $\vec{B}$ with $\vec{r}'$,

$$\left(\vec{j}_0 \cdot \nabla \right)\vec{r}' = \vec{j}_0 \tag{3.25}$$

and this combined with the first of the vector identities gives

$$\oint_{S'} \vec{r}' \left(\vec{j}_0 \cdot d\vec{a} \right) = \int_{V'} \left[\vec{j}_0 + \vec{r}' \left(\nabla \cdot \vec{j}_0 \right) \right] d^3 r' = 0. \tag{3.26}$$

We can set the above expression equal to zero since the currents comprising the source vanish at our (arbitrarily large) boundary surface. Combining these relations gives

$$\int_{V'} \vec{j}_0 \, d^3 r' = -\int_{V'} \vec{r}' \left(\nabla \cdot \vec{j}_0 \right) d^3 r' = -i\omega \int_{V'} \rho_0(\vec{r}')\vec{r}' d^3 r' = -i\omega\vec{\wp}_0. \tag{3.27}$$

The new vector introduced here is the electric dipole moment of the charge distribution. The vector potential for electric dipole radiation is thus also given by

$$\vec{A}(\vec{r},\, t) = -\frac{i\mu_0 \omega \vec{\wp}_0}{4\pi r}\, e^{i(kr-\omega t)}.$$

Example. Consider a dipole located at the origin of a spherical coordinate system, oriented along the z-axis (see figure 3.2).

The dipole moment is given by

$$\vec{\wp}_0 = \wp_0\, \hat{z} = \wp_0(\cos\theta\, \hat{r} - \sin\theta\, \hat{\theta}) \tag{3.28}$$

For a harmonically varying field, our usual assumption,

$$\frac{d\vec{\wp}}{dt} = -i\omega\vec{\wp},$$

where $\vec{\wp} = \vec{\wp}_0 e^{-i\omega t}$ and thus the vector potential can be written

$$\vec{A} = \frac{\mu_0 e^{i(kr-\omega t)}}{4\pi r}(-i\omega\wp_0)\, \hat{z} \quad \Longrightarrow \quad \vec{A} \text{ is in the same direction as the dipole.}$$

Using the coordinate transformation given in equation (3.28),

$$\vec{A} = \frac{-i\omega\mu_0}{4\pi r}\wp_0\, e^{i(kr-\omega t)}(\cos\theta\, \hat{r} - \sin\theta\, \hat{\theta}). \tag{3.29}$$

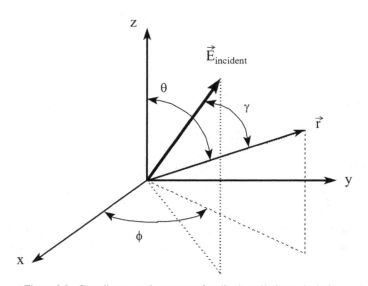

Figure 3.2. Coordinates and geometry for dipole radiation calculation.

From here we can directly calculate the fields $\vec{B}$ and $\vec{E}$:

$$\vec{B} = \nabla \times \vec{A}$$

$$= \frac{\hat{r}}{r \sin \theta} \left[\frac{\partial}{\partial \theta} (\sin \theta \, A_\phi) - \frac{\partial A_\theta}{\partial \phi} \right] \tag{3.30}$$

$$+ \frac{\hat{\theta}}{r} \left[\frac{1}{\sin \theta} \frac{\partial A_r}{\partial \phi} - \frac{\partial}{\partial r} (r A_\phi) \right]$$

$$+ \frac{\hat{\phi}}{r} \left[\frac{\partial}{\partial r} (r A_\theta) - \frac{\partial A_r}{\partial \theta} \right]. \tag{3.31}$$

Since the vector potential has only $\hat{r}$ and $\hat{\theta}$ components and no ϕ dependence, this simplifies considerably. We find as the final result

$$\vec{B} = -\frac{\mu_0 k^2 \omega \wp_0}{4\pi} \sin \theta \left[\frac{1}{kr} + \frac{i}{(kr)^2} \right] e^{i(kr - \omega t)} \, \hat{\phi}. \tag{3.32}$$

We can also calculate $\vec{E}$ using the relation (this step is left as a problem for the reader):

$$\vec{E} = \frac{ic}{k} \nabla \times \vec{B}$$

$$= \frac{ic}{k} \left\{ \frac{\hat{r}}{r \sin \theta} \frac{\partial}{\partial \theta} (\sin \theta \, B_\phi) - \frac{\hat{\theta}}{r} \left[\frac{\partial}{\partial r} (r \, B_\phi) \right] \right\}$$

$$= -\frac{k^3 \wp_0}{4\pi \varepsilon_0} \left\{ \left[\frac{2i}{k^2 r^2} - \frac{2}{k^3 r^3} \right] \cos \theta \, \hat{r} \right. \tag{3.33}$$

$$\left. + \left[\frac{1}{kr} + \frac{i}{k^2 r^2} - \frac{1}{k^3 r^3} \right] \sin \theta \, \hat{\theta} \right\} e^{i(kr - \omega t)}.$$

Note that the $\vec{B}$-field is always transverse, that is, the direction of $\vec{B}$ is orthogonal to that of the observation direction $\hat{r}$. In addition, the angular dependence of $\vec{B}$ on $\hat{\phi}$ alone means that the B-field lines are circles surrounding the dipole (which is along the z-axis). We can imagine this by making a comparison to magnetostatics. If we have a current flowing along the z-axis, the magnetic field lines will surround the current. Here we have an oscillating charge, which is a current and it gives rise to a similar magnetic field pattern, oscillating in time. The electric field, on the other hand, has a transverse component (in the $\hat{\theta}$-direction) and in addition there is a longitudinal component (i.e. in the $\hat{r}$-direction).

We want to find the time-averaged Poynting vector, which is

$$\langle \vec{S} \rangle = \frac{1}{2\mu_0} Re(\vec{E} \times \vec{B}^*), \tag{3.34}$$

or more directly, the power propagating through a spherical surface a distance R away from the source, which is given by

$$P = \int \langle \vec{S} \rangle \cdot d\vec{a}.$$

Before going through this calculation, we note that terms of order $1/r^2$ or higher in the fields will vanish in the far-field (i.e. $r \to \infty$) limit. Thus we can write the fields in this approximation as

$$\vec{B} \simeq -\frac{\mu_0 k \omega \wp_0}{4\pi r} \sin \theta \; e^{i(kr - \omega t)} \; \hat{\phi} \tag{3.35}$$

$$\vec{E} \simeq -\frac{k^2 \wp_0}{4\pi \epsilon_0 r} \sin \theta \; e^{i(kr - \omega t)} \; \hat{\theta}. \tag{3.36}$$

Notice that in this limit both fields are transverse, which is not true in general. Essentially this means that the emitted radiation looks like plane waves if one is far enough away from the source.

Using the above definition for the Poynting vector we find

$$\langle \vec{S} \rangle = \frac{\mu_0 \omega^4 \, |\wp_0|^2}{32\pi^2 c r^2} \sin^2 \theta \; \hat{r} \tag{3.37}$$

and

$$P = \int_0^\pi \int_0^{2\pi} \left[\frac{\mu_0 \omega^4 \, |\wp_0|^2}{32\pi^2 c R^2} \sin^2 \theta \right] R^2 \sin \theta \; d\theta \; d\phi \tag{3.38}$$

$$= \frac{\mu_0 \omega^4 \, |\wp_0|^2}{12\pi c}. \tag{3.39}$$

By looking at the expression for $\langle \vec{S} \rangle$ we see that an observer looking along the axis of the dipole would see no radiation at all. In figure 3.3(a) the radiation pattern is illustrated in a spherical coordinate plot. The z-axis in a rectangular coordinate system pierces the 'hole' of the doughnut-shaped radiation pattern. The radiation is independent of the azimuthal angle (ϕ-direction), therefore the pattern is symmetric about the z-axis. The Poynting vector maximum occurs for $\theta = \pi/2$, i.e. in the x–y plane. Figure 3.3(b) illustrates this effect, showing a cross section in the y–z plane of the radiation pattern.

Example: antenna radiation
Now we take a look at the problem referred to above, namely an antenna which is driven by a given current and which then radiates. In this case the source size is some multiple of the wavelength, but in any case the approximation made in equation (3.19) is no longer valid. Now we must begin by writing

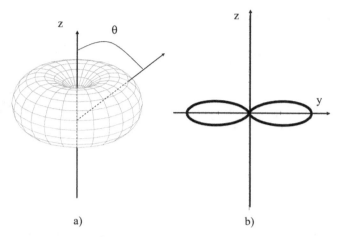

Figure 3.3. (a) Spherical plot of the $\sin^2 \theta$ dependence of the radiation from a dipole. The z-axis, along which the radiating dipole is oriented, pierces the 'hole' in the doughnut. (b) Plot of the dipole radiation in two dimensions, in the y-z plane.

$$\vec{A}(\vec{r}, t) = \frac{\mu_0}{4\pi} e^{-i\omega t} \int d^3 r' \frac{\vec{j}(\vec{r}')}{R} e^{ikR}.$$

In the denominator we can use the same expansion for R as before,

$$\frac{1}{R} \simeq \frac{1}{r}\left(1 + \frac{r'}{r} \cos \Theta\right) \simeq \frac{1}{r},$$

since we are still assuming that the observation point is far away from the source point. In the exponential term we must be more careful, since the variation of phase can lead to interference effects for spatial path differences on the order of a wavelength. Thus we use equation (3.18) and can now write

$$\vec{A}(\vec{r}, t) = \frac{\mu_0}{4\pi r} e^{i(kr-\omega t)} \int d^3 r' \, \vec{j}(\vec{r}') e^{-ikr' \cos \Theta}. \tag{3.40}$$

We will assume for the form of the current used to drive the antenna of length d the following:

$$\vec{j}(\vec{r}') = I \sin\left(\frac{kd}{2} - k\,|\,z'|\right)\delta(x')\delta(y')\,\hat{z}. \tag{3.41}$$

That is, the antenna is driven from the center with a current which has a sinusoidal distribution in the $\pm z$-direction and vanishes at the ends of the antenna, and is confined to the z-axis only (the δ-functions do this mathematically). Substituting this current density into equation (3.40) we have

$$\vec{A}(\vec{r}, t) = \frac{\mu_0 I}{4\pi r} e^{i(kr-\omega t)} \int_{-d/2}^{d/2} d^3 r' \, \sin\left(\frac{kd}{2} - k\,|z'|\right)\delta(x')\delta(y')e^{-ikr' \cos \theta} \,\hat{z}.$$

Carrying out the integrations (left for a homework problem) yields

$$\vec{A}(\vec{r},\,t) = \frac{2\mu_0 I}{4\pi k r}e^{+i(kr-\omega t)}\left[\frac{\cos\left(\frac{kd}{2}\cos\theta\right) - \cos\frac{kd}{2}}{\sin^2\theta}\right]\hat{z}.$$

Again we can calculate the $\vec{B}$-field from $\vec{B} = \nabla \times \vec{A}$ (again the details are left as an exercise). In the radiation zone (keeping only terms of order $1/r$),

$$\vec{B} = -\frac{\mu_0 I}{2\pi r}ie^{i(kr-\omega t)}\left(\frac{\cos\left(\frac{kd}{2}\cos\theta\right) - \cos\frac{kd}{2}}{\sin\theta}\right)\hat{\phi}.$$

3.5 Scattering

A first step toward understanding how electromagnetic radiation propagates through matter is to consider the scattering of radiation by individual particles. For an electromagnetic wave incident on a system of charged particles, the electric and magnetic fields of the wave will exert a force on the particles and set them in motion. Since the wave is periodic, we expect the motion of the charged particles to be periodic and this in turn implies that energy will be absorbed from the incident wave by the charged particles. A significant fraction of the absorbed energy will then be re-radiated, since all oscillating (accelerating) charges radiate. We call this process *scattering*. For example, in the simplest classical model of refraction, we say that an incident wave induces the atoms of a transmitting material to form oscillating dipoles which then re-radiate at the same frequency, but with a phase lag. These secondary 'scattered' waves along with the incident wave form the 'refracted' wave. It is the phase lag which, when the scattered wave and the incident wave are combined, leads to the bending of the wave front or what we know of as refraction. However, we are not ready to look at refraction in any detail until chapter 5. Actually, refraction, reflection, and diffraction are just a form of scattering. The nature of the scattered wave depends upon the nature of the scatterers, their positions, the arrangement of the scattering objects, and possible interactions among the scatterers. Obviously, a detailed accounting of such an approach can become quite complicated. Fortunately, for the most part, the basic concepts can be developed by using fairly simple examples.

Suppose we have a beam of light of power P_0 which is incident upon a region of matter, e.g. a collection of particles as in a gas as shown in the sketch below (figure 3.4). As the beam encounters the scatterers, some power, P_e, is removed from the beam so that the power seen by a detector in the forward direction would be $P_0 - P_e$. The beam is thus said to have undergone some *extinction*. The power lost from the beam consists of that which is scattered in various directions by the irradiated particles and that which is lost due to the conversion of energy into heat by some internal mechanism of the scatterers. The power lost due to heat is generally called *absorption*, although this convention is by no means universal. Some authors

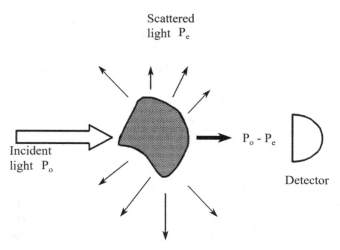

Figure 3.4. Scattering geometry.

refer to the entire extinction process as absorption. This is a potential source of confusion and one should always be aware when reading the literature of what the author means by absorption. We have chosen the definition of extinction and absorption found in standard texts such as *Principles of Optics* by Born and Wolf [5].

We now consider the scatterers to be atoms in a dilute gas. As pointed out above, each atom will become an oscillating dipole and re-radiate secondary waves. If the atoms are far apart compared to the wavelength of light, the secondary waves from one atom will have little effect on another atom compared to the incident driving electromagnetic field. In this case, we can regard each atom as a single scatterer so that if we have N atoms, the total scattered power would just be N times the single scattered power.

When could we not use this single scattering model? This occurs in the case where the scattering medium is dense. By dense we mean *optically dense*, which means there are many atoms within the dimensions of a wavelength of the incident light. For example, a gas at one atmosphere is optically dense for visible light, but it would not be optically dense for x-rays. For an optically dense medium, the secondary waves generated by one atom will now have an effect on atoms that are close to it. This secondary field superimposed with the incident field form what is called an internal field which is now the driving field for the induced dipoles. To treat a dense medium from a strictly scattering viewpoint is quite complicated and the entire subject of propagation in a dense medium is handled much more easily by the macroscopic Maxwell equations, which are the topic of chapter 5.

3.5.1 Scattering by a dipole

In this section we will examine how the concept of scattering applies to a dipole and how the simple model of an oscillating dipole, driven by an incident electromagnetic field, can explain much of the behavior of light propagating through matter. We will also introduce here some of the basic definitions associated with the scattering of light. At the end of this section we will be able to tackle a more detailed example of great interest, namely, a calculation of the scattering of unpolarized light by molecules. This phenomenon helps to explain why the sky is blue, why the setting sun becomes progressively more red in color, and why skylight is polarized.

If the incident field on an atom is given by $\vec{E} = \vec{\mathcal{E}}e^{-i\omega t}$, then the linear response of the induced dipole, $\vec{\wp}$, will also oscillate as $e^{-i\omega t}$ according to

$$\vec{\wp} = \alpha \vec{\mathcal{E}}e^{-i\omega t}, \tag{3.42}$$

where α is the polarizability of the particle or atom. The polarizability is a coefficient associated with the dipole and provides a measure of the strength of response of the atom or molecule to the applied electric field. From equation (3.37) the scattered irradiance (recall from chapter 1 that the irradiance is defined as the magnitude of the time-averaged Poynting vector, and has units of W m^{-2}) will be

$$I_S = \frac{\omega^4 \mu_0}{32\pi^2 c} \frac{|\alpha\vec{\mathcal{E}}|^2 \sin^2\gamma}{r^2}, \tag{3.43}$$

where γ is the angle between the direction of polarization of the incident electric field and the vector $\vec{r}$ to the observation point of the scattered radiation as shown in figure 3.5. Replacing the spherical polar angle θ with the more general angle γ will allow us to look at problems for which the incident polarization is not necessarily along the z-direction.

There are a couple of different ways by which we may characterize the properties of light scattered by a single dipole. At this point, it is useful to

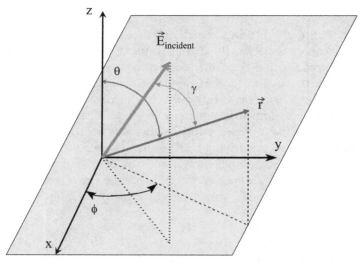

Figure 3.5. Coordinate geometry for induced dipole scattering. Since the induced dipole will be along the polarization direction of the incident electric field, we are now treating a more general case than in our previous example in which the dipole was along the z-axis.

introduce the idea of a scattering cross section. We have until this point usually considered the intensity of radiation fields (units of W m^{-2}) rather than the power (units of watts). Converting from one to the other then requires a factor corresponding to an area. We postulate that the total time-averaged scattered power can be written as

$$\langle P_{\text{scat}} \rangle \equiv \sigma_{\text{scat}} I_0, \tag{3.44}$$

where I_0 is the irradiance of the incident beam and σ_{scat} is called the scattering cross section; clearly σ_{scat} has the units of area.

In a similar manner, we can introduce an absorption cross section according to

$$\langle P_{\text{abs}} \rangle \equiv \sigma_{\text{abs}} I_0 \tag{3.45}$$

to represent incident radiation absorbed by a medium and *not* re-radiated (leading to heating, for example) and an extinction cross section,

$$\langle P_{\text{ext}} \rangle \equiv \sigma_{\text{ext}} I_0. \tag{3.46}$$

By definition, $\sigma_{\text{ext}} = \sigma_{\text{scat}} + \sigma_{\text{abs}}$. In general, these cross sections will depend on the frequency and polarization of the incident wave. We will return to the a discussion of the relations between these quantities in chapter 4.

One often speaks as well of a quantity called the differential scattering cross section, which is a measure of the relative irradiance scattered in a particular angular direction given by the angular coordinates θ and ϕ as shown in figure 3.6. The time-averaged power scattered into the solid angle $d\Omega$ in the angular direction (θ, ϕ) is defined as

$$\left\langle \frac{dP}{d\Omega} \right\rangle \equiv I_0 \left[\frac{d\sigma(\theta, \phi)}{d\Omega} \right], \tag{3.47}$$

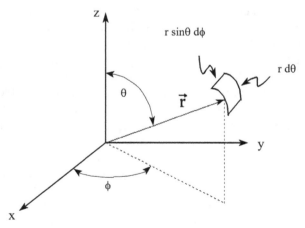

Figure 3.6. In the spherical coordinate system we can define the 'solid angle' subtended by an area $r \sin \theta d\phi \times r d\theta$ located at a distance r from the origin.

which can be rearranged to yield

$$\frac{d\sigma}{d\Omega} = \frac{\text{Scattered power per unit solid angle at } (\theta, \phi)}{\text{Incident power per unit area}},$$

where $d\sigma/d\Omega$ is called the differential scattering cross section per unit solid angle. Equation (3.47) says that the differential scattering cross section is the ratio of the time-averaged power scattered per unit solid angle to the incident irradiance. The relationship between the time-averaged scattered power per unit solid angle at (θ, ϕ) and the scattered irradiance at the observation point $\vec{r}$, which is just the time-averaged scattered power per unit area at (θ, ϕ), using figure 3.6, can be defined as

$$I_S = \frac{1}{r^2}\left\langle \frac{dP}{d\Omega} \right\rangle, \tag{3.48}$$

where $r^2 d\Omega$ is the area at a distance r of the solid angle element $d\Omega$. The scattered irradiance at a point (r, θ, ϕ) is thus

$$I_S = I_0 \frac{[d\sigma/d\Omega]}{r^2}. \tag{3.49}$$

Now from equations (3.43) and (3.49)

$$\frac{\omega^4 \mu_0 \sin^2 \gamma \, |\alpha|^2 |\mathcal{E}|^2}{32\pi^2 c r^2} = \frac{I_0}{r^2} \frac{d\sigma}{d\Omega} \tag{3.50}$$

and from the fact that $I_0 = |\mathcal{E}|^2/2\mu_0 c$, we obtain the following expression for the differential scattering cross section of an induced dipole of polarizability α,

$$\left[\frac{d\sigma}{d\Omega}\right]_{\text{scat}} = \frac{\mu_0^2 \omega^4 |\alpha|^2 \sin^2 \gamma}{16\pi^2} = \frac{\omega^4 |\alpha|^2 \sin^2 \gamma}{16\pi^2 \epsilon_0^2 c^4}. \tag{3.51}$$

The above result, referred to as 'Rayleigh scattering', is the answer to the question, 'Why is the sky blue?' The scattering cross section gives us information about the relative probability that light of various frequencies will be scattered by dipoles (i.e. molecules in which oscillations are induced by the incident field). Here we see that for unpolarized light, the probability increases very rapidly with frequency. Thus, when we look overhead, in a direction away from the Sun, the light we tend to see is that of higher frequency (blue instead of red). For the same reason, when we observe the setting sun low on the horizon, the light has passed through a great deal of the atmosphere and thus had ample opportunity to be scattered by air molecules, and we see only what remains after the higher frequency blue or green light is gone: orange or red. Recall that our starting point for this exercise was to consider a single small ($\ll \lambda$) particle with an induced dipole moment. The phenomenon of Rayleigh scattering appears for a dilute gas of very small molecules; if the scatterers are larger, e.g. water droplets, or at higher density, the analysis of the scattering problem becomes significantly more difficult. For example, in a medium with higher density (more than about one molecule per cubic wavelength), it can be shown that random fluctuations in the index of refraction caused by density fluctuations will also lead to a similarly strong frequency dependence for scattering.

For scattering of light from larger particles, of dimensions comparable to the radiation wavelength, the amount of scattering becomes independent of wavelength, the regime of so-called 'Mie scattering'. One example of this is the scattering of light from clouds or from pockets of air in a mass of whipped egg whites. The scattered light has the same spectral characteristics as the incident light, i.e. it is white light. More details on the topic of scattering can be found in [5].

3.6 Polarization of Rayleigh scattered light

In this section we will treat in more detail a specific case of scattering and look at the polarization characteristics of initially unpolarized light scattered by molecules. We will assume propagation along the z-axis; we can see from figure 3.7 that it is possible to define the angle γ (still defined as the angle between the *polarization* direction of the incident driving E-field and the scattering (observation) direction, $\hat{r}$) in terms of the scattering angles θ and ϕ. θ and ϕ are now the usual spherical angles measured relative to the direction of the incident beam, indicated in figure 3.7 by the wavevector $\vec{k}_0$. From figure 3.7 we see that

$$\cos \gamma = \sin \theta \cos \phi,$$

so that

$$\sin^2 \gamma = 1 - \sin^2 \theta \cos^2 \phi.$$

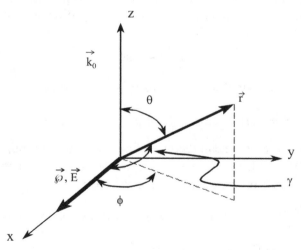

Figure 3.7. Coordinates for light propagation in the z-direction, with an induced dipole along the direction of the electric field (in the x–y plane).

Thus the differential cross section becomes

$$\left[\frac{d\sigma(\theta,\phi)}{d\Omega}\right]_{\text{scat}} = \frac{\omega^4|\alpha|^2}{16\pi^2\epsilon_0^2 c^4}(1 - \sin^2\theta\cos^2\phi). \tag{3.52}$$

We should remind ourselves that what we have just determined is the differential scattering cross section for a linearly polarized incident beam. Again, recall that $\frac{d\sigma(\theta,\phi)}{d\Omega}$ measures scattered power per unit solid angle per incident irradiance as opposed to σ_{scat} which measures the total scattered power per incident irradiance.

Continuing with this example, let us find the differential scattering cross section for an *unpolarized* incident beam. To do this we must average over the angle ϕ since, for unpolarized light, the orientation of $\vec{E}$ is random, and ϕ is the variable which measures the orientation of $\vec{E}$. The average of $\cos^2\phi$ yields a factor of 1/2 so that

$$\left[\frac{d\sigma}{d\Omega}\right]_{\text{unpol}} = \frac{\omega^4|\alpha|^2}{16\pi^2\epsilon_0^2 c^4}\left(1 - \frac{1}{2}\sin^2\theta\right) = \frac{\omega^4|\alpha|^2}{16\pi^2\epsilon_0^2 c^4}\frac{1}{2}(1 + \cos^2\theta). \tag{3.53}$$

The total cross section is found by integrating over the entire solid angle, $\sigma_{\text{scat;unpol}} = \int\left[\frac{d\sigma}{d\Omega}\right]_{\text{unpol}} d\Omega$ with the result being

$$\sigma_{\text{scat;unpol}} = \frac{\omega^4|\alpha|^2}{16\pi^2\epsilon_0^2 c^4}\frac{1}{2}\int_0^\pi\int_0^{2\pi}(1 + \cos^2\theta)\sin\theta\, d\theta\, d\phi = \frac{\omega^4\mu_0^2|\alpha|^2}{6\pi}. \tag{3.54}$$

3.6.1 Polarization of scattered light

Now we will take a step back and look at the electric field of the scattered light. For propagation in the z-direction, we can consider the incident unpolarized light in a slightly different way. We can consider the incident light to consist of two components, one in the x-direction and one in the y-direction. We will make use of the following unit vector identities:

$$\hat{z} = \hat{r}\cos\theta - \hat{\theta}\sin\theta \tag{3.55}$$

$$\hat{y} = \hat{r}\sin\theta\sin\phi + \hat{\phi}\cos\phi + \hat{\theta}\cos\theta\sin\phi \tag{3.56}$$

$$\hat{x} = \hat{r}\sin\theta\cos\phi - \hat{\phi}\sin\phi + \hat{\theta}\cos\theta\cos\phi. \tag{3.57}$$

For the x-direction dipole we can therefore find the vector potential $\vec{A}$, and then the fields $\vec{B}$ and $\vec{E}$. We will keep only terms of order $1/r$ in our calculations since we are interested in the radiation zone only. For $\vec{A}$ we have

$$\vec{A}_1 = \frac{-i\omega\mu_0}{4\pi r}\wp_0 e^{i(kr-\omega t)}(\hat{r}\sin\theta\cos\phi - \hat{\phi}\sin\phi + \hat{\theta}\cos\theta\cos\phi), \tag{3.58}$$

which, using equation (3.31), leads to

$$\vec{B}_1 = \frac{k\omega\mu_0}{4\pi r}\wp_0 e^{i(kr-\omega t)}(-\sin\phi\hat{\theta} + \cos\theta\cos\phi\hat{\phi}),$$ (3.59)

where, again, we keep only terms of order $1/r$. Equation (3.33) then gives us

$$\vec{E}_1 = \frac{k\omega\mu_0 c}{4\pi r}\wp_0 e^{i(kr-\omega t)}(\cos\theta\cos\phi\hat{\theta} - \sin\phi\hat{\phi}).$$ (3.60)

Before moving on to calculate the field for a dipole oriented in the y-direction, we check our result with the intuition gained earlier in the chapter. For an x-aligned dipole, we would expect to see zero radiation when looking along the x-axis, which in this case means (see figure 3.7) setting $\theta = \pi/2$ and $\phi = 0$ or $\phi = \pi$. Using these values, we do indeed find that the field is zero. We also see that the maximum of equation (3.60) occurs along the z-axis, where $\theta = 0$ and $\phi = \pi/2$.

For a dipole oriented along the y-axis we jump right to the final result,

$$\vec{E}_2 = \frac{k\omega\mu_0 c}{4\pi r}\wp_0 e^{i(kr-\omega t)}(\cos\theta\sin\phi\hat{\theta} + \cos\phi\hat{\phi}).$$ (3.61)

Our actual interest in this problem is motivated by the question of the polarization of scattered skylight. If we imagine light propagating in the z-direction and that we are observers in the x–z plane, corresponding to $\phi = 0$, we may define a degree of polarization by

$$P(\theta) = \frac{I_\perp - I_\parallel}{I_\perp + I_\parallel},$$ (3.62)

where $I_\perp$ is the intensity of all scattered light with polarization perpendicular to the x–z plane (i.e. in the y-direction) and $I_\parallel$ is the intensity of scattered light with polarization parallel to the x–z plane. To calculate these quantities we use the definitions of the unit vectors above and calculate

$$I_\perp \equiv I_y = |\hat{y}\cdot\vec{E}_1|^2 + |\hat{y}\cdot\vec{E}_2|^2.$$ (3.63)

For $\phi = 0$ we arrive at the result

$$I_\perp = \left(\frac{k\omega\mu_0 c}{4\pi r}\right)^2$$ (3.64)

since $E_{1,y} = 0$.

To calculate $I_\parallel$ we must find both the x- and z-components and with $\phi = 0$. We find the following results,

$$E_{1,x} = \left(\frac{k\omega\mu_0 c}{4\pi r}\right)\cos^2\theta$$ (3.65)

$$E_{2,x} = 0$$ (3.66)

$$E_{1,z} = -\left(\frac{k\omega\mu_0 c}{4\pi r}\right)\cos\theta\sin\theta \tag{3.67}$$

$$E_{2,z} = 0. \tag{3.68}$$

Now we use these to find the degree of polarization

$$P(\theta) = \frac{1 - \cos^2\theta\sin^2\theta - \cos^4\theta}{1 + \cos^2\theta\sin^2\theta + \cos^4\theta} = \frac{\sin^2\theta}{1 + \cos^2\theta}. \tag{3.69}$$

The function $P(\theta)$ has a maximum at $\theta = \pi/2$ and is zero for $\theta = 0$. We can map out this function roughly with a polaroid sheet. If we look at the sky at right angles to the Sun, we should find fairly strong polarization (ideally, from the above, the light would be completely polarized). However, the other limit, looking directly at the Sun, or nearly so, we will find that the sunlight is unpolarized. We can understand the $\theta = \pi/2$ limit by our simple dipole picture. For dipoles along the x- and y-directions, with our observation being in the x–z plane, the parallel component of radiation comes from the x dipole only, but here we are looking along its axis, and therefore see no parallel component. That is, the scattered light is then totally due to the y-direction dipole.

We conclude by summarizing our results of this section. We have seen that the simple model of a molecular dipole being forced to oscillate and thus scattering incident radiation leads to an explanation of two commonly observed sky phenomena. First, blue scattered light predominates when white light is incident on the molecular dipoles, known as Rayleigh scattering. Second, skylight may be polarized depending on the direction of observation with respect to the direction the sunlight is incident on the scattering molecules. These phenomena represent our first real model of the interaction between light and matter, albeit at a very simplified and microscopic level.

3.7 Radiation in the Coulomb gauge

We mentioned at the beginning of this chapter that there is more than one possibility for defining the divergence of the vector potential. In this section we work through the so-called 'Coulomb gauge' in which $\nabla \cdot \vec{A} = 0$. We will make no further use of this derivation, but present it here for the sake of completeness.

Substituting equations (3.5) and (3.6) into the first and fourth Maxwell equations gives

$$-\nabla^2\Phi - \nabla\cdot\left(\frac{\partial\vec{A}}{\partial t}\right) = \rho/\epsilon_0 \tag{3.70}$$

and

$$\nabla(\nabla\cdot\vec{A}) - \nabla^2\vec{A} + \frac{1}{c^2}\frac{\partial}{\partial t}\nabla\Phi + \frac{1}{c^2}\frac{\partial^2\vec{A}}{\partial t^2} = \mu_0\vec{j}. \tag{3.71}$$

These are the field equations in terms of the potentials Φ and $\vec{A}$ which result from the charge and current distributions ρ and $\vec{j}$, respectively. These field equations, as they stand, are rather complicated due to the fact that Φ and $\vec{A}$ are mixed together in the same equation. However, we can manipulate Φ and $\vec{A}$ to obtain a significant simplification. It turns out that equations (3.5) and (3.6) do not completely determine Φ and $\vec{A}$ since these potentials can be varied in certain ways without changing the fields $\vec{E}$ and $\vec{B}$. The fields $\vec{E}$ and $\vec{B}$ are the measurable fields, so it does not matter how we transform Φ and $\vec{A}$ if they still give the same results for $\vec{E}$ and $\vec{B}$. A transformation of the potentials in this manner is called a gauge transformation.

Now we turn again to the gauge transformations discussed at the beginning of the chapter. Recall that we suppose that Φ' and $\vec{A}'$ are solutions to the field equations, equations (3.70) and (3.71), for a given ρ and $\vec{j}$. We can perform the gauge transformations on Φ' and $\vec{A}'$ to obtain a new Φ and $\vec{A}$ (repeating equations (3.7) and (3.8)),

$$\Phi = \Phi' + \frac{\partial \Psi}{\partial t} \tag{3.72}$$

and

$$\vec{A} = \vec{A}' - \nabla\Psi, \tag{3.73}$$

where Ψ is some scalar function. Note, from equations (3.5) and (3.6), that neither $\vec{B}$ nor $\vec{E}$ is changed by this transformation. Now suppose the scalar quantity Ψ is chosen such that

$$\nabla \cdot \vec{A} = 0.$$

This is called the Coulomb gauge. In order for this to hold, Ψ must be chosen such that

$$\nabla^2\Psi = \nabla \cdot \vec{A}'$$

The choice of the Coulomb gauge results in the following simplification for equations (3.70) and (3.71):

$$-\nabla^2\Phi = \rho/\epsilon_0 \tag{3.74}$$

$$-\nabla^2\vec{A} + \frac{1}{c^2}\frac{\partial}{\partial t}\nabla\Phi + \frac{1}{c^2}\frac{\partial \vec{A}}{\partial t} = \mu_0\vec{j}. \tag{3.75}$$

The advantage of the Coulomb gauge is that Φ becomes easy to calculate, but $\vec{A}$ now becomes very difficult to calculate, whereas in the Lorentz gauge it was the vector potential $\vec{A}$ that formed the basis for arriving at all other results of interest (primarily the electric field $\vec{E}$). Equation (3.74) is Poisson's equation of electrostatics and therefore Φ can be obtained from the usual techniques employed in electrostatics.

Notice that equation (3.75) still has Φ and $\vec{A}$ mixed in the same equation. This can be further simplified by using a result due to Helmholtz which says that any vector field can be written as the sum of two terms, one of which has a curl of zero and the other of which has a divergence of zero. In this case, consider the current density $\vec{j}$. The term which has zero curl we will designate as $\vec{j}_{\parallel}$, called the longitudinal component, and the zero divergence term is $\vec{j}_{\perp}$, called the transverse (or solenoidal, or irrotational) component. Thus $\vec{j}$ is written as

$$\vec{j} = \vec{j}_{\parallel} + \vec{j}_{\perp}, \tag{3.76}$$

where

$$\nabla \times \vec{j}_{\parallel} = 0 \tag{3.77}$$

and

$$\nabla \cdot \vec{j}_{\perp} = 0. \tag{3.78}$$

Because of equation (3.77), $\vec{j}_{\parallel}$ can be expressed in terms of a scalar potential which we can designate as χ so that

$$\vec{j}_{\parallel} = \nabla \chi.$$

In addition, the continuity equation becomes

$$\nabla \cdot \vec{j}_{\parallel} + \frac{\partial \rho}{\partial t} = 0$$

and from Poisson's equation (equation (3.74)) we have

$$\chi = \epsilon_0 \frac{\partial \Phi}{\partial t}.$$

This then leads to the following expression for $\vec{j}_{\parallel}$,

$$\vec{j}_{\parallel} = \epsilon_0 \nabla \frac{\partial \Phi}{\partial t}, \tag{3.79}$$

which then, using the fact that $\epsilon_0 \mu_0 = 1/c^2$, simplifies equation (3.75) to be

$$-\nabla^2 \vec{A} + \frac{1}{c^2} \frac{\partial^2 \vec{A}}{\partial t^2} = \mu_0 \vec{j}_{\perp}. \tag{3.80}$$

The solution to equation (3.80), which can be found in standard texts on electromagnetic theory [6] is

$$\vec{A}(\vec{r}, t) = \frac{\mu_0}{4\pi} \int \frac{\vec{j}_{\perp} d^3 r'}{|\vec{r} - \vec{r}'|}, \tag{3.81}$$

where again, $t' = t - \left|\frac{\vec{r} - \vec{r}'}{c}\right|$ and $\vec{r}'$ stands for the coordinates of the source point, and $\vec{r}$ is the coordinate of the observation point, called the field point, see figure 3.1. Recall that the time t' is called the retarded time and it reflects the fact that the vector potential at the point $\vec{r}$ and the time t is due to the effect of a current distribution at $\vec{r}'$ and the earlier time t'. It takes a finite time, $\left|\frac{\vec{r} - \vec{r}'}{c}\right|$, for the effects at $\vec{r}'$ and t' to propagate to $\vec{r}$.

Again, the Helmholtz theorem says that any vector can be written as a sum of two terms, longitudinal and transverse, and therefore the electric field can be written as

$$\vec{E} = \vec{E}_{\parallel} + \vec{E}_{\perp}, \tag{3.82}$$

where by definition $\nabla \cdot \vec{E}_{\perp} = \nabla \times \vec{E}_{\parallel} = 0$. From $\vec{E} = -\nabla\Phi - \frac{\partial \vec{A}}{\partial t}$ and $\nabla \times \nabla\Phi = 0$, and using the Coulomb gauge condition, we have

$$\vec{E}_{\parallel} = -\nabla\Phi \tag{3.83}$$

$$\vec{E}_{\perp} = -\frac{\partial \vec{A}}{\partial t}. \tag{3.84}$$

What the above two equations illustrate is that the Coulomb gauge allows us to associate the longitudinal field with the scalar potential and the transverse field with the vector potential. Furthermore, we see that the field equation for the radiated field is determined from $\vec{A}$, equation (3.81), and that the third and fourth Maxwell equations are concerned with the transverse E-field as seen below:

$$\nabla \times \vec{E}_{\perp} = -\frac{\partial \vec{B}}{\partial t} \tag{3.85}$$

$$\nabla \times \vec{B} = \mu_0 \vec{j}_{\perp} + \frac{1}{c^2}\frac{\partial \vec{E}_{\perp}}{\partial t}. \tag{3.86}$$

The first Maxwell equation would now be written,

$$\nabla \cdot \vec{E}_{\parallel} = \rho/\epsilon_0. \tag{3.87}$$

Thus we see that the longitudinal field equations are the field equations associated with fields from charges as in the case of electrostatics, whereas the transverse field equations are the field equations associated with electromagnetic waves.

This final section of the chapter has presented an alternative method for calculating electric fields for radiating systems. Both the Lorentz and Coulomb gauge approaches have their advantages and disadvantages, as we have mentioned; in the end, of course, calculations done in either gauge representation must lead to the same final results for quantities that we observe physically, such as the radiation patterns discussed earlier in the chapter.

In the next chapter we will take our treatment of the interaction between radiation and matter one step further and look at the propagation of light through a dilute sample of atoms or molecules.

3.8 Problems

1. Show that the force on an electric dipole in a nonuniform electric field is
$\vec{F} = (\vec{p} \cdot \nabla)\vec{E}$.
2. Use the appropriate Maxwell equation to calculate the electric field due to an infinitely long cylindrical conductor of radius R which as a charge per unit length λ. Also determine the scalar potential.
3. Use the appropriate Maxwell equation to calculate the magnetic field at a distance r_0 from an infinitely long straight wire which carries a current I.
4. Show that a sphere charged in spherical symmetry and oscillating purely radially will not radiate in the far-field.
5. Consider a center-fed antenna of total length $\lambda/4$ with a current density

$$\vec{j}\,(\vec{r}) = j_0 \delta(x)\delta(y) \cos \frac{2\pi z}{\lambda} \hat{z}, \quad \frac{-\lambda}{8} \leqslant z \leqslant \frac{\lambda}{8}.$$

Calculate the far-field irradiance corresponding to the electric dipole moment.
6. Find σ_{scat} for a dipole driven by a linearly polarized beam of incident light.
7. Find the expression for the differential scattering cross section for the cases of a circularly polarized plane wave. Repeat for an elliptically polarized plane wave scattered by an oscillating induced dipole. Hint: Start with

$$I_s = I_0 \frac{[d\sigma/d\Omega]}{r^2}$$

and determine what I_s is for elliptically polarized light.
8. Find the dipole moment for the following charge distribution,

$$\rho = e[\delta(x - \zeta \cos \omega t - x_0)\delta(y - \eta \cos \omega t - y_0)\delta(z - \xi \cos \omega t - z_0)$$
$$- \delta(x - x_0)\delta(y - y_0)\delta(z - z_0)].$$

9. A thin linear antenna of length d is excited in such a way that the sinusoidal current makes a full wavelength of oscillation, i.e.

$$I = I_0 \sin \left(\frac{kd}{2} - k\,|z| \right) e^{-i\omega t}.$$

The antenna is also surrounded by a spherical shell of static charge of total charge Q_0 distributed over the surface of the shell.
   (a) Find the total electric field associated with this arrangement of charge and current.
   (b) Find the far-field irradiance.
   (c) Find the total power radiated.

10. Find the far-field electric field and irradiance for the charge distribution

$$\rho = e\delta(x)\delta(y)\delta(z - a\cos\omega_0 t) - e\delta(x)\delta(y)\delta(z + a\cos\omega_0 t).$$

   Hint: Calculate the dipole moment of the charge distribution first.
11. An electric dipole $\vec{\wp}(t) = \wp_0\hat{z}\cos\omega t$ is placed at the origin.
    (a) Determine the fields and the radiation along the axis of the dipole. Use the far-field approximation.
    (b) Repeat part (a) for the direction normal to the dipole.
    (c) Give a sketch of the radiation as a function of Θ.
12. Consider two electric dipoles placed at the origin in the x–y plane:

$$\vec{\wp}_1(t) = \wp_0\hat{x}\cos\omega t \quad \text{and}$$
$$\vec{\wp}_2(t) = \wp_0\big(\sin\phi_0\hat{x} + \cos\phi_0\hat{y}\big)\sin\omega t.$$

   Determine the polarization of the radiation in the following cases:
   (a) $\theta = 90°$, (b) $\theta = 0$ and $\phi_0 = \pi/2$, and (c) $\theta = 0$, π and $\phi_0 \neq \pi/2$, where θ is the standard spherical angle.
13. A linear antenna of length d carries a standing current distribution of amplitude I_0 and frequency ω. The antenna has m current half-waves and the ends of the antenna are nodes.
    (a) Write an expression for the current in the antenna.
    (b) Determine the electric dipole moment per unit length. Hint: recall the relation between current density and dipole moment.
    (c) Determine the magnetic field produced by the antenna in the far-field radiation zone.
    (d) Determine the total power radiated by the antenna.
14. Consider an electric dipole of charges q and $-q$ placed at a distance d from each other. The magnitude of each of the charges varies sinusoidally with time at a frequency ω.
    (a) Determine the dipole moment of the charges.
    (b) Determine the current flowing between them. Hint: recall the relation between current density and the dipole moment.
    (c) Determine the radiation fields of the charges using the far-field approximation. What are these fields along the line joining the charges?
15. An electric dipole, located at the origin, is rotating in the x–y plane with an angular frequency ω.
    (a) Decompose the dipole into two oscillating dipoles along the x- and y- axes.
    (b) What is the phase difference between the two components of the dipole?
    (c) What is the angular distribution of the power radiated by the dipole?
16. Express the total scattering cross section in terms of the index of refraction, n, and the extinction coefficient, κ for the case of a dilute gas of N molecules.
17. Find an expression for the scattered irradiance of an induced dipole driven by elliptically polarized light.

References

[1] Wikipedia: Heinrich Hertz https://en.wikipedia.org/wiki/Heinrich_Hertz (Accessed: 11 Jan. 2021)

[2] Wikipedia: John William Strutt, 3rd Baron Rayleigh https://en.wikipedia.org/wiki/John_William_Strutt,_3rd_Baron_Rayleigh (Accessed: 11 Jan. 2021)

[3] Wikipedia: John Tyndall https://en.wikipedia.org/wiki/John_Tyndall (Accessed: 11 Jan. 2021)

[4] Jackson J D 1998 *Classical Electrodynamics* 3rd edn (New York: Wiley)

[5] Born M and Wolf E 1999 *Principles of Optics* 7th edn (New York: Macmillan)

[6] Griffiths D 2017 *Introduction to Electrodynamics* 4th edn (Cambridge: Cambridge University Press)

IOP Publishing

Optical Radiation and Matter

Robert J Brecha and J Michael O'Hare

Chapter 4

Absorption and line broadening

In this chapter we will use some of the intuition gained from chapter 3 and apply it to a first specific example of the interaction of light and matter. We have seen how incident radiation may induce a response in an atomic oscillator dipole and how that resulting oscillator will re-radiate energy, becoming itself a (nearly) point source. As this model is the basis for all other interactions, at least in the classical physics regime (i.e. non-quantum mechanics), we will consider it here in more detail.

Before moving on to treat crystals and other solids in later chapters, we will introduce in this chapter the principles of radiation interacting with a dilute atomic medium. This will serve as well as a first, very abbreviated, introduction to concepts commonly encountered in atomic and molecular physics.

4.1 Historical introduction

One unifying theme of the present chapter is the interaction of radiation and dilute matter, such as a gas. More specifically, we may say that 'spectroscopy' is the focus of our interest—understanding properties of matter by using light (or electromagnetic radiation of frequencies other than that in the visible or near-visible spectrum) to probe a sample. The origin of spectroscopic investigation may be credited to Joseph Fraunhofer (1787–1826), who started as an apprentice lens maker [1]. His early (although not the first) observation that light from the sun, suitably spread by a prism into a spectrum, is characterized by a series of dark lines. He further investigated these features and found that they exist in light reflected from the Moon and planets, that other stars show a different set of lines, and that one set of lines appear at wavelengths at which a sodium flame produces light. It can be concluded that the cause of these 'Fraunhofer lines' is absorption, or removal, of light emitted by the Sun as that light traverses the atmosphere of the Sun itself. It was quickly realized that absorption and emission lines might be used as a 'fingerprint' to identify different elements, a concept still commonly used in various forms today.

doi:10.1088/978-0-7503-2624-7ch4

Naturally, the next step beyond realizing a tool that may be used to identify what we now call atoms or molecules is to gain an understanding of *why* each element emits or absorbs light at the particular frequencies that make up its signature. A model based on classical physics concepts was a first attempt in that direction and appears in Hendrik Antoon Lorentz's (1853–1928) *Theory of Electrons* [2]. We will consider this model in great deal in chapter 6; for now we simply note that it describes atomic and molecular systems in terms of a driven, damped simple harmonic oscillator. It is well known that such systems display resonant responses for certain well-defined driving frequencies, and those resonant frequencies will be taken to correspond to the absorption frequencies of atoms and molecules. Although the classical model succeeds in a qualitative way in describing real molecules, it is always left with undetermined constants, such as the strength of absorption or emission, or the exact resonant frequency, both of which can only be supplied by comparison with experiment.

The next stage in this highly abbreviated description of spectroscopic history was a formulaic description of the resonant frequencies of the hydrogen atom at the end of the nineteenth century. Although there was still no model available that explained exactly why certain frequencies absorbed light, the formulas of Lyman, Balmer, and Paschen had a certain degree of predictive power, at least for hydrogen, and with some extensions to the alkali atoms. With the advent of quantum mechanics in the late 1920s, a tremendous breakthrough was made in understanding the spectra of atoms and molecules. To go further into these developments would take us too far afield; it suffices to say that quantum mechanical descriptions of microscopic systems provide, with only a relatively few very general assumptions, highly accurate predictions for absorption and emission frequencies of atomic and molecular gases, and even of solid-state systems.

By way of summary, we will essentially return to the pre-quantum mechanical formulation of the interaction of light and matter by investigating the Lorentz harmonic oscillator model in some detail. This will be sufficient for our purposes, namely to gain a qualitative understanding of absorption characteristics of dilute atomic gases. In later chapters we will expand the model to include systems that cannot be considered dilute, and to consider the optical properties of metals as well.

4.2 Extinction by a dipole

As a starting point we take a more detailed look at the mechanics of the radiation-induced dipole. We will investigate the extinction of radiation by a dipole and will be concerned with only one degree of freedom, that being the vibrational mode of the dipole.

Consider an atom or molecule which is charge neutral and whose dimensions are small compared to the wavelength of the incident light. Assume that only one valence electron is affected by the incident field and that its response to the field gives us the oscillating dipole. Since the nucleus is so much more massive than the electron, we will only need to consider the motion of the electron relative to the nucleus. The electron, in a classical picture, can thus be regarded as bound to the nucleus by a spring of spring constant ζ. Let x be the displacement of the electron from equilibrium. We consider a one-dimensional case here; the generalization to three dimensions is straightforward to carry out. The displacement from equilibrium is what gives rise to a 'dipole moment', defined by $\wp = ex$. The forces on the atom or molecule can be summarized using Newton's second law according to

$$m\ddot{x} = -\zeta x - m\Gamma\dot{x} + eE, \tag{4.1}$$

where m is the electron mass (again assuming only the electron moves) and e the electron charge, the term $-\zeta x$ is the restoring force of the spring (Hooke's law), the term $-m\Gamma\dot{x}$ is a damping term (reasonably assuming that damping is proportional to the velocity $\dot{x}$ with Γ being the damping constant), and the term eE is the force on the dipole due to the external electric field.

This last term arises from the following considerations: the potential of a charge distribution in an E-field can be written as a multipole expansion, the first term being the monopole, the second the dipole, and higher order terms, etc. In our example, there is no monopole term so that the dipole is the quantity which interacts with the field according to the potential $V = -\vec{\wp} \cdot \vec{E}$. The force is then found from the expression $\vec{F} = -\nabla V$, which in our example would just be eE, since $\vec{\wp} = e\vec{r}$.

We may write the field incident on the atom as

$$\vec{E} = \vec{\mathcal{E}}_0 e^{-i\omega t}. \tag{4.2}$$

We expect that x will be synchronous with the driving E-field in steady-state, i.e. $x = x_0 e^{-i\omega t}$. In addition, since we will from now on consider only one direction, we will drop the vector notation; we should keep in mind, however, that the more general case is three-dimensional. Using this ansatz we have, from equation (4.1),

$$\ddot{x} = -\omega_0^2 x - \Gamma\dot{x} + (e/m)E \tag{4.3}$$

$$-\omega^2 x_0 e^{-i\omega t} = -\omega_0^2 x_0 e^{-i\omega t} + i\omega x_0 \Gamma e^{-i\omega t} + \frac{e\mathcal{E}_0}{m}e^{-i\omega t}, \tag{4.4}$$

where x_0 is the amplitude of the displacement and $\omega_0^2 = \zeta/m$ is the usual definition of the resonant frequency of a harmonic oscillator. Solving for x_0 yields

$$x_0 = \frac{(e/m)\mathcal{E}_0}{(\omega_0^2 - \omega^2) - i\Gamma\omega}. \tag{4.5}$$

Therefore, the induced dipole moment is

$$\wp = ex = \frac{(e^2/m)\mathcal{E}_0}{(\omega_0^2 - \omega^2) - i\Gamma\omega}e^{-i\omega t}. \tag{4.6}$$

From this expression and equation (3.42) the polarizability of the dipole (defined as $\wp = \mathcal{E}$) is

$$\alpha = \frac{(e^2/m)}{(\omega_0^2 - \omega^2) - i\Gamma\omega}. \tag{4.7}$$

Now let us investigate the mechanism of extinction, or removal of energy from the incident electric field. The coupling of the electromagnetic field to the coordinate x results in energy being removed from the incident field by the dipole, since energy is needed to excite the motion of the dipole. This energy loss is accounted for mathematically by the damping mechanism represented by Γ. Γ can represent any mechanism which is responsible for removing energy from the incident field. Of course, if there is more than one mechanism responsible for removing energy from the incident field, then there would be a damping coefficient for each mechanism. To obtain a quantitative expression for power extinction, we look at the energy stored in a dipole, and more explicitly, at the power needed to force the oscillation of the dipole in steady-state. For any mechanical oscillator this power is given by

$$\frac{dW}{dt} = \vec{F} \cdot \vec{v}. \tag{4.8}$$

For our case, the force on the electron exerted by the driving field is $\vec{F} = e\vec{E}$. Combining these, along with the definition of velocity, we can write

$$\frac{dW}{dt} = \vec{E} \cdot \frac{d\vec{\wp}}{dt}, \tag{4.9}$$

where we have used the definition of the dipole moment.

We are interested in finding an expression relating the power expended in driving the dipole to the input intensity of the electromagnetic field. To this end, we rewrite the expression for the dipole moment, equation (4.6), so that we may look at only the physically relevant real part,

$$\wp = \frac{(e^2/m)\mathcal{E}_0}{(\omega_0^2 - \omega^2)^2 + \Gamma^2\omega^2}(\omega_0^2 - \omega^2 + i\Gamma\omega)(\cos \omega t - i \sin \omega t), \tag{4.10}$$

from which we take the real part,

$$\Re[\wp] = \frac{(e^2/m)\mathcal{E}_0}{(\omega_0^2 - \omega^2)^2 + \Gamma^2\omega^2}((\omega_0^2 - \omega^2) \cos \omega t + \Gamma\omega \sin \omega t). \tag{4.11}$$

To calculate the power we need the time derivative of this expression, arriving finally at

$$\frac{dW}{dt} = \frac{(e^2/m)\mathcal{E}_0^2}{(\omega_0^2 - \omega^2)^2 + \Gamma^2\omega^2}[-\omega(\omega_0^2 - \omega^2) \sin \omega t + \Gamma\omega^2 \cos \omega t]. \tag{4.12}$$

The two terms in square brackets represent motion of the dipole in-quadrature (i.e. shifted in phase by $\pi/2$) and in-phase with the driving field. As usual, we are interested in the time-averaged behavior, for which the first term goes to zero and the second averages to 1/2. Now we may write the time-averaged power taken from the field as

$$P_{\text{ext}} \equiv \left\langle \frac{dW}{dt} \right\rangle = \frac{\mu_0 c e^2 \Gamma \omega^2}{m[(\omega_0^2 - \omega^2)^2 + \Gamma^2\omega^2]}\left(\frac{\mathcal{E}_0^2}{2\mu_0 c}\right). \tag{4.13}$$

Written in this way we identify the term in parentheses as the average intensity and define the extinction cross section to be

$$\sigma_{\text{ext}} = \frac{\mu_0 c e^2 \Gamma \omega^2}{m[(\omega_0^2 - \omega^2)^2 + \Gamma^2 \omega^2]}. \tag{4.14}$$

A simple dimensional analysis shows that the units of the cross section are m²; in some approximate sense this quantity represents the 'size' of the dipole as 'seen' by the incident electromagnetic field. We note as well that the extinction cross section is proportional to the damping constant. A sketch of σ_{ext} as a function of frequency is shown in figure 4.1. This sketch shows that σ increases as ω approaches the resonant frequency ω_0.

We will manipulate the expression for the extinction cross section one final time to cast it in a form that will be useful later in this chapter. In general we will be looking at the response of the electronic oscillator at frequencies reasonably close to the resonance frequency, ω_0. In that case we may write $\omega_0^2 - \omega^2 \simeq 2\omega_0(\omega_0 - \omega)$. With that substitution the extinction cross section becomes

$$\sigma_{\text{ext}} = \frac{e^2}{2\pi m \varepsilon_0 c} \left[\frac{\Gamma/2\pi}{(\omega - \omega_0)^2 + (\Gamma/2)^2} \right]. \tag{4.15}$$

The term in square brackets defines the frequency response of the atoms and is known as the 'lineshape'. As we shall see later in this chapter, as defined above the lineshape is normalized, i.e. the integral of the expression in brackets over all frequencies is unity—which is why we went to the effort to write the expression in this particular form.

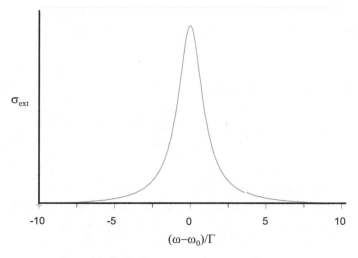

Figure 4.1. Extinction cross section versus frequency.

It is also interesting to return and take a look at the scattering cross section for this damped harmonic oscillator. From equation (3.54) the total scattering cross section for unpolarized light is

$$\sigma_{\text{scat}} = \frac{\omega^4 \mu_0^2 \, |\alpha|^2}{6\pi},$$

which, when combined with equation (4.7), gives

$$\sigma_{\text{scat}} = \frac{\omega^4}{6\pi\varepsilon_0^2 c^4} \frac{e^4/m^2}{|(\omega_0^2 - \omega^2) - i\Gamma\omega|^2}. \tag{4.16}$$

Suppose that scattering is the only mechanism for dissipating energy. Then σ_{scat} and σ_{ext} should be equal. From the above expressions, the only way the two expressions could be equal is if the decay constant Γ were dependent on frequency, such that

$$\Gamma = \frac{\omega_0^2 e^2}{6 \, m\pi\varepsilon_0 c^3}, \tag{4.17}$$

which gives the dependence on frequency of the classical electron oscillator. However, this type of comparison begins to expose some of the limitations to the strictly classical approach that we have used here, since we are making somewhat ad hoc assumptions about how well we can relate the various models. To pursue this matter any further we would need to treat the scattering and absorption problems quantum mechanically, which is beyond the scope of this text. A thorough discussion of some of the different quantities describing decay rates and the strength of absorption (extinction) for both classical models and a quantum mechanical formulation is given in [3]. In the same journal issue there is a more mathematically sophisticated discussion of absorption, interference, and 'stimulated emission' (another atom-field interaction process that is important for lasers) within the framework of the classical oscillator model, along with some consideration of the limits of validity of that model [4].

4.3 Field from a sheet of dipoles

We extend our model of a material medium by one step, before we move on in the next section to propagation through a dilute medium. We wish to start here with what one might think of as a single thin layer of a medium in which our electromagnetic field is propagating. Effectively we will take the result from this section and then (mathematically) we will stack up a series of these sheets to arrive at results for a finite-thickness medium. Consider an infinite sheet of dipoles in the x–y plane, all oscillating in phase and driven by an incident field propagating in the $+z$-direction, as shown in figure 4.2.

A detailed calculation shows that the field radiated by these dipoles is proportional to a dipole on the z-axis, but shifted in phase by 90°, and is given by

$$\vec{E}_{\text{radiated}} = i\frac{k\eta}{2\varepsilon_0}\alpha\vec{\mathcal{E}}e^{i(kz-\omega t)},$$

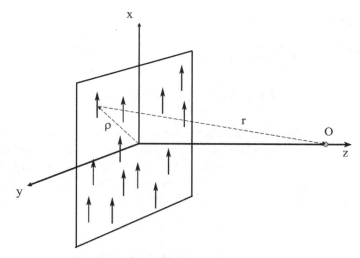

Figure 4.2. Radiation from a sheet of dipoles.

where k is the wavevector and η is the number of dipoles per unit area. Note that this result is obtained by averaging over the orientations of the dipoles on the sheet, assuming that the locations of the dipoles are fixed. We now wish to present a more complete derivation of this result following reference [5].

If we assume a dipole oscillating in the $\hat{x}$-direction we can write the dipole moment as

$$\vec{p} = \hat{x}\wp_0 e^{-i\omega t}.$$

We need to write the radiated field in the form appropriate for this orientation of the dipole, which looks more complicated than the result we found previously, for which we assumed the dipole to be oriented along the z-axis. Here we just give the result [6]

$$\vec{E}_{\text{radiated}} = \frac{\wp_0}{4\pi\varepsilon_0 r} e^{i(kr-\omega t)} \left\{ \frac{1-ikr}{r^2}[-\hat{x} + 3(\hat{x}\cdot\hat{r})\hat{r}] + k^2 (\hat{r}\times\hat{x})\times\hat{r} \right\}. \qquad (4.18)$$

We will work in Cartesian coordinates to carry out the integration over the entire sheet. The unit vector $\hat{r}$ is given by

$$\hat{r} = \frac{x\hat{x} + y\hat{y} + z\hat{z}}{r},$$

with $r^2 = x^2 + y^2 + z^2$. The vector products in equation (4.18) can be calculated; the results are

$$(\hat{x}\cdot\hat{r})\,\hat{r} = \frac{x}{r}\,\hat{r} \quad (\hat{r}\times\hat{x})\times\hat{r} = \left(\frac{x^2}{r^2} + \frac{y^2}{r^2}\right)\hat{x} + \frac{1}{r^2}(xy\,\hat{y} - xz\,\hat{z}).$$

Substituting these into equation (4.18) yields

$$\vec{E}_{\text{radiated}} = \frac{\wp_0}{4\pi\varepsilon_0 r}e^{i(kr-\omega t)}\left(\hat{x}\left[\frac{1-ikr}{r^2}\left(-1+3\frac{x^2}{r^2}\right)+k^2\frac{x^2+y^2}{r^2}\right]\right)$$

$$+\frac{\wp_0}{4\pi\varepsilon_0 r}e^{i(kr-\omega t)}\times\left(\hat{y}\left[3\frac{1-ikr}{r^2}\frac{xy}{r^2}+k^2\frac{xy}{r^2}\right]+\hat{z}\left[3\frac{1-ikr}{r^2}\frac{xz}{r^2}-k^2\frac{xz}{r^2}\right]\right). \tag{4.19}$$

As a first step in evaluating the above, we notice that all terms that involve odd powers of x or y will vanish when we integrate over the x–y plane. We can proceed either by following the clever method shown in [5], or by using a symbolic manipulation package such as Maple©, Matlab©, or Mathematica©. In either case, we find the result for the radiated field to be

$$\vec{E}_{\text{radiated}} = \hat{x}\eta\frac{i\wp_0 k}{2\varepsilon_0}e^{i(kz-\omega t)},$$

with η being the area density of dipoles on the sheet. The incident field is

$$\vec{E}_I = \hat{x}\mathcal{E}e^{i(kz-\omega t)} \tag{4.20}$$

so that the total field at the point z can be written as the sum of two pieces,

$$\vec{E}_T = \hat{x}\mathcal{E}e^{i(kz-\omega t)}+\hat{x}i\frac{k\eta}{2\varepsilon_0}\alpha\mathcal{E}e^{i(kz-\omega t)}, \tag{4.21}$$

where we have again used the relation $\wp = \alpha\mathcal{E}$. Note that the re-radiated field is shifted in phase by $\pi/2$ with respect to the incident field, since the factor 'i' is equivalent to $e^{i\pi/2}$. Another way of saying this is that the re-radiated field is proportional to the *velocity* of the charges rather than the position.

Thus the total field is the superposition of two contributions: the initial field and another out-of-phase piece proportional to the density and polarizability of the dipoles. We will make use of this result in the next section when we begin to look at the index of refraction of dilute media (figure 4.3).

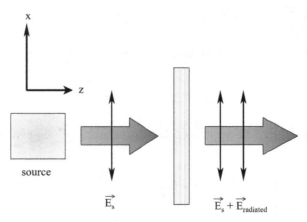

Figure 4.3. Illustration of the superposition of fields: the incident field and the re-radiated field from the sheet of dipoles add to give an apparent phase-delayed field.

4.4 Propagation in a dilute medium

In this section we are interested in propagation in a dilute medium, such that the molecules are far enough apart so that the fields produced by the molecules do not effect each other. To find the cumulative effects of all molecules in the medium on absorption and scattering, we may then simply add the contributions of the individual molecules.

In addition to scattering light and absorbing light, an assembly of induced dipoles affects the velocity of light as it passes through the medium. As pointed out earlier, the field at any point in the medium is the sum of the external (incident) field and all of the fields re-radiated by the induced dipoles. The phase of the re-radiated dipolar fields is shifted with respect to the incident field and thus the phase of the total field is affected by this superposition of two fields. A change of phase in the total wave is equivalent to a change in the phase velocity of the wave. Therefore, we find that the fact that the (phase) velocity of light in matter differs from the velocity of light in vacuum, a consequence of scattering and leads to the concept of the refractive index. This will be shown mathematically below.

We will follow an approach due to Feynman [7] to find an expression for the phase delay due to the index of refraction and for the attenuation of the beam caused by a thin dielectric plate. The derivation of the exact form for the phase shifted field due to an infinite sheet of dipoles we will leave for a section at the end of this chapter.

In chapter 3 we saw that the source of the electromagnetic waves we consider in this course is fundamentally an oscillating charge. If an electric field (in the form of a laser beam, for concreteness) is incident on a thin glass plate, for example, the electric field of the laser causes in the molecules of the medium a separation of the negatively charged electron cloud from the positively charged nucleus. The direction of the separation changes sign every half-cycle of the incident field. Thus we have created oscillating and therefore radiating dipoles in the medium. If we ask about the electric field on the opposite side of the plate from the incoming laser beam, we realize that it must be a superposition of the incident field and that of the radiating dipoles. Again, it is this superposition which leads finally to the effective phase delay of the field and the index of refraction.

We expect that the light will slow down somewhat as it travels through the plate, due to the excitation and reradiation of the dipoles, which takes some amount of time. We can write down an expression for the extra time delay

$$\Delta t = t_{\text{additional}} = (n - 1)\frac{\Delta z}{c},$$

where Δz is the thickness of the plate, c is the speed of light in vacuum and we introduce the factor n to account for the extra propagation time. With our previously determined plane wave form we can write the field after the plate as

$$\hat{x}\, E_{\text{after}} = \hat{x}\, E_s \exp\left[i\left\{ \frac{\omega}{c}z - \omega(t - \Delta t) \right\} \right]$$

$$= \hat{x}\, E_s \exp\left[i\left\{ \frac{\omega}{c}z - \omega\left(t - (n-1)\frac{\Delta z}{c} \right) \right\} \right] \tag{4.22}$$

$$= \hat{x}\, E_s \exp\left[i\left\{ \frac{\omega}{c}z - \omega t \right\} \right] \exp\left[i\frac{\omega}{c}(n-1)\Delta z \right],$$

where t is the propagation time for the distance z in the absence of the plate and E_s is the source electric field. If we consider that the phase delay due to the plate is small, the second exponential can be approximated by the first two terms of a Taylor series expansion ($e^{ix} \simeq 1 + ix$), giving us for the field

$$\hat{x}\, E_{\text{after}} = \underbrace{\hat{x}\, E_s \exp\left[i\left\{ \frac{\omega}{c}z - \omega t \right\} \right]}_{\text{source wave}}$$

$$+ \underbrace{i\, \hat{x}\, E_s \frac{\omega}{c}(n-1)\Delta z \exp\left[i\left\{ \frac{\omega}{c}z - \omega t \right\} \right]}_{\text{wave radiated by the dipoles in the plate}}. \tag{4.23}$$

Both waves propagate with a phase velocity c, as can be seen from the above expressions. The net result of adding both waves, one of which is phase shifted by $\pi/2$, is an effective change of the phase velocity of the wave as was described qualitatively above.

We can return to our harmonic oscillator model of the atoms in the dielectric thin plate and write the expression for the amplitude of the position as a function of frequency, equation (4.5):

$$x_0 = \frac{(e/m)\mathcal{E}_0}{(\omega_0^2 - \omega^2) - i\Gamma\omega}. \tag{4.24}$$

If we consider for a moment the situation for which the incident laser frequency, ω is far from the resonance frequency ω_0, the imaginary part of the denominator may be ignored and the position of the electron as a function of time can be written as

$$x(t) = x_0 e^{-i\omega t} \simeq \frac{(e/m)\mathcal{E}_0 \, e^{-i\omega t}}{(\omega_0^2 - \omega^2)}. \tag{4.25}$$

As we saw in the previous section, the field radiated by the thin sheet of dipoles is proportional to the velocity of the charges according to

$$\vec{E}_{\text{radiated}} = -\frac{\eta e}{2\varepsilon_0 c} \times \{\text{velocity of charges}\}_{\text{ret}} \, \hat{x}, \tag{4.26}$$

where 'ret' means that the field is evaluated at the 'retarded' time $t - z/c$ and η is the area density of dipoles. This result takes into account the fact that a field seen by an observer at time t was radiated by the dipole at an earlier time $t - z/c$, where z is the separation between the source and observer and c is the speed of light in a vacuum.

The velocity of the charges can be evaluated easily from (4.25) and substituted into the expression for the radiated field $\vec{E}_{\text{radiated}}$:

$$\vec{E}_{\text{radiated}} = -\hat{x}\frac{\eta e}{2\varepsilon_0 c}\{\dot{x}\}_{\text{ret}} = -\hat{x}\frac{\eta e}{2\varepsilon_0 c}\frac{e\mathcal{E}_0}{m(\omega_0^2 - \omega^2)}(-i\omega)e^{-i\omega(t-z/c)}.$$

We can rewrite this slightly by realizing that η, the number of dipoles per area, is related to the volume density of dipoles, N_v, by $\eta = N_v \cdot \Delta z$, and by using x_0 again. This yields

$$\vec{E}_{\text{radiated}} = -\hat{x}\frac{N_v\Delta z\ e}{2\varepsilon_0 c}(-i\omega)x_0 e^{-i\omega(t-z/c)}.$$

Our final step is to compare this result to that found above in equation (4.23). The two results can agree only if we have

$$n - 1 = \frac{N_v e^2}{2\varepsilon_0 m(\omega_0^2 - \omega^2)}. \tag{4.27}$$

This result gives us the index of refraction as a function of frequency, known as 'dispersion'. In addition, we have arrived at a self-consistent result that confirms our initial guess that the result of the sheet of material would cause an apparent time delay in the field propagating through. Recall that the result in equation (4.27) is valid only far from resonance.

We can look at the absorption due to a thin sheet of dipoles in a phasor diagram as well. The incident and radiated fields are shifted in phase by 90° as shown above. In addition we assume that the radiated field E_{radiated} is a small contribution (in magnitude). Thus we can represent the phasors as shown in figure 4.4(b). The resultant field $\vec{E}_{\text{after}}$ is barely changed in magnitude but is shifted in phase by 90°, since the radiated field is proportional to the *velocity* of the charges, which is cosinusoidal if the

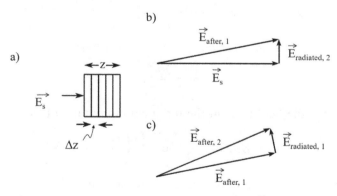

Figure 4.4. Phasor diagram illustrating the addition of the incident field, $\vec{E}_S$ and the radiated field, $\vec{E}_{\text{radiated}}$, which has a relative phase shift of 90° with respect to the incident field.

driving field is sinusoidal. If we were to consider a medium of some finite thickness, as illustrated in figure 4.4(a), the same technique could be applied successively to a series of thin sheets. At each step $\vec{E}_{after}$ from one piece becomes the input field to the next infinitesimal sheet. Essentially we would be performing an integration over the finite thickness. The rate of accumulation of the phase shift due to the medium is given by the factor $\exp\left(i\frac{n\omega}{c}z\right)$ (see equation (4.22)). In figure 4.4(c) the process is carried out one step further using the phasor representation.

After the light has passed through the block of material (we still consider the case of zero absorption), the electric field can be written, starting with equation (4.22):

$$\vec{E}_{after} = \hat{x}\,\mathcal{E}_0 \exp\left[i\left\{\frac{\omega}{c}z - \omega\left(t - (n-1)\frac{z}{c}\right)\right\}\right]$$

$$= \hat{x}\,\mathcal{E}_0 \exp\left[i\left\{\frac{\omega}{c}z + \frac{\omega}{c}(n-1)z - \omega t\right\}\right] \qquad (4.28)$$

$$= \hat{x}\,\mathcal{E}_0 \exp\left[i\left\{\frac{n\omega}{c}z - \omega t\right\}\right] = \vec{E}_S \exp\left[i\{kz - \omega t\}\right],$$

where we have used $\Delta z = z$ and define a more general wavenumber $k = n\omega/c$. This is the form of a plane wave propagating through a medium of index of refraction n. As we discussed in chapter 1, we can consider this to be a wave of frequency ω propagating with a phase velocity given by $v = c/n$.

Now we wish to consider the possibility that the electric field, and therefore the intensity of the light, is attenuated while propagating through the medium. From our discussion earlier in this chapter, we know that the polarizability of the driven damped oscillator/dipole is a complex quantity. In addition we have just seen that in the case in which damping can be neglected (our approximation that $\omega - \omega_0 \gg \Gamma$), the index of refraction is closely related to the real part of the polarizability. We will assume therefore that we can define a *complex* index of refraction $\tilde{n} = n + i\kappa$. If we use this in our expression for $\vec{E}_{after}$, equation (4.28), we can write

$$\vec{E}_{after} = \vec{E}_S \exp\left[-\frac{\kappa\omega}{c}z\right]\exp\left[i\left\{\frac{n\omega}{c}z - \omega t\right\}\right].$$

If we define an *intensity absorption coefficient* γ such that

$$\gamma = \frac{2\omega}{c}\kappa,$$

we can then write the field $\vec{E}_{after}$ in the more commonly used form

$$\vec{E}_{after} = \vec{E}_S \exp\left[-\frac{\gamma}{2}z\right]\exp\left[i\left\{\frac{n\omega}{c}z - \omega t\right\}\right].$$

The imaginary part of the complex index of refraction is thus proportional to the absorption in the medium.

We can return to our phasor diagram representation and look at absorption as well. If we examine equation (4.24) we see that on resonance ($\omega = \omega_0$) the position and velocity of the electron are given by

$$x(t) = \frac{(e/m)\mathcal{E}_0}{-i\Gamma\omega_0}e^{-i\omega t}$$

$$\dot{x}(t) = -i\omega_0\frac{(e/m)\mathcal{E}_0}{-i\Gamma\omega_0}e^{-i\omega t} = \frac{e\mathcal{E}_0}{m\Gamma}e^{-i\omega t}.$$

The position of the electron is shifted in phase by $90° = \pi/2$ with respect to the driving field, while the velocity is exactly in phase with the incident field. Since the field from the dipole sheet is proportional to the charge velocity (equation (4.26)), we conclude that the radiated field is exactly out of phase with the driving field ($180°$ shift). This means that the radiated field subtracts from the driving field, i.e. the light is attenuated. In the phasor diagram shown in figure 4.5(a) this is represented by a small arrow pointing in the opposite direction of the incident field phasor.

If we assume a thin slab (thickness Δz) of absorptive material we can write, for $\gamma\Delta z \ll 1$,

$$\vec{E}_{after} = \vec{E}_S\left\{1 - \tfrac{\gamma}{2}\Delta z\right\}\exp\left[i\left\{\tfrac{\omega_0}{c}\Delta z - \omega_0 t\right\}\right]$$

$$= \hat{x}\mathcal{E}_0\exp\left[i\left\{\tfrac{\omega_0}{c}\Delta z - \omega_0 t\right\}\right] - \hat{x}\tfrac{\gamma}{2}\Delta z\mathcal{E}_0\exp\left[i\left\{\tfrac{\omega_0}{c}\Delta z - \omega_0 t\right\}\right],$$

which shows more directly the subtraction of the source field and the radiated field. We can also consider this to be an interference effect: absorption is simply partial destructive interference of the incident wave with the radiated wave. Using

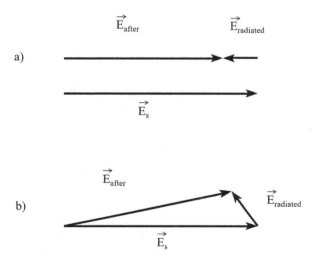

Figure 4.5. Phasor diagrams for (a) pure absorption and (b) a general case in which both absorption and dispersion are present.

equation (4.26) we can find an expression for the attenuation coefficient γ in terms of the electron oscillator parameters:

$$\gamma = \frac{N_v e^2}{m\Gamma\varepsilon_0 c}.$$

The units of γ are inverse distance, m^{-1}.

For a general case in which both absorption and dispersion are present, the simple harmonic oscillator model gives a complex index of refraction (the details are left for chapter 6)

$$\tilde{n} = n + i\kappa = 1 + \frac{N_v e^2}{2m\varepsilon_0\left(\omega_0^2 - \omega^2 - i\Gamma\omega\right)}.$$

We can separate the real and imaginary parts of the above expression to yield

$$n = 1 + \frac{N_v e^2\left(\omega_0^2 - \omega^2\right)}{2m\varepsilon_0\left(\left(\omega_0^2 - \omega^2\right)^2 + \Gamma^2\omega^2\right)}$$

$$\kappa = \frac{N_v e^2 \Gamma\omega}{2m\varepsilon_0\left(\left(\omega_0^2 - \omega^2\right)^2 + \Gamma^2\omega^2\right)}.$$

These reduce to the special cases we considered above for $\Gamma = 0$ and for $\omega = \omega_0$. In figure 4.5(b) the corresponding phasor diagram is shown in which both dispersion and absorption are present. Plots of the real and imaginary parts of the complex index of refraction $\tilde{n}$ are shown in figures 4.6 and 4.7. Note the region of 'anomalous dispersion' near the scaled frequency $\omega/\omega_0 = 1$. This will be discussed later in more

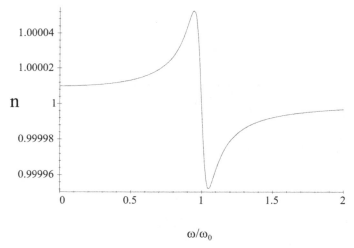

Figure 4.6. Real part of the complex index of refraction, n, plotted as a function of the scaled frequency ω/ω_0. The plot is for parameters $N_v e^2/2m\varepsilon_0\omega_0^2 = 10^{-5}$ and $\Gamma/\omega_0 = 0.1$. The dispersive lineshape is characteristic.

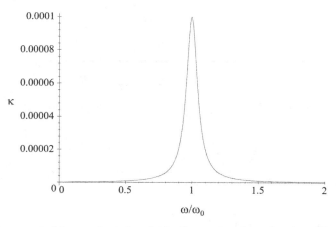

Figure 4.7. Imaginary part of the complex index of refraction, κ, plotted as a function of the scaled frequency ω/ω_0. The plot is for parameters $N_v e^2/2m\varepsilon_0\omega_0^2 = 10^{-5}$ and $\Gamma/\omega_0 = 0.1$. The absorption coefficient, γ, as defined in the text is $(2\omega_0/\Gamma c) \times \kappa$.

detail. In addition, note that the index of refraction, n can take on values less than unity near an absorption feature. In terms of the phase velocity, $v_p = c/n$, this implies velocities greater than the speed of light. We mentioned briefly in chapter 1, while discussing phase and group velocities, why this does not contradict the common wisdom that 'nothing can go faster than the speed of light'.

4.5 Beer's law

One illustration of the extinction cross section we discussed in the previous chapter is 'Beer's law of absorption' and follows immediately from the results we derived there. Consider a volume of thickness dz (the incident wave propagates in the z-direction) and area A as shown in figure 4.8.

Suppose also that the volume is filled with molecules of number density N_v, the number of molecules per unit volume. The total number of atoms in the volume is then $N_v A dz$. The total effective area blocked by the molecules is $\sigma_{\text{ext}} N_v A dz$, which is the 'area' of each molecule multiplied by the total number of molecules. Therefore the amount of power removed from the beam is

$$P_{\text{ext}} = \sigma_{\text{ext}}(N_v A dz)I.$$

If we call ΔI the irradiance removed from the incident beam, we have

$$\Delta I = -\frac{P_{\text{ext}}}{A} = -(\sigma_{\text{ext}} N_v dz)I,$$

which leads to, for a short pathlength increment ($\Delta I \to dI$),

$$\frac{dI}{dz} = -\gamma I \tag{4.29}$$

A

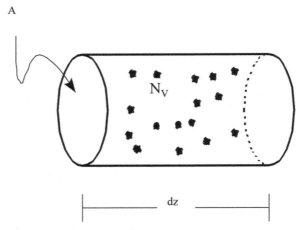

N_V

dz

Figure 4.8. Geometry for absorption by a collection of atoms.

with the 'attenuation' or 'absorption' coefficient given by $\gamma = \sigma_{ext} N_v$. Integrating this equation leads to 'Beer's Law', written in the following form,

$$I = I_0 e^{-\gamma z}.$$

Note, we do not call γ the extinction coefficient because a different quantity which we will introduce soon is commonly referred to as the extinction coefficient. Here again we have an example of the ambiguity that exists over the definition of extinction and absorption.

4.6 Broadening

From the sketch of σ_{ext}, figure 4.1, we see that the scattering is a maximum at resonance, $\omega = \omega_0$. However, as seen in the figure, the transition to resonance scattering is not a sharp one, i.e. there is a finite width to the curve of σ about ω_0. This is commonly called a linewidth and it is expressed quantitatively by its width at half maximum, which in figure 4.1 is Γ. There are several mechanisms which may be responsible for the spread in frequency (called line broadening) of the light scattered from any source [8]. These mechanisms are, first of all, the radiative decay process itself; the quantum mechanical process by which light (photons) are emitted from a source, which is the quantum mechanical analog to the classical oscillating dipole model. To treat this *natural broadening* or *radiative broadening* rigorously involves a treatment of the quantum mechanical concept of spontaneous emission which leads to a so-called Lorentzian frequency linewidth. We will treat a model for spontaneous emission in terms of the damped harmonic oscillator which reveals the most important features of the linewidth. The natural linewidth is in most situations the *minimum possible* spread of a particular emission line. Other mechanisms which may contribute to the linewidth are due to the motion of the atoms or molecules. A spread in the velocities of the atoms or molecules leads to what is called *Doppler broadening. Collision broadening*, as the name implies, is due to random collisions of atoms. Collision broadening also leads to a shift in the resonance frequency ω_0. If we probe an atomic sample with a laser, there is another broadening mechanism to consider, namely *power broadening*. This occurs when the atom is effectively driven so hard that it is forced to emit light in a time shorter than its natural lifetime (stimulated emission). Treatment of this latter effect is beyond the scope of this text, but is extremely important for understanding lasers [9]. Another effect we will not consider here is *time-of-flight broadening* which is due to an atom flying through a probe laser beam so fast that it effectively sees a short pulse of light with a frequency spread larger than its own natural linewidth. We will consider the natural linewidth first and then the collision and Doppler broadening mechanisms in some detail, along with a discussion of combinations of these, which are typically present in a real sample.

4.6.1 Natural line broadening

In discussing the natural lineshape of an emission or absorption line, we need to consider the lifetime of the emission. In a quantum mechanical picture, this is just the lifetime of the excited state which ultimately decays to a lower state with the emission of a photon. The lifetime of the excited quantum state corresponds in a classical picture to the duration of the optical wave train emitted by an excited harmonic oscillator. The finite temporal extent of the wave train has a Fourier decomposition which includes frequencies other than ω_0. This gives what is called lifetime or natural broadening whose lineshape is also Lorentzian. The lifetime, τ, of the state is the inverse of the damping factor, i.e. $\tau = 1/(\Gamma_N/2)$. As stated previously, the minimum line broadening mechanism is usually associated with the natural linewidth.

We can consider the harmonic oscillator model of an atom, now without the driving term. The equation of motion is given by

$$\ddot{x} + \frac{\Gamma_N}{2}\dot{x} + \omega_0^2 x = 0,$$

where, as usual, $\omega_0^2 = \zeta/m$, with ζ being the spring constant. We are interested in the situation in which oscillator is started at some maximum amplitude and then released. Damping will cause the oscillator to eventually return to a state of zero motion ($x(t) \rightarrow 0$; $\dot{x}(t) \rightarrow 0$). This corresponds to starting with an atom in an excited state, with energy E_b and (in the quantum mechanical picture) allow it to decay spontaneously to the ground state (energy E_a). The energy difference between the two states is given by $E_b - E_a = \hbar\omega_0$.

The general solution to the above differential equation is given by

$$x(t) = x_0 e^{-(\Gamma_N/2)\,t}\left[\cos \omega t + \left(\frac{\Gamma_N}{2\omega}\right)\sin \omega t\right]. \qquad (4.30)$$

Where the initial conditions $x(0) = x_0$ and $\dot{x}(0) = 0$ have been used. Here

$$\omega = \sqrt{\omega_0^2 - \Gamma_N^2/4}\,.$$

In nearly any case we might consider, the frequencies of interest are on the order of $10^{13} - 10^{16}$ s^{-1}, whereas the damping constants might range from $10^6 - 10^{10}$ s^{-1}. Therefore we can simplify equation (4.30) by dropping the second term in brackets; in addition we can set $\omega \approx \omega_0$. This leaves the result we are after,

$$x(t) = x_0 e^{-(\Gamma_N/2)t} \cos \omega_0 t.$$

We can look at this result another way. The damping of the oscillatory motion means that the resulting electromagnetic wave is no longer exactly monochromatic. The *Fourier transform* of a time-varying function gives the set of frequency components which make up the oscillatory motion. We can define the motion $x(t)$ in terms of the different frequency components as

$$x(t) = \frac{1}{\sqrt{2\pi}} \int_0^\infty A(\omega)\, e^{i\omega t} d\omega.$$

Thus the oscillatory motion is made up of a 'sum' of oscillations at frequency ω, each with an amplitude of vibration $A(\omega)$. The goal is to determine the amplitudes $A(\omega)$, which follow from the Fourier transform relation

$$A(\omega) = \frac{1}{\sqrt{2\pi}} \int_0^\infty x(t) e^{-i\omega t} dt$$
$$= \frac{1}{\sqrt{2\pi}} \int_0^\infty \left(x_0\, e^{-\frac{\Gamma_N}{2}t} \cos \omega_0 t \right) e^{-i\omega t} dt.$$

This integration can be easily carried out, since we have made the approximation that $\omega \approx \omega_0$. The intensity, which is the variable of interest for a measurement, is proportional to $A^*(\omega) \cdot A(\omega)$; if we make the approximation that frequencies of interest to us are relatively near resonance, so that $(\omega - \omega_0) \ll \omega_0$, the final result is

$$I(\omega - \omega_0) = C \cdot \frac{1}{(\omega - \omega_0)^2 + \left(\frac{\Gamma_N}{2}\right)^2},$$

where C is a constant, often chosen such that the intensity is normalized. This characteristic intensity profile is called a *Lorentzian lineshape*.

To find the normalized lineshape function we integrate over all frequencies,

$$\int_{-\infty}^{+\infty} d\omega I(\omega - \omega_0) = \frac{2\pi}{\Gamma_N}.$$

Thus we can write the general Lorentzian lineshape function as

$$g_L(\omega - \omega_0) = \frac{\Gamma/2\pi}{(\omega - \omega_0)^2 + \left(\frac{\Gamma}{2}\right)^2},$$

where $\Gamma = \Gamma_N$ for natural broadening, but might represent other mechanisms in a more general treatment. The Lorentzian lineshape is illustrated in figure 4.9 and compared to the Gaussian lineshape to be derived in the next section.

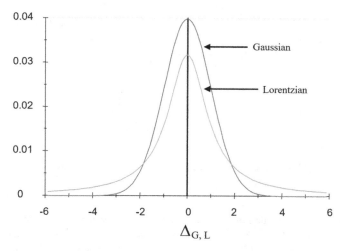

Figure 4.9. Plot of Lorentzian and Gaussian lineshapes. The two are plotted here such that the full-width at half-maximum (FWHM) is the same in each case. The curves are normalized as well. The detuning parameters for the frequency offset, scaled by the width of the curve, are given by $\Delta_L = (\omega - \omega_0)/(\Gamma_L/2)$ and $\Delta_G = (\omega - \omega_0)/(\Gamma_D/2)$.

4.6.2 Doppler broadening

The fact that atoms and molecules are not at rest can often produce observable optical effects. This motion is generally due to random thermal agitation, but in addition there may be organized motion of a part of the system containing many atoms or molecules such as in the case of turbulence. Suppose an atom at rest with respect to an observer emits a wave at the frequency ω_0. If the atom emitting the wave is moving with a velocity $\vec{v}$ relative to the observer, the frequency is Doppler shifted according to the expression, $\omega' = \omega - \vec{k} \cdot \vec{v}$. Applications of this Doppler shift are numerous, particularly in the field of astronomy where it is used to determine the velocities of distant galaxies. Or by the police when measuring the speed of your car. For the case of the Doppler shift in gases, the net effect when many atoms are involved in chaotic motion is a broadening of the shape of the spectral line in a way that can help reveal the temperature of the source.

To consider the Doppler shift for a many particle system such as a gas of molecules, one needs to take note of the fact that not all of the molecules are moving at the same velocity. In addition, for light incident along the z-axis, only the velocity component along z, i.e. v_z, contributes to the broadening. We may take the velocity distribution of the molecules to be given by a Maxwellian distribution, which says that the probability $\rho(v_z)dv_z$ of finding molecules in the velocity interval between v_z and $v_z + dv_z$ is given by [10],

$$\rho(v_z)dv_z = \left[\frac{m}{2\pi k_B T}\right]^{1/2} e^{-mv_z^2/2k_B T} dv_z, \tag{4.31}$$

where k_B is Boltzmann's constant (not to be confused with the other use of k, namely the wavevector), T is the Kelvin temperature, and m is the mass of the molecule.

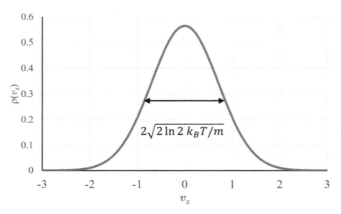

Figure 4.10. Maxwellian distribution of velocities, $\rho(v_z)$. The full-width at half-maximum (FWHM) of the distribution is given by $\Delta = 2\sqrt{2 \ln 2k_BT/m}$. For convenience here, $m/2k_BT$ is set to unity.

The range of velocities of the particles is determined by the half-width, Δ, of the Maxwellian distribution. It can be shown that Δ is equal to $\sqrt{2 \ln 2k_BT/m}$. The velocity distribution is shown in figure 4.10.

Using equation (4.13) we are now able to write down the power extinction from a light beam incident along the z-axis on a dilute gas of N molecules, taking into account the different velocities of the molecules and the Doppler shift of the scattered light simply by integrating over the normalized velocity distribution,

$$\langle P_{\text{ext}} \rangle = \frac{N_v \, \omega^2}{2} \frac{|eE|^2 \, \Gamma}{m} \int_{-\infty}^{\infty} \frac{\rho(v_z)}{|[\omega_0^2 - (\omega - kv_z)^2] - i\Gamma(\omega - kv_z)|^2} dv_z \qquad (4.32)$$

$$\equiv \sigma_{\text{ext}} I_0,$$

where $I_0 = |E^2|/2\mu_0 c$ and the extinction cross section is defined as earlier in the chapter. We have replaced ω in equation (4.13) by $\omega - kv_z$, except in the numerator where we use the fact that $\omega \gg kv_z$. Note equation (4.13) is the power extinction for a single molecule. The factor N_v appears in equation (4.32) because there are N_v molecules per unit volume. Generally, Γ and kv_z will be small compared to ω and ω_0, so that the denominator in the integral above becomes (dropping terms of order v_z^2),

$$\omega_0^2 - \omega^2 + 2kv_z\omega - i\Gamma\omega \simeq 2\omega(\omega_0 - \omega) + 2kv_z\omega - i(2\omega)\frac{\Gamma}{2}. \qquad (4.33)$$

We can now write the denominator as

$$\simeq 4\omega^2[(\omega_0 - \omega + kv_z)^2 + (\Gamma/2)^2].$$

With these approximations the expression for the extinction cross section for N molecules becomes

$$\sigma_{\text{ext}} = \frac{N_v e^2 \Gamma}{4 \, m\varepsilon_0 c} \sqrt{\frac{m}{2\pi k_BT}} \int_{-\infty}^{\infty} \frac{e^{-mv_z^2/2k_BT}}{[(\omega_0 - \omega + kv_z)^2 + (\Gamma/2)^2]} dv_z. \qquad (4.34)$$

In the above expression for σ_{ext}, a modified Lorentzian lineshape function can be seen and can be written for our present purposes as

$$g_L(kv_z - (\omega - \omega_0)) \rightarrow \frac{\Gamma/2\pi}{(kv_z - (\omega - \omega_0))^2 + (\Gamma/2)^2},$$

from which we can write the expression for the extinction cross section as

$$\sigma_{\text{ext}} = \frac{N_v e^2}{4\,m\varepsilon_0 c}\sqrt{\frac{m}{2\pi k_B T}}\int_{-\infty}^{\infty} e^{-mv_z^2/2k_B T}\,2\pi\,g_L[kv_z - (\omega - \omega_0);\Gamma]dv_z. \tag{4.35}$$

We are often interested in the limiting case for which the spread in energies of the molecules is large compared to the energy spread corresponding to the natural linewidth. If $k\Delta \gg \Gamma$, as might often be the case in the visible region of the spectrum, the variation of the Gaussian function will be very slight compared to that of the Lorentzian function which means that we can essentially pull the exponential term outside the integral. To do so we assume the exponential function takes on the value it has at the center of the Lorentzian, where $kv_z = \omega - \omega_0$. Substituting this value for v_z, we arrive at the following expression for the extinction cross section, since the Lorentzian function left inside the integral is normalized:

$$\sigma_{\text{ext}} = \frac{N_v \pi e^2 \omega_0}{2\,m\varepsilon_0 c^2}\left[\frac{c}{\omega_0}\sqrt{\frac{m}{2\pi k_B T}}\,\exp\left[-\frac{mc^2}{2k_B T}\frac{(\omega - \omega_0)^2}{\omega_0^2}\right]\right]. \tag{4.36}$$

The expression in square brackets is called a Gaussian or Doppler-broadened lineshape, $g_D(\omega - \omega_0)$, and is also referred to as an inhomogeneously broadened line profile, in reference to the fact that different velocity sub-groups of atoms experience different frequency shifts. Figure 4.9 shows an example of this lineshape. The width of the line, the full-width at half-maximum (FWHM) is given by $\Gamma_D = 2\omega_0(2k_B T \ln 2/mc^2)^{1/2}$. We may use this expression to write the lineshape in a slightly different form,

$$g_D(\omega - \omega_0) = \frac{2}{\Gamma_D}\sqrt{\frac{\ln 2}{\pi}}\,e^{-4\ln 2\left[(\omega - \omega_0)/\Gamma_D\right]^2}. \tag{4.37}$$

For oxygen molecules (atomic mass $M = 32$) at room temperature, $T = 300$ K, and considering the near-infrared transition at $\lambda = 762$ nm ($\omega_0 = 2.5 \times 10^{15}$ s^{-1}), we find a Doppler width of $\frac{\Gamma_D}{2} = 2\pi \times 426$ MHz. Common spectroscopic units for linewidths are 'wavenumbers', $\Delta\tilde{\nu}_D = \Gamma_D/c = 426$ MHz/3×10^{10} cm s^{-1} = 0.0142 cm^{-1}.

A second limiting case is relevant in the far infrared part of the spectral region. In this case, the Maxwellian distribution becomes a delta function,

$$\sqrt{\frac{m}{2\pi k_B T}}\,\exp\left[-\frac{mv_z^2}{2k_B T}\right] \simeq \delta(v_z). \tag{4.38}$$

The above equation for the delta function can be found in most texts which treat the properties of delta-functions. The expression for the extinction cross section becomes

$$\sigma_{\text{ext}} = \frac{N_v \pi e^2}{2m\varepsilon_0 c} \int_{-\infty}^{\infty} g_L((kv_z - (\omega - \omega_0)))\delta(v_z)\, dv_z, \tag{4.39}$$

which again is straightforward to integrate, yielding

$$\sigma_{\text{ext}} = \frac{N_v \pi e^2}{2m\varepsilon_0 c} g_L(\omega - \omega_0). \tag{4.40}$$

The Lorentzian lineshape is called a homogeneously broadened line because essentially all atoms or molecules experience the same frequency shifts.

4.6.3 Collision broadening

We will now look at the mechanism of collision broadening and for simplicity assume that the other two broadening mechanisms are not present. Consider an atom radiating light of frequency ω_0. One can imagine a wave train of electromagnetic radiation constantly emanating from the atom until the atom experiences a collision. During a collision the energy 'levels' of the radiating atom are shifted by the interaction forces of the two colliding atoms. Thus the radiating wave train is interrupted during the duration of the collision. After the collision, the radiation of frequency ω_0 is resumed with the identical characteristics of the previous wave train except that the phase is totally unrelated to the phase of the wave before collision, as shown in figure 4.11. If the duration of the collision is

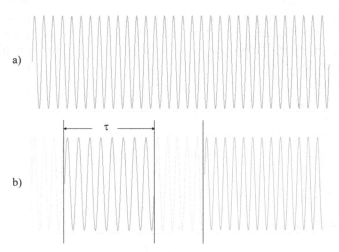

Figure 4.11. Two examples of the wave emitted by a harmonic oscillator. In (a) we see an undisturbed oscillator which emits radiation at a frequency ω_0, whereas in (b) the emission is shown from an oscillator that is interrupted by collisions. The emission frequency is still always ω_0, but the phase is randomized by the interactions with the other atoms or molecules.

sufficiently brief, it is possible to ignore any radiation emitted during the collision while the frequency is shifted from ω_0. The collision broadening effect can then be represented by a model in which each excited atom always radiates at the frequency ω_0, but with random changes in the phase of the radiated wave each time a collision occurs. The apparent spread in the emitted frequencies arises because the wave is chopped up into finite sections whose Fourier decompositions include frequencies other than ω_0. It can be shown (see reference [11] for example) that the broadening due to collisions is homogeneous or Lorentzian in shape.

As shown in figure 4.11 we can write the emitted field amplitude as

$$E(t) = E_0 \exp(-i\omega_0 t + i\phi)$$

during the time τ for any one of the (collision-shortened) wave trains. Note that we assume the amplitude to always be the same, as is the frequency. To gain some information about the time intervals τ between collisions, we turn to the kinetic theory of gases [10] and find a probability distribution that can be written as

$$p(\tau) = \frac{1}{\tau_0} \exp(-\tau/\tau_0)d\tau.$$

The mean collision time τ_0 can be related to molecular parameters, as well as to the gas density. For most purposes it is adequate to assume that the collision rate can be described by a function of the form

$$\frac{1}{\tau_0} = a + bP, \tag{4.41}$$

where P is the gas pressure, which for an ideal gas is proportional to the density.

We are interested in a spectrum, which is proportional to the modulus-squared of the Fourier transform of the electric field. For one atom the averaged intensity over one cycle is thus

$$I(\omega) \propto |E(\omega)|^2 \tag{4.42}$$

$$= \left| \frac{1}{2\pi} \int_{t_0}^{t_0+\tau} E_0 e^{-i(\omega_0 t - \phi)} e^{i\omega t} dt \right|^2 \tag{4.43}$$

$$= \left(\frac{E_0}{\pi} \right)^2 \frac{\sin^2((\omega_0 - \omega)\tau/2)}{(\omega_0 - \omega)^2}. \tag{4.44}$$

To find the total intensity emitted by a collection of N interacting atoms, with a mean time between collisions of τ_0, we average the previous result weighted by the collision probability distribution, yielding finally a Lorentzian lineshape again:

$$I_{\text{total}} \propto \frac{1}{(\omega_0 - \omega)^2 + (1/\tau_0)^2}. \tag{4.45}$$

We can write this in terms of a linewidth, $\Gamma/2 = 1/\tau_0$, such that the linewidth is related to the inverse of the mean time between collisions:

$$g_L = \frac{\Gamma/2\pi}{(\omega_0 - \omega)^2 + (\Gamma/2)^2}.$$

Collision broadening coefficients (the factor b in equation (4.41)) vary widely depending on molecular species and on the transitions involved. A rough estimate might give $b = 0.05 \text{ cm}^{-1}/\text{atm} = 2 \text{ MHz/Torr}$, taken for near-infrared molecular oxygen transitions.

4.6.4 Voigt profile

From the discussion in the previous sections it should be clear that for an actual atomic or molecular system one must somehow combine the various linewidth broadening mechanisms. For example, an atomic gas might have a natural linewidth of $\Gamma_N = 2\pi \times 10 \text{ MHz}$, a Doppler broadening of $\Gamma_D = 2\pi \times 500 \text{ MHz}$, and pressure broadening of similar or larger magnitude. In general it is not possible to find an analytic expression for the combined lineshape. The mathematical procedure used for finding the resultant of two functions is called a convolution and in our case takes the form

$$g(\omega - \omega_0) = \int_{-\infty}^{+\infty} d\omega' g_1(\omega' - \omega) g_2(\omega_0 - \omega').$$

Effectively, one multiplies every frequency ω' of the first lineshape by the second lineshape, and then integrates over all possible frequencies of that first lineshape. There are some simple cases, such as the convolution of two Lorentzians or the convolution of two Gaussians, in which the result can be obtained analytically. However, the more common situation is probably the convolution of a Gaussian Doppler-broadened line with a Lorentzian pressure-broadened, and this requires a numerical solution.

In figure 4.12 the lineshape components due to Doppler broadening and collision broadening are shown as a function of pressure. We assume that the sample temperature, and thus the Doppler contribution, is constant.

4.7 Absorption spectroscopy experiment

As an example to illustrate some of the ideas presented thus far in this chapter we would like to consider a concrete experiment in which absorption of light by molecular oxygen is measured. The basic schematic of the experiment is shown in figure 4.13. Molecular oxygen has a well-known set of absorption peaks in the near-infrared portion of the spectrum, spanning several nanometers about a center wavelength of 762 nm. The frequency corresponding to this wavelength is commonly given in terms of wavenumbers or cm^{-1}, in this case 13 123 cm^{-1}. These wavelengths are conveniently reached using semiconductor diode lasers, solid-state titanium sapphire (Ti:Al$_2$O$_3$) lasers, or liquid dye lasers. The light source can be fixed at one frequency or can be scanned in frequency if one is interested in measuring the absorption as a function of frequency.

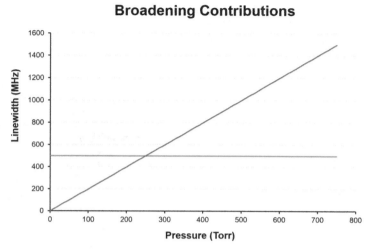

Figure 4.12. Lineshape components due to Doppler and pressure broadening as a function of gas pressure. In the low- and high-pressure extremes the lineshape would be adequately described by a Gaussian or a Lorentzian function, respectively.

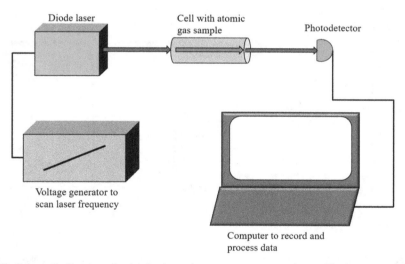

Figure 4.13. Schematic diagram of a simple absorption spectroscopy experiment. The laser output is incident on a sample cell, and the amount of light transmitted through the sample is monitored with a photodetector and recorded for further data processing.

A first experiment would be to vary the pathlength of the sample and measure the relative transmitted power as a function of the pathlength. A possible set of data might look as shown in figure 4.14. Along with the data points we have included an exponential fit to the data, with the result displayed on the plot. The plot is conveniently made such that 100% transmission is 1 mW on the plot, a somewhat artificial choice. Often one would normalize the power output to the input power so

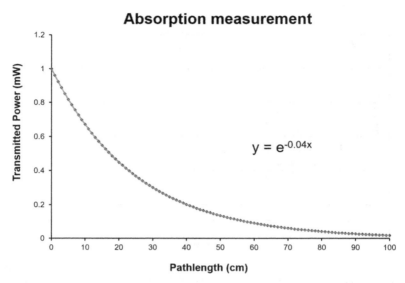

Figure 4.14. Plot of transmitted power (in mW) as a function of sample pathlength in centimeters. An exponential fit to the data allows one to extract relevant molecular parameters.

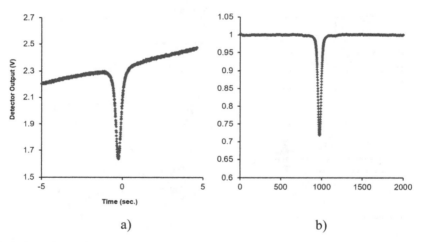

Figure 4.15. Plot of transmitted power (in terms of photodiode signal voltage) as a function of laser frequency. In (a) the raw data are shown, with the abscissa representing the scan time in seconds. In (b) the transmitted power has been normalized and the 'frequency' scale is still uncalibrated. Data collected by the author.

that '1' represents total transmission. The exponential fit function is precisely Beer's law as discussed earlier in this chapter; it is possible to extract from this fit to the data the attenuation coefficient; if the number density of molecules is known (from the pressure and temperature, which are easily measured) then the extinction coefficient can also be found from the data.

Next we look at the data taken as the frequency of the laser is swept over a range of about 1.5 cm^{-1} or 45 GHz. In figure 4.15(a) we show the raw data. The vertical scale corresponds to the photodiode voltage, while the horizontal axis is the sweep time in seconds. To eliminate the need to worry about absolute powers, it is convenient to normalize the data, usually to an identical scan, but taken with an empty cell. Next the time scale must be converted to frequency. In figure 4.15(b) we have presented the data only in terms of the number of points taken, i.e. this scan is a record of 2000 individual points on the digital oscilloscope. A further step is necessary, namely to calibrate the number of scan points per frequency interval. In the case of oxygen this can be done by tuning the laser to a pair of absorption lines with a well-known separation and using them as a calibration for the laser scan. There are many subtleties involved in correctly calibrating frequencies, but these are beyond the scope of what we want to cover in this text.

If we perform the same absorption spectroscopy experiment using oxygen at different pressures, always scanning the same frequency range, we might find the results shown in figure 4.16, in which several absorption curves are shown for a weak probe laser propagating through a 6 m pathlength of oxygen at various pressures. For this oxygen transition the natural linewidth is completely negligible at only 2 mHz. We see that the absorption lineshape changes qualitatively as the pressure increases from 15 Torr through intermediate pressures up to nearly one atmosphere of oxygen pressure. As discussed in section 4.6, at low pressures the lineshape is represented best by a Gaussian function, with width determined by the temperature of the sample. For the highest pressures the lineshape is well-represented by a Lorentzian function, with the linewidth determined by the pressure of the gas.

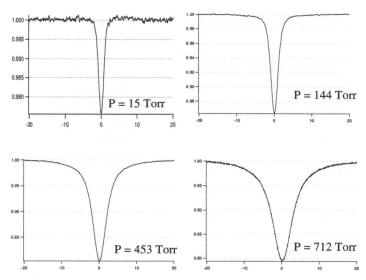

Figure 4.16. Examples of different lineshapes for absorption by molecular oxygen at different pressures. At the lowest pressure the lineshape is essentially Gaussian (Doppler) broadened. At the highest pressures a Lorentzian lineshape describes the spectrum adequately, since pressure broadening dominates. For intermediate pressures the Voigt profile or more complicated expression is required to fit a theoretical curve to the data. The absorption curves are normalized, and the horizontal scale is in GHz.

The presence of a pressure-broadened component is signaled by the long 'tails' of the lineshape; the Lorentzian distribution falls off as a function of frequency much more slowly than the Gaussian distribution. When the pressure is between these two extremes the lineshape is a combination of the limiting profiles and described by the convolution of Gaussian and Lorentzian lineshapes known as the Voigt profile. Note as well that as the pressure is increased, the absorption increases initially (there are more molecules available to scatter light) and the linewidth broadens. Eventually the amount of peak absorption saturates as the linewidth broadens.

To extract molecular parameters from these data we must perform a nonlinear least-squares fit to the data, taking the Voigt profile as our fitting function. This procedure can be performed analytically; an example is shown in figure 4.17. The best fit curve cannot even be distinguished on this plot, as it is buried in the data points. To help determine the quality of the fit, it is useful to include the 'residuals' (upper part of figure 4.17). The residuals are simply the difference at each data point between the least-squares fit function and the actual data point. It can be seen from figure 4.17 that near the absorption peak some non-random deviations from zero occur in the residuals. These are a sign that even the Voigt function might not be sufficient to describe the lineshape perfectly. To go beyond this point it is necessary to look in more detail at the types of collisions that are taking place and is well beyond the scope of this text [8].

With this example we finish our discussion of propagation of light in a dilute medium, extinction of the electromagnetic radiation by atomic or molecular dipoles, and line broadening due to spontaneous emission, temperature, and motion, and to collisions.

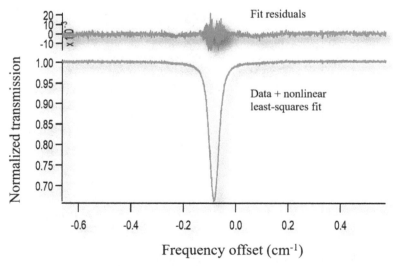

Figure 4.17. An example of the fitting to an absorption profile using the Voigt function. The vertical scale shows the normalized absorption, while the abscissa is the relative frequency, given in cm^{-1}. The fit function is not distinguishable, being buried in the data. The residuals plotted at the top represent the difference between the fit function and the actual data.

4.8 Problems

1. Find σ_{abs} for a dilute gas of N dipoles.
2. Show that the index of refraction of a dilute gas of N_V atoms/volume, with small damping, is given by

$$n \sim 1 + \frac{N_v e^2}{2\epsilon_0 m_e(\omega_0^2 - \omega^2)}.$$

3. Consider a dilute gas of N_V oscillators/volume. Find an approximate expression for the attenuation coefficient, γ, in Beer's law for ω near ω_0.
4. Determine how k'' is related to the extinction cross section.
5. Show by making a plot that the approximate expression and the exact expression for the extinction cross section and for the real part of the index of refraction are essentially identical. Assume optical transitions. Of what order are the neglected terms in the approximation? Give a numerical estimate.
6. Show that for the normalized Lorentzian function given in the text

$$\int_{-\infty}^{\infty} g_L(x)dx = 1.$$

7. Find the total scattering cross section by an induced dipole when,
   (a) $\omega \gg \omega_0$
   (b) $\omega \sim \omega_0$
   (c) $\omega \ll \omega_0$.
8. Show that the full-width at half-maximum of the extinction cross section, figure 4.1, is Γ.
9. Show that two independent homogeneous (Lorentzian) sources of broadening with rates Γ_1 and Γ_2 relevant for the same molecular system will lead to an overall Lorentzian lineshape with a full-linewidth of $\Gamma = \Gamma_1 + \Gamma_2$.
10. The pressure broadening coefficient for alkali atom vapors is approximately 0.25 cm^{-1}/atm. Find the Doppler width for a room temperature vapor of rubidium, with a transition at 780 nm. For what pressure (roughly) might a pure Lorentzian lineshape provide an adequate description of the absorption coefficient? For what temperature does the Doppler width no longer dominate the natural linewidth of 5 MHz?

References

[1] Wikipedia: Joseph von Fraunhofer https://en.wikipedia.org/wiki/Joseph_von_Fraunhofer
[2] Wikipedia: Hendrik Lorentz https://en.wikipedia.org/wiki/Hendrik_Lorentz
[3] Hilborn R C 1982 Einstein coefficients, cross sections, f values, dipole moments, and all that *Am. J. Phys.* **50** 982–6
[4] Cray M, Shih M-L and Milonni P W 1982 Stimulated emission, absorption, and interference *Am. J. Phys.* **50** 1016–21
[5] Sargent M, Scully M O and Lamb W 1974 *Laser Physics* (Reading, MA: Addison-Wesley)

[6] Jackson J D 1998 *Classical Electrodynamics* 3rd edn (New York: Wiley)
[7] Feynman R, Leighton R B and Sands M 1977 *The Feynman Lectures on Physics* (Boston, MA: Addison-Wesley)
[8] Demtröder W 1988 *Laser Spectroscopy* 1st edn (New York: Springer)
[9] Davis C 2014 *Lasers and Electro-optics* 2nd edn (Cambridge: Cambridge University Press)
[10] Reif F 1965 *Fundamentals of Statistical and Thermal Physics* (New York: McGraw-Hill)
[11] Loudon R 1973 *The Quantum Theory of Light* (Oxford: Clarendon Press)

IOP Publishing

Optical Radiation and Matter

Robert J Brecha and J Michael O'Hare

Chapter 5

Macroscopic electrodynamics

In the previous chapter we considered propagation in dilute matter; we now wish to extend these results to propagation in dense matter. For dilute matter we looked at the field due to a sheet of dipoles and spoke of averages over molecular orientations. In dense matter we also consider spatial averages, which have the effect of reducing the medium to a continuum. Propagation in a dense medium is then treated by 'so-called' macroscopic electrodynamics, which indeed regards the medium as a continuum with spatially averaged fields. We will begin by reviewing the relations between microscopic and macroscopic characterizations of matter, including the polarization density and the dielectric constant. The electromagnetic wave equation for fields in matter will be derived and we will consider the form taken by plane-wave solutions. One of the key points of this chapter is to understand the processes of reflection and refraction at the boundaries between dielectric media and at a dielectric–metal interface. Finally, we will consider an application of these results in which one can gain information about the dielectric and absorption properties of a material by analyzing the polarization of light reflected from the surface of that material.

5.1 Historical introduction

The phenomenon of refraction, with which we will be concerned in this chapter, was already known to and studied by the ancient Greeks. The first mathematical treatment of refraction is due, however, to Willebrord Snel (1591–1626) [1] who never published his result. It was through the work of René Descartes [2] that the law of refraction actually became known. Even without a mathematical understanding of refraction, eyeglasses had been in use for some four centuries and telescopes for several decades, based purely on empirical results and experimentation.

In the early optics experiments of Isaac Newton (1642–1727) [3] his focus was on the description of the apparent dependence of refraction on wavelength. We tend to take for granted the fact that white light is composed of wavelengths corresponding

doi:10.1088/978-0-7503-2624-7ch5

to the complete visible spectrum, but it was not until this work by Newton that careful and definitive measurements were made of the light dispersed (differentially refracted) by a prism. Newton discovered that violet light is refracted more than red light. It is interesting that all of Newton's correct conclusions were based on his corpuscular theory of light, whereas nearly all of the results presented in this text are based on the wave theory of light.

Closely related to the phenomenon of refraction are the polarization experiments mentioned at the beginning of chapter 2. Malus realized the possibility of producing polarized light by reflection from a dielectric surface and David Brewster (1781–1868) [4] derived the for the first time the relation that the polarizing angle, or Brewster's angle as we usually call it, is given by $\theta_B = \tan^{-1}\left(\frac{n_1}{n_2}\right)$ with n_1 (n_2) the index of refraction of the first (second) medium and $n_1 < n_2$.

5.2 The local field

Macroscopic electrodynamics takes as its focus the polarizations $\vec{P}$ and $\vec{M}$ which are the average electric dipole moment per unit volume and the average magnetic dipole moment per unit volume, respectively. The field that induces the dipole in a single molecule is the *molecular* or *local field*, $\vec{E}_{\text{loc}}$. The local field is the incident field plus all of the fields due to all *other* dipoles, permanent and induced, in the medium. Thus the local field is the microscopic field in the medium seen by an individual molecule. For a dilute medium, the local field is simply the incident field, since 'dilute' implies that all other molecules are far enough away that our molecule of interest does not 'see' them. The actual electric field at any point in a dense medium is a rapidly fluctuating spatial quantity. Obviously a detailed calculation of such a quantity would be difficult and of questionable value. We are interested in the so-called macroscopic field which averages over all of the various sources of field in the medium, including the external field, and is thus a smoothly varying spatial and temporal quantity. Note that, with this definition, the macroscopic field includes the field of the molecule in question and is thus smaller than the local field (the field produced by the induced dipole in the molecule is opposite to the applied field). In a dilute medium, the macroscopic field, local field, and incident field are all the same quantity. In a dense medium the local field consists of the macroscopic field plus the effects of an internal field; the internal field being the polarization field of the molecule in question and the polarization fields produced by close-by neighboring molecules. Therefore

$$\vec{E}_{\text{loc}} = \vec{E} + \vec{E}_i.$$

In the above expression $\vec{E}$ stands for the macroscopic field and $\vec{E}_i$ is the internal field. The internal field consists of two contributions, the microscopic field contribution of molecules close to the molecule in question minus the contributions from those same molecules and the molecule in question that were included in the continuum average to obtain the macroscopic field. Various models exist for calculating the internal field, but these are outside the scope of this discussion. From [5] the internal field can be written,

$$\vec{E}_i = \left(\frac{1}{3\varepsilon_0} + s\right)\vec{P},$$
(5.1)

where s is called a structure factor and $\vec{P}$ is the macroscopic polarization per unit volume given by

$$\vec{P} = N_v\langle\vec{\wp}\rangle \quad \text{and} \quad \langle\vec{\wp}\rangle = \alpha(\vec{E} + \vec{E}_i),$$
(5.2)

where N_v is the number of dipoles per unit volume, $\vec{\wp}$ is the average dipole moment, and, as in chapter 4, α is the polarizability of the molecules. The details of the structure factor s are not of particular interest at this point and we will take it to be zero. Therefore, from equation (5.1) the polarization is

$$\vec{P} = N_v\alpha\left(\vec{E} + \frac{\vec{P}}{3\varepsilon_0}\right).$$
(5.3)

The polarization is defined in terms of the macroscopic field through the electric susceptibility, χ_e, given by

$$\vec{P} \equiv \varepsilon_0\chi_e\vec{E}.$$
(5.4)

Similar definitions exist for a magnetic susceptibility which will be summarized in the next section. From equations (5.3) and (5.4) above χ_e is found to be

$$\chi_e = \frac{N_v\alpha/\varepsilon_0}{1 - \frac{N_v\alpha}{3\varepsilon_0}}.$$
(5.5)

We will not be concerned with the details of equation (5.5), the reason for working through it is to gain insight into its nature; which is that it gives us the relationship between the two macroscopic quantities $\vec{P}$ and $\vec{E}$, thus allowing us to treat propagation in dense matter in terms of macroscopic fields. Equation (5.5) relates the polarizability α, a microscopic parameter, to the susceptibility χ_e, which is a macroscopic parameter. With results such as these the world at the atomic level, not directly observable, can be related to measurable properties of bulk matter.

5.3 The macroscopic Maxwell equations

Our study of the interaction of light and matter continues with a review of the basic elements of macroscopic electromagnetic theory. This review will be formulated in terms of the macroscopic field variables, $\vec{E}$, $\vec{B}$, $\vec{D}$, and $\vec{H}$ and the charge and current densities. In microscopic theory there are no $\vec{D}$ and $\vec{H}$ fields. In MKS units the macroscopic Maxwell equations are

$$\nabla \cdot \vec{D} = \rho_f$$
(5.6)

$$\nabla \cdot \vec{B} = 0 \tag{5.7}$$

$$\nabla \times \vec{E} = -\frac{\partial \vec{B}}{\partial t} \tag{5.8}$$

$$\nabla \times \vec{H} = \vec{j}_f + \frac{\partial \vec{D}}{\partial t}. \tag{5.9}$$

In these equations $\vec{E}$ and $\vec{H}$ are called the electric and magnetic field vectors, respectively, while $\vec{D}$ and $\vec{B}$ are the electric displacement and the magnetic induction, respectively. The term $\frac{\partial \vec{D}}{\partial t}$ in equation (5.9) is called the 'Maxwell displacement current density'. Often we are concerned with materials (insulators for example) in which the free charge density, ρ_f, and the free current density, $\vec{j}_f$, are zero. In such cases equations (5.6) and (5.7) become

$$\nabla \cdot \vec{D} = 0 \tag{5.10}$$

$$\nabla \times \vec{H} = \frac{\partial \vec{D}}{\partial t}. \tag{5.11}$$

In addition to Maxwell's equations we require two 'constitutive relations' between macroscopic quantities. The first of these is given by

$$\vec{D} = \varepsilon_0 \vec{E} + \vec{P} = \varepsilon_0(1 + \chi_e)\vec{E} = \varepsilon_0 \varepsilon_r \vec{E} = \varepsilon \vec{E}, \tag{5.12}$$

where $\varepsilon_r = (1 + \chi_e)$ is the relative permittivity, ε is the dielectric constant, and χ_e is the electric susceptibility. The permittivity of free space is $\varepsilon_0 = 8.854\,188 \times 10^{-12}$ F m^{-1}.

We can make use immediately of these relations and look back at our result from the previous section relating the susceptibility and polarizability. With the constitutive relation $\varepsilon_r = 1 + \chi_e$ we can derive a relation for the polarizability in terms of the relative permittivity, which, as we will see later, is a quantity to which we have experimental access. The result is

$$\alpha(\omega) = \frac{3\varepsilon_0}{N_v} \frac{\varepsilon_r(\omega) - 1}{\varepsilon_r(\omega) + 2}. \tag{5.13}$$

This equation is known as the Lorentz–Lorenz formula or, in another context, the Clausius–Mosotti relation.

The second constitutive relation, for the magnetic field and magnetic induction is

$$\vec{B} = \mu_0(\vec{H} + \vec{M}) = \mu_0(1 + \chi_m)\vec{H} = \mu_0 \mu_r \vec{H} = \mu \vec{H} \tag{5.14}$$

in which we introduce the magnetization $\vec{M}$, the magnetic dipole moment per unit volume. In the above we also use the relative magnetic permeability, $\mu_r \equiv 1 + \chi_m$. The permeability of free space, $\mu_0 = 4\pi \times 10^{-7}$ N/A^{-2} and the magnetic susceptibility, χ_m, is defined through $\vec{M} = \chi_m \vec{H}$.

Finally, we can also relate the current density to the electric field vector by

$$\vec{j}_f = \sigma\vec{E}, \tag{5.15}$$

where σ is the electric conductivity. As we shall see in the next chapter, σ may be a complex number. Although it is not obvious in this form, this is an alternative form of Ohm's law from circuit theory.

For the sake of completeness, we re-introduce as well the Poynting vector in terms of macroscopic fields, given by

$$\vec{S} = \vec{E} \times \vec{H}. \tag{5.16}$$

In these first two sections we have a summary of all of the fundamental macroscopic field equations and auxiliary relationships which will be needed throughout this course. When reviewing the macroscopic form of Maxwell's equations, it is important to remember that $\vec{E}$ and $\vec{B}$ are the fundamental fields. The fields $\vec{D}$ and $\vec{H}$ are introduced as a means of taking into account, in an average way, the contributions of charge densities and current densities of the atomic charges and currents in the medium.

5.4 The polarization density and constitutive relation

Before considering the wave equation derived from the macroscopic Maxwell equations it is important to investigate the constitutive relations for dielectrics a bit further. The constitutive relations given above hold strictly only in the limit of a non-dispersive, non-absorbing medium. Another way of saying this is that we tacitly assumed a medium that can respond instantaneously to an applied field. As this is not the case for a real system, we need to at least see what implications arise out of our simplifying assumption. We will ignore the corresponding magnetic constitutive relations as they play little role in most optics applications.

We have already seen that the response of the material medium can be characterized by the polarization density vector $\vec{P}$. Recall that we can relate this quantity to the number (per unit volume) of oscillating dipoles induced by an incident electromagnetic field. The oscillating collection of dipoles described by $\vec{P}$ radiates its own electric field and gives rise to an altered net field, which we can then detect outside the material (at least as a time-averaged quantity). We will be concerned with primarily a linear response to the external field, but in general we could express the polarization density as a power series

$$\vec{P}(t) = \vec{P}^{(0)}(t) + \vec{P}^{(1)}(t) + \vec{P}^{(2)}(t) + \cdots.$$

The first term is independent of the applied field and represents a permanent dipole moment. For example a collection of polar molecules, such as water, would be described by this polarization density. The last two terms shown above are linear and quadratic in dependence on the applied field, respectively. The second-order terms (and all higher orders) are what lead to nonlinear effects such as the conversion of light of one frequency to another frequency. These effects are beyond the scope of this course for the most part. In our investigation of electro-optics and acousto-optics we will touch on nonlinear processes briefly, but most of the detailed study will be left for another course.

We have also seen that the response of a medium depends on the frequency of the incident field, through the absorption coefficient and the index of refraction. Since these are in turn related to the dielectric constant, and thus to the susceptibility, as defined in the previous section, we need to consider the frequency dependence of the polarization density more carefully as well. The frequency dependence of the classical oscillator implies that there is a time response as well. For a mass on a spring we can imagine this by thinking of a heavy versus a light mass. Intuitively we realize that the heavier mass–spring system will not respond as quickly to a given applied force. We can describe this mathematically by the relation

$$\vec{P}^{(1)}(t) = \varepsilon_0 \int_{-\infty}^{\infty} d\tau \, \bar{R}^{(1)}(t - \tau) \cdot \vec{E}(\tau)$$
$$= \varepsilon_0 \int_{-\infty}^{\infty} d\tau' \bar{R}^{(1)}(\tau') \cdot \vec{E}(t - \tau'), \tag{5.17}$$

where $\bar{R}^{(1)}(\tau)$ is the time response function for the (linear) polarization. We can interpret the above equation by thinking in terms of the material and how it responds to an applied field. Any real system will take some time to respond to the oscillations of the external field. Because of this, the polarization at a given time t is due to the field a time τ' earlier, with the response function $\bar{R}^{(1)}(\tau)$ giving the details of how the material reacts. We have assumed in the foregoing that all of this takes place at one space point, that is, the polarization at a point $\vec{r}$ depends on the field at the same point $\vec{r}$. In addition, $\bar{R}^{(1)}(\tau)$ must be identically zero for τ' less than zero to preserve causality.

To discuss the frequency dependence of χ, the susceptibility, we must look at the frequency dependence of the electric fields, which can be done mathematically through use of the Fourier transform. We can write

$$\vec{E}(t) = \int_{-\infty}^{\infty} d\omega \, \vec{E}(\omega)e^{-i\omega t},$$

where $\vec{E}(\omega)$ is defined by

$$\vec{E}(\omega) = \int_{-\infty}^{\infty} d\tau \, \vec{E}(\tau)e^{+i\omega\tau}.$$

We will not discuss the details of the validity of these relations; more can be found in any mathematical physics or engineering math text. If we substitute for $\vec{E}(t - \tau)$ in equation (5.17) using the Fourier transform relation we find

$$\vec{P}^{(1)}(t) = \varepsilon_0 \int_{-\infty}^{\infty} d\tau \, \bar{R}^{(1)}(\tau) \int_{-\infty}^{\infty} d\omega \, \vec{E}(\omega)e^{-i\omega(t-\tau)}$$
$$= \varepsilon_0 \int_{-\infty}^{\infty} d\omega \, \bar{\bar{\chi}}^{(1)}(\omega)\vec{E}(\omega)e^{-i\omega t}$$

with

$$\bar{\chi}^{(1)}(\omega) = \int_{-\infty}^{\infty} d\tau \, \bar{R}^{(1)}(\tau) e^{+i\omega\tau}. \tag{5.18}$$

This latter result should be taken as the definition of the linear susceptibility. The convergence of the integral is guaranteed by the conditions that $\tau > 0$ and that we take the integration to be in the upper half of the complex plane $\omega = x + iy$, so that as $\tau \to \infty$, $e^{+i\omega\tau} \to 0$. Since $\vec{E}(t)$ is real, as is $\bar{R}^{(1)}(\tau)$, we have the auxiliary conditions

$$\left\{ \vec{E}(\omega) \right\}^* = \vec{E}(-\omega^*)$$

$$\left\{ \bar{\chi}^{(1)}(\omega) \right\}^* = \bar{\chi}^{(1)}(-\omega^*).$$

5.4.1 Monochromatic waves

In the case most commonly treated in this text, that of the monochromatic plane wave, the electric field can be written as

$$\vec{E}(t) = \frac{1}{2} \left\{ \vec{E}_{\omega_0} e^{-i\omega_0 t} + \vec{E}_{-\omega_0} e^{+i\omega_0 t} \right\},$$

which can be written in the frequency domain as

$$\vec{E}(\omega) = \frac{1}{2} \left\{ \vec{E}_{\omega_0} \delta(\omega - \omega_0) + \vec{E}_{-\omega_0} \delta(\omega + \omega_0) \right\},$$

where we use the subscript ω_0 to emphasize the fact that we are dealing with the fields for one frequency only. The Dirac delta-functions $\delta(\omega \mp \omega_0)$ take on values of unity if $\omega = \pm\omega_0$, respectively, and zero for any other value of ω. (See the appendix for further information about the δ-function.) Using this expression for the frequency components of the electric field, we find the polarization to be

$$\vec{P}^{(1)}(t) = \varepsilon_0 \int_{-\infty}^{\infty} d\omega \, \bar{\chi}^{(1)}(\omega) \, \vec{E}(\omega) e^{-i\omega t}$$

$$= \varepsilon_0 \int_{-\infty}^{\infty} d\omega \, \bar{\chi}^{(1)}(\omega)$$

$$\times \frac{1}{2} \left\{ \vec{E}_{\omega_0} \delta(\omega - \omega_0) + \vec{E}_{-\omega_0} \delta(\omega + \omega_0) \right\} e^{-i\omega t}$$

$$= \frac{1}{2} \varepsilon_0 \left\{ \bar{\chi}^{(1)}(\omega_0) \cdot \vec{E}_{\omega_0} \delta(\omega - \omega_0) + \bar{\chi}^{(1)}(-\omega_0) \cdot \vec{E}_{-\omega_0} \delta(\omega + \omega_0) \right\}.$$

Thus the constitutive relation can be rewritten as

$$\vec{D}(t) = \varepsilon_0 \vec{E}(t) + \vec{P}(t)$$

$$\frac{1}{2} \left\{ \vec{D}_{\omega_0} e^{-i\omega_0 t} + \vec{D}_{-\omega_0} e^{i\omega_0 t} \right\} = \frac{1}{2} \varepsilon_0 \left\{ \vec{E}_{\omega_0} e^{-i\omega_0 t} + \vec{E}_{-\omega_0} e^{i\omega_0 t} \right\}$$

$$+ \frac{1}{2} \varepsilon_0 \left\{ \bar{\chi}^{(1)}(\omega_0) \cdot \vec{E}_{\omega_0} e^{-i\omega_0 t} + \bar{\chi}^{(1)}(-\omega_0) \cdot \vec{E}_{\omega_0} e^{i\omega_0 t} \right\}$$

in which the subscripts indicate the appropriate frequency components. Picking out, for example, only those with frequency $+\omega_0$, we can write

$$\vec{D}_{\omega_0} = \varepsilon_0 \vec{E}_{\omega_0} + \varepsilon_0 \overleftrightarrow{\chi}^{(1)}(\omega_0) \cdot \vec{E}_{\omega_0}$$

$$= \varepsilon_0 \left(1 + \chi^{(1)}(\omega_0) \right) \cdot \vec{E}_{\omega_0}$$

$$= \varepsilon_0 \overleftrightarrow{\varepsilon}_r(\omega) \cdot \vec{E}_{\omega_0}.$$

These are the constitutive relations written in terms of frequency domain functions, i.e. they are valid at the given frequency ω_0 only.

Again we want to point out that the constitutive relations written down at the beginning of this chapter are really only valid for a (fictional) medium which responds instantaneously to an external field. If the response of a real medium is significantly faster than the oscillation frequency of the optical electric field, the instantaneous version of the constitutive relations can be used. After having gone through the development above, we can, however, make the statement that for most of what follows in this text the 'instantaneous approximation' will be assumed valid.

5.5 Dielectric and impermeability tensors

The relative dielectric 'constant', $\overleftrightarrow{\varepsilon}_r$, presented above should be given a bit more attention. The overbar denotes the fact that $\overleftrightarrow{\varepsilon}_r$ is a tensor of rank two. For our purposes this means essentially that the relative dielectric constant is to be represented in general by a 3×3 matrix. Thus, if we look at the relation between the displacement field and the electric field we have

$$\vec{D}_\omega = \overleftrightarrow{\varepsilon} \cdot \vec{E}_\omega$$

$$\begin{pmatrix} D_x \\ D_y \\ D_z \end{pmatrix} = \begin{pmatrix} \varepsilon_{11} & \varepsilon_{12} & \varepsilon_{13} \\ \varepsilon_{21} & \varepsilon_{22} & \varepsilon_{23} \\ \varepsilon_{31} & \varepsilon_{32} & \varepsilon_{33} \end{pmatrix} \cdot \begin{pmatrix} E_x \\ E_y \\ E_z \end{pmatrix}.$$

We will be discussing the implications of this type of relation in a great deal more detail in chapter 7. For now, it is important to note simply that a component of the displacement vector $\vec{D}$, say D_x, can depend on components of the electric field in directions other than x. That is, the two vectors $\vec{D}$ and $\vec{E}$ are not necessarily parallel to one another.

In optics one often encounters in addition to the dielectric tensor the *impermeability tensor*, $\overleftrightarrow{\eta}(\omega)$, defined through the relationship

$$\vec{E}_\omega = \frac{1}{\varepsilon_0} \overleftrightarrow{\eta}(\omega) \cdot \vec{D}_\omega,$$

i.e. $\overleftrightarrow{\eta}(\omega) = \left(\overleftrightarrow{\varepsilon}_r(\omega) \right)^{-1}$. Recall that these quantities are tensors or, mathematically, matrices, so that the inverse operation may not be completely trivial.

The impermeability tensor will be useful when considering propagation in anisotropic media, the subject of chapters 7–9. Mathematically we can describe the advantage of using the impermeability instead of the relative dielectric tensor as being due to the possibility of reducing wave propagation problems to eigenvalue problems for different polarizations of the electromagnetic field.

5.6 The electromagnetic wave equation

The electromagnetic wave equation in a medium is readily obtained from Maxwell's equations and the auxiliary equations given in the previous section. Taking the curl of equation (5.8) gives

$$
\begin{aligned}
\nabla \times (\nabla \times \vec{E}) &= -\frac{\partial}{\partial t}(\nabla \times \vec{B}) \\
&= -\mu_r \mu_0 \frac{\partial}{\partial t}(\nabla \times \vec{H}) \\
&= -\mu_r \mu_0 \frac{\partial}{\partial t}\left(\vec{j}_f + \frac{\partial \vec{D}}{\partial t}\right) \\
&= -\mu_r \mu_0 \sigma \frac{\partial \vec{E}}{\partial t} - \mu_r \mu_0 \frac{\partial^2 \vec{D}}{\partial t^2}.
\end{aligned}
\tag{5.19}
$$

Using a standard vector identity on the left-hand side of the above equation as we did in chapter 1, we obtain

$$
\nabla(\nabla \cdot \vec{E}) - \nabla^2 \vec{E} = -\mu_r \mu_0 \sigma \frac{\partial \vec{E}}{\partial t} - \mu_r \mu_0 \frac{\partial^2 \vec{D}}{\partial t^2}.
$$

Using the first Maxwell equation,

$$
\nabla \cdot \vec{D} = \rho_f.
$$

Here we will assume that there are no free charges, $\rho_f = 0$. This does not mean in general that $\nabla \cdot \vec{E} = 0$, however. Because the relation between $\vec{E}$ and $\vec{D}$ is not a simple linear one for anisotropic materials (see equation (5.12)) $\vec{E}$ is not necessarily orthogonal to the vector $\vec{k}$, which would be implied by the vanishing divergence of $\vec{E}$. We will return to this situation in chapter 7. We can, however, simplify the above wave equation with the assumption that $\nabla \cdot \vec{E} \approx 0$ implying that any differences in index of refraction (indirectly, in components of the tensor ε) are very small. This is a valid approximation for most optics problems. Using in addition the fact that for nonmagnetic materials $\mu_r = 1$, we obtain the following form for the wave equation,

$$
\nabla^2 \vec{E} = \mu_0 \sigma \frac{\partial \vec{E}}{\partial t} + \mu_0 \frac{\partial^2 \vec{D}}{\partial t^2}.
\tag{5.20}
$$

Finally, using the fact that $\mu_0\epsilon_0 = 1/c^2$, we obtain the wave equation in the form

$$\nabla^2 \vec{E} = \mu_0\sigma\frac{\partial \vec{E}}{\partial t} + \frac{\epsilon_r}{c^2}\frac{\partial^2 \vec{E}}{\partial t^2}. \tag{5.21}$$

Equation (5.21) represents the wave equation for the macroscopic electric field for an electromagnetic wave in a dense medium.

5.7 Plane waves in dense matter

We will continue to represent the electromagnetic wave by a plane wave with the only addition being to relate the complex propagation vector to the macroscopic field variables introduced in chapter 1. As we pointed out in chapter 2, the complex vector amplitude is a notational device to designate the polarization of the wave, whereas the complex wavevector allows us to take into account attenuation of the wave. This form of the plane wave is

$$\vec{E}(\vec{r}, t) = \Re \vec{\mathcal{E}}(\vec{K}, \omega)e^{i(\vec{K}\cdot\vec{r} - \omega t)}. \tag{5.22}$$

In equation (5.22) $\vec{\mathcal{E}}$ and $\vec{K}$ both denote *complex* vectors. We have already seen that the complex propagation vector, $\vec{K}$, can be written as

$$\vec{K} = \vec{k}' + i\vec{k}''$$

so that the plane-wave expression for the field vector, $\vec{E}$, becomes

$$\vec{E} = \vec{\mathcal{E}}e^{-\vec{k}''\cdot\vec{r}}e^{i(\vec{k}'\cdot\vec{r} - \omega t)}. \tag{5.23}$$

From this expression for the field vector, we see that $\vec{k}'$ is a vector which is normal to *surfaces of constant phase*, while the vector $\vec{k}''$ is a vector which is normal to *surfaces of constant amplitude*.

If equation (5.22) is substituted into the wave equation, we find the following expression for the square of the wavevector $\vec{K}$ (not to be confused with $|\vec{K}|^2 = (\vec{K}\cdot\vec{K}^*)$):

$$\vec{K}\cdot\vec{K} = \vec{K}^2 = \frac{\omega^2}{c^2}\left(\epsilon_r + \frac{i\mu_0\sigma}{\omega}c^2\right)$$

$$= \frac{\omega^2}{c^2}\left(\epsilon_r + \frac{i\sigma}{\omega\epsilon_0}\right) \equiv \frac{\omega^2}{c^2}\tilde{n}^2, \tag{5.24}$$

where $\tilde{n}$ is the complex index of refraction and is given by

$$\tilde{n}^2 = \epsilon_r + \frac{i\sigma}{\omega\epsilon_0}. \tag{5.25}$$

If we express the complex index of refraction as a complex number, i.e.

$$\tilde{n} = n + i\kappa, \tag{5.26}$$

where n is the ordinary index of refraction and κ is the extinction coefficient and, if we square the expression for $\tilde{n}$ and equate real and imaginary parts to equation (5.25), we find

$$n^2 - \kappa^2 = \varepsilon_r \tag{5.27}$$

$$2n\kappa = \sigma/\omega\varepsilon_0. \tag{5.28}$$

Solving for n^2 and κ^2 yields

$$n^2 = \frac{1}{2}\left(\left(\varepsilon_r^2 + \frac{\sigma^2}{\omega^2\varepsilon_0^2}\right)^{\frac{1}{2}} + \varepsilon_r\right) \tag{5.29}$$

$$\kappa^2 = \frac{1}{2}\left(\left(\varepsilon_r^2 + \frac{\sigma^2}{\omega^2\varepsilon_0^2}\right)^{\frac{1}{2}} - \varepsilon_r\right). \tag{5.30}$$

Note that σ is the conductivity at the optical frequency concerned and is not generally equal to the dc low frequency conductivity. For the case of $\sigma \to 0$, $\kappa = 0$ and $n^2 = \varepsilon_r$.

The expression above suggest an alternative and entirely equivalent approach, which is to *define* a complex relative dielectric constant, $\tilde{\varepsilon}_r$, where

$$\tilde{\varepsilon}_r = \varepsilon_1 + i\varepsilon_2 = \tilde{n}^2. \tag{5.31}$$

Therefore,

$$\varepsilon_1 = n^2 - \kappa^2 \quad \text{and} \quad \varepsilon_2 = 2n\kappa, \tag{5.32}$$

which give equivalent forms for n^2 and κ^2

$$n^2 = \frac{1}{2}\left(\left(\varepsilon_1^2 + \varepsilon_2^2\right)^{1/2} + \varepsilon_1\right) \tag{5.33}$$

$$\kappa^2 = \frac{1}{2}\left(\left(\varepsilon_1^2 + \varepsilon_2^2\right)^{1/2} - \varepsilon_1\right). \tag{5.34}$$

5.8 Classification of wave types

The expression $\vec{K}^2 = \frac{\omega^2}{c^2}\tilde{n}^2$ does not completely determine $\vec{K}$ even if n and κ are known. There are many possible values of $\vec{k}'$ and $\vec{k}''$ which can satisfy this relation depending on how the wave enters into the medium. Waves can be classified as either homogeneous or inhomogeneous according to the orientation of $\vec{k}'$ and $\vec{k}''$ relative to each other. This classification is as follows:

1. Homogeneous waves: $\vec{k}'$ is parallel to $\vec{k}''$ or $\vec{k}'' = 0$.
2. Inhomogeneous waves: $\vec{k}'$ and $\vec{k}''$ are not parallel.

5.8.1 Homogeneous and inhomogeneous waves

Consider, again, the expression for the complex wavevector,

$$\vec{K} = \vec{k}' + i\vec{k}'',$$

and consider the case of propagation in a *non-attenuating* medium. If the medium is non-attenuating, the square of the wavevector will be real. (Why is this so?) Therefore,

$$\vec{K} \cdot \vec{K} = k'^2 - k''^2 + 2i\vec{k}' \cdot \vec{k}'' = \frac{\omega^2}{c^2}\tilde{n}^2. \tag{5.35}$$

The middle term in equation (5.35) is real if

$$\vec{k}' \cdot \vec{k}'' = 0 \quad \text{(non-attenuating)},$$

which leads to the fact that either $\vec{k}'\perp\vec{k}''$ or $\vec{k}'' = 0$. If $\vec{k}'' = 0$, the wave is homogeneous. If the wave is inhomogeneous then $\vec{k}''$ must be perpendicular to $\vec{k}'$ if the medium is non-attenuating.

Next consider an attenuating medium for which

$$\vec{K} \cdot \vec{K} = k'^2 - k''^2 + 2ik'k''\cos\Theta = \frac{\omega^2}{c^2}\big((n^2 - \kappa^2) + 2in\kappa\big). \tag{5.36}$$

This equation says that if we know n, κ, and Θ, we can solve for k' and k''. For a homogeneous wave $\Theta = 0$ and equation (5.36) gives $k' = \frac{\omega}{c}n$ and $k'' = \frac{\omega}{c}\kappa$. In general, however, it is not so simple to separate the n and κ contributions to the complex propagation vector.

5.8.2 Transverse electric and magnetic waves

For inhomogeneous waves it is possible to define two distinct types of linear polarization. These are called transverse electric and transverse magnetic. To examine these polarizations, let $\vec{\mathcal{E}}'$ and $\vec{\mathcal{E}}''$ be the real and imaginary parts of $\vec{\mathcal{E}}$, respectively; and let $\vec{\mathcal{H}}'$ and $\vec{\mathcal{H}}''$ be the real and imaginary parts of $\vec{\mathcal{H}}$.
  - *Transverse electric (TE) polarization.*
    We define TE polarization by the conditions $\vec{\mathcal{E}}'\|\vec{\mathcal{E}}''$ and both are $\perp$ to the plane of k' and k''. Thus we can set $\vec{\mathcal{E}} = \vec{\mathcal{E}}_0 e^{i\phi_E}$ where $\vec{\mathcal{E}}_0$ is just the real vector amplitude of the wave and ϕ_E is a phase factor.
  - *Transverse magnetic (TM) polarization.*
    $\vec{\mathcal{H}}'\|\vec{\mathcal{H}}''$ and both are $\perp$ to the plane of k' and k''. Thus we can set $\vec{\mathcal{H}} = \vec{\mathcal{H}}_0 e^{i\phi_M}$ where $\vec{\mathcal{H}}_0$ is just the real vector amplitude of the wave and ϕ_M is a phase factor.

Here we have employed the properties of a complex number z given by

$$z = x + iy = re^{i\phi},$$

where $r = \sqrt{x^2 + y^2}$ and $\phi = \tan^{-1}(y/x)$.

Note: For homogeneous waves the distinction between TE and TM is lost. Can you formulate a simple explanation of why this is the case?

Question: Why is it possible to write $\vec{E}(\vec{H})$ in the above simple form for TE (TM) waves?

5.8.3 Additional properties of plane waves

Again consider our complex expressions for the macroscopic electric and magnetic fields of a plane EM wave,

$$\vec{E} = \vec{\mathcal{E}} \exp(i(\vec{K} \cdot \vec{r} - \omega t)) = \left(\vec{\mathcal{E}}' + i\vec{\mathcal{E}}''\right)e^{i(\vec{K}\cdot\vec{r}-\omega t)}$$

$$\vec{H} = \vec{\mathcal{H}} \exp(i(\vec{K} \cdot \vec{r} - \omega t)) = \left(\vec{\mathcal{H}}' + i\vec{\mathcal{H}}''\right)e^{i(\vec{K}\cdot\vec{r}-\omega t)}.$$

From the third Maxwell equation,

$$\nabla \times \vec{E} = -\mu_0 \mu_r \frac{\partial \vec{H}}{\partial t},$$

we obtain

$$\vec{K} \times \vec{\mathcal{E}} = \mu_0 \mu_r \omega \vec{\mathcal{H}}, \tag{5.37}$$

which, to satisfy the cross product vector relationship of this expression, leads to the following result:

$$\vec{\mathcal{E}} \perp \vec{\mathcal{H}} \quad \text{and} \quad \vec{K} \perp \vec{\mathcal{H}},$$

as shown in figure 5.1(a). We can rewrite the above relation as

$$|\vec{K}| \, \hat{k} \times |\vec{\mathcal{E}}| \, \hat{e} = \mu_0 \mu_r \omega \, |\vec{\mathcal{H}}| \, \hat{h}$$

$$\mathcal{H} = \frac{K\mathcal{E}}{\mu_0 \mu_r \omega} \tag{5.38}$$

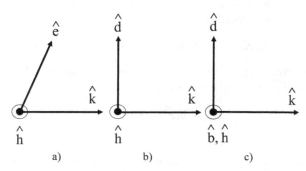

Figure 5.1. Relations between the electromagnetic field vectors and the wavevector. In (a) the consequence of the third Maxwell equation is shown, as described in the text. The first Maxwell equation is illustrated in (b) for a charge-free region, while the divergenceless B-field relation is shown in (c).

$$\hat{k} \times \hat{e} = \hat{h}. \tag{5.39}$$

The latter two relations give the relative magnitudes and directions of the vectors $|\vec{K}|$, $|\vec{\mathcal{E}}|$, and $|\vec{\mathcal{H}}|$. We have written the fields in terms of magnitudes and unit vectors $\hat{k}$, $\hat{e}$, and $\hat{h}$ for the directions of the wavevector, electric field, and magnetic field, respectively. Since $\mu_r = 1$ for a nonmagnetic medium, and if we consider a non-attenuating medium for which $k = n\omega/c$, we can write

$$\mathcal{H} = \frac{n\omega/c}{\mu_0\omega}\mathcal{E} = \frac{n}{\sqrt{\mu_0/\varepsilon_0}}\mathcal{E} = n\frac{\mathcal{E}}{377\Omega}, \tag{5.40}$$

where 377Ω is the so-called impedance of free space.

Also, from the fourth Maxwell equation

$$\nabla \times \vec{H} = \varepsilon_0\varepsilon_r\frac{\partial\vec{E}}{\partial t} + \sigma\vec{E} \tag{5.41}$$

we obtain

$$i\vec{K} \times \vec{\mathcal{H}} = -i\varepsilon_0\varepsilon_r\omega\vec{\mathcal{E}} + \sigma\vec{\mathcal{E}} = (\sigma - i\varepsilon_0\varepsilon_r\omega)\vec{\mathcal{E}},$$

which leads to the fact that $\vec{K} \perp \vec{\mathcal{E}}$ *if the medium is isotropic,* i.e. if ε_r can be treated as a scalar. These results show that the electromagnetic wave is a transverse wave, that is, the E and H vectors are perpendicular to the direction of propagation given by the K-vector. How does this apply to the real and imaginary part of these vectors?

Finally, if we continue with this approach we can obtain an analytical expression for the relationship between K, $\mathcal{E}$, and $\mathcal{H}$. From the third Maxwell equation

$$\vec{K} \times \vec{\mathcal{E}} = \mu_0\mu_r\omega\vec{\mathcal{H}} \tag{5.42}$$

and forming the cross product of both sides of the equation with $\vec{K}$ gives

$$\vec{K} \times (\vec{K} \times \vec{\mathcal{E}}) = \mu_0\mu_r\omega\vec{K} \times \vec{\mathcal{H}}$$

and with the help of a vector identity for the double cross product, the left-hand side of the above expression becomes

$$\vec{K}(\vec{K} \cdot \vec{\mathcal{E}}) - \vec{\mathcal{E}}(\vec{K} \cdot \vec{K}) = \mu_0\mu_r\omega(\vec{K} \times \vec{\mathcal{H}}).$$

This allows us to obtain the following expression for the complex amplitude of the macroscopic E-field of the wave in terms of K and $\mathcal{H}$

$$\vec{\mathcal{E}} = \frac{-\mu_0\mu_r\omega(\vec{K} \times \vec{\mathcal{H}})}{\vec{K} \cdot \vec{K}} \tag{5.43}$$

since $\vec{K}$ is perpendicular to $\vec{\mathcal{E}}$ and thus $\vec{K} \cdot \vec{\mathcal{E}} = 0$. The first Maxwell equation $\nabla \cdot \vec{D} = 0$ (for a charge-free region) leads immediately to

$$\hat{k} \cdot \hat{d} = 0,$$

where $\hat{d}$ is the unit vector in the direction of the electric displacement vector. This result, together with that for $\hat{k}$, $\hat{e}$, and $\hat{h}$ is shown in figure 5.1(b). Finally, from the second Maxwell equation, $\nabla \cdot \vec{B} = \mu_0 \nabla \cdot \vec{H} = 0$, we find

$$\hat{k} \cdot \hat{h} = \hat{k} \cdot \hat{b} = 0,$$

the consequences of which are illustrated in figure 5.1(c). Here $\hat{b}$ is the unit vector in the direction of the magnetic induction.

If we return to the fourth Maxwell equation we can determine the *dispersion relation*, the relation between the wavevector k and the frequency ω. For zero free current density, $\vec{j}_f = 0$, we have

$$\nabla \times \vec{H} = \frac{\partial \vec{D}}{\partial t}$$

$$\rightarrow iK \hat{k} \times \mathcal{H}\hat{h} = -i\omega \mathcal{D} \hat{d}.$$

Therefore we have $\hat{k} \times \hat{h} = \hat{d}$ as already seen and further

$$k = \omega \frac{\mathcal{D}}{\mathcal{H}} = \omega \frac{\varepsilon_0 n^2(\omega) \mathcal{E}}{\frac{k}{\mu_0 \omega} \mathcal{E}},$$

which then reduces to

$$k = \frac{n\omega}{c}. \tag{5.44}$$

5.9 Reflection and transmission at an interface

By way of introduction to the topic of reflection and refraction at a surface, let us first consider the more basic example of an interface between two dielectric media. After arriving at our desired results we will see that an extension to the more complicated interface between a dielectric and a conducting medium requires only a straightforward generalization. The rigorous details of propagation in a conducting medium, including the possibility of propagation from one conductor to another, will be left for the end of this chapter.

5.9.1 Nature of reflected and transmitted waves

A wave incident on the surface of a boundary between two materials with different optical properties will break into a reflected and a transmitted wave. The measurement of the optical properties of a solid material will depend on the characteristics of these waves. A description of the field at one point in space, $\vec{r}_1$, may be given in terms of the field at another point in space, $\vec{r}_2$, and the vector displacement between the two spatial points; thus, in the same medium, we may write

$$\vec{E}(at\ \vec{r}_2) = \vec{E}(at\ \vec{r}_1)\ e^{i\vec{k}\cdot(\vec{r}_2-\vec{r}_1)} = \tilde{\rho}\vec{E}(at\ \vec{r}_1). \tag{5.45}$$

The complex quantity indicates the manner in which $\vec{E}(at\ \vec{r_2})$ is altered from $\vec{E}(at\ \vec{r_1})$. The character of the reflected and transmitted waves will be determined from the geometry of this situation.

Our objective in this section is to determine how a wave, incident on an interface, is transmitted and how it is reflected. We will write the incident, reflected, and transmitted waves as follows:

$$\vec{E}_i = \vec{\mathcal{E}}_i e^{i(\vec{K}_i \cdot \vec{r} - \omega_i t)} \quad \text{(incident wave)} \tag{5.46}$$

$$\vec{E}_r = \vec{\mathcal{E}}_r e^{i(\vec{K}_r \cdot \vec{r} - \omega_r t)} \quad \text{(reflected wave)} \tag{5.47}$$

$$\vec{E}_t = \vec{\mathcal{E}}_t e^{i(\vec{K}_t \cdot \vec{r} - \omega_t t)} \quad \text{(transmitted wave).} \tag{5.48}$$

If these are the only waves in the medium then they must 'match' at all positions $\vec{r}$ on the surface of the boundary at all times. This 'matching' is required by boundary conditions which flow from Maxwell's equations. These matching conditions will assume that a number, $\tilde{t}$, will exist such that

$$\vec{E}_t = \tilde{t}\vec{E}_i \tag{5.49}$$

and a number $\tilde{r}$ exists such that

$$\vec{E}_r = \tilde{r}\vec{E}_i. \tag{5.50}$$

The boundary conditions at an interface require that the tangential components of both $\vec{E}$ and $\vec{H}$ be continuous across the interface. This must be true for any time and any point on the interface, which in turn is only possible if the phases of $\vec{E}_i$, $\vec{E}_r$, and $\vec{E}_t$ are all the same. That is

$$(\vec{K}_i \cdot \vec{r} - \omega_i t) = (\vec{K}_r \cdot \vec{r} - \omega_r t) = (\vec{K}_t \cdot \vec{r} - \omega_t t) \tag{5.51}$$

for any t and $\vec{r}$ on the boundary. Thus the waves can match only if

$$\omega_i = \omega_r = \omega_t$$

and

$$\begin{aligned}\vec{k}_i' \cdot \vec{r} = \vec{k}_r' \cdot \vec{r} = \vec{k}_t' \cdot \vec{r} \\ \vec{k}_i'' \cdot \vec{r} = \vec{k}_r'' \cdot \vec{r} = \vec{k}_t'' \cdot \vec{r}.\end{aligned} \tag{5.52}$$

Since the boundary conditions are valid for any $\vec{r}$, the origin is chosen such that $\vec{r}$ is parallel to the surface. Then $\vec{k}_i'$, $\vec{k}_r'$, and $\vec{k}_t'$ are coplanar as would be $\vec{k}_i''$, $\vec{k}_r''$, and $\vec{k}_t''$. In addition, we consider first the case of a non-attenuated wave, $\vec{k}'' = 0$, for all waves. It can readily be shown that the above relations correspond to the 'the law of reflection', that is,

$$\theta_r = \theta_i \tag{5.53}$$

and 'Snell's law',

$$n_i \sin \theta_i = n_t \sin \theta_t. \tag{5.54}$$

To demonstrate these latter relations we choose a coordinate system such that the x-axis lies along the interface. Then we can write the scalar products $\vec{k}' \cdot \vec{r}$ as

$$\left(k'_{ix}\,\hat{x} - k'_{iy}\,\hat{y}\right) \cdot (x\,\hat{x} + y\,\hat{y}) = k'_i\,x \sin \theta_i - k'_i\,y \cos \theta_i$$

$$= \frac{n_i \omega}{c}(x \sin \theta_i - y \cos \theta_i)$$

$$\left(k'_{rx}\,\hat{x} + k'_{ry}\,\hat{y}\right) \cdot (x\,\hat{x} + y\,\hat{y}) = \frac{n_r \omega}{c}(x \sin \theta_r + y \cos \theta_r)$$

$$\left(k'_{tx}\,\hat{x} - k'_{ty}\,\hat{y}\right) \cdot (x\,\hat{x} + y\,\hat{y}) = \frac{n_t \omega}{c}(x \sin \theta_t - y \cos \theta_t),$$

where we have used the sketch of the incident, reflected, and transmitted waves in figure 5.2. Since the incident and reflected waves are in the same medium, $n_i = n_r = n_1$. Thus if we compare the x coefficients of the first two of the above relations, and use the translational invariance property, we find

$$\frac{n_1 \omega}{c} \sin \theta_i = \frac{n_1 \omega}{c} \sin \theta_r$$

$$\theta_i = \theta_r,$$

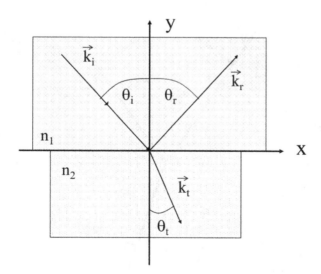

Figure 5.2. Illustration of the k-vectors for the incident, reflected, and transmitted waves at an interface between two media with indices n_1 and n_2. Note that the angles of incidence, reflection, and transmission are all defined with respect to the normal to the surface.

which is the law of reflection. Similarly, if we compare the incident and transmitted waves, with $n_t = n_2$, we have

$$\frac{n_1\omega}{c}\sin\theta_i = \frac{n_2\omega}{c}\sin\theta_t$$

$$n_1\sin\theta_i = n_2\sin\theta_t,$$

which is Snell's law.

5.9.2 Amplitude of transmitted and reflected waves (Fresnel's equations)

In considering the reflection and transmission of optical fields at an interface we must take into account two different situations, namely waves with E-vectors perpendicular to the plane of incidence, and those with E lying in the plane of incidence. If we know the behavior of polarizations perpendicular and parallel to the plane of incidence, the principle of superposition says that we can determine what happens to any other wave, considered as a superposition of the two 'basis' waves.

TE waves
Referring to figure 5.3 we can see the situation for the former case, called transverse electric (TE) or *s*-polarized waves. Considering this case first, the boundary conditions for the continuity of the tangential parts of $\vec{E}$ and $\vec{H}$ lead to the following relations:

$$E_i + E_r = E_t \tag{5.55}$$

$$H_i\cos\theta_i - H_r\cos\theta_i = H_t\cos\theta_t, \tag{5.56}$$

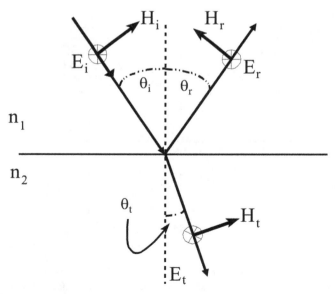

Figure 5.3. Sketch of the boundary conditions for TE or '*s*-polarized' waves.

where we have already included the fact that $\theta_r = \theta_i$. Using Maxwell's equations we have that $\vec{k}' \times \vec{E} = \mu_0 \omega \vec{H}$, and since, for example, $k_t'/k_i' = n_2/n_1$,

$$n_1 E_i \cos \theta_i - n_1 E_r \cos \theta_i = n_2 E_t \cos \theta_t. \tag{5.57}$$

We can find the *reflection coefficient* by eliminating E_t from equation (5.57) using equation (5.55):

$$r_s = \left(\frac{E_r}{E_i} \right)_s = \frac{n_1 \cos \theta_i - n_2 \cos \theta_t}{n_2 \cos \theta_t + n_1 \cos \theta_i}. \tag{5.58}$$

If we define β_s as

$$\beta_s = \frac{n_2 \cos \theta_t}{n_1 \cos \theta_i} = \frac{\sqrt{n_2^2 - n_1^2 \sin^2 \theta_i}}{n_1 \cos \theta_i} \tag{5.59}$$

then the reflection coefficient becomes

$$r_s = \frac{1 - \beta_s}{1 + \beta_s}. \tag{5.60}$$

By the same technique we can find the field *transmission coefficient*

$$t_s = \left(\frac{E_t}{E_i} \right)_s = \frac{2n_1 \cos \theta_i}{n_2 \cos \theta_t + n_1 \cos \theta_i} = \frac{2}{1 + \beta_s}. \tag{5.61}$$

We can write the field reflection and transmission coefficients in an alternative form by using trigonometric relations; the details are left for an exercise. The results are

$$r_s = -\frac{\sin (\theta_i - \theta_t)}{\sin (\theta_i + \theta_t)} \tag{5.62}$$

$$t_s = \frac{2 \sin \theta_t \cos \theta_i}{\sin (\theta_i + \theta_t)}. \tag{5.63}$$

TM waves
We next want to determine what fraction of the amplitude of a wave incident upon an interface is reflected and what fraction of the amplitude is transmitted when the waves are transverse magnetic (TM) or *p*-polarized. This means that $\vec{H}$ is perpendicular to the plane of incidence and that $\vec{E}$ lies in the plane of incidence. Referring to figure 5.4 the boundary conditions for continuity of $\vec{E}$ and $\vec{H}$ yield

$$H_i + H_r = H_t \tag{5.64}$$

$$E_i \cos \theta_i - E_r \cos \theta_i = E_t \cos \theta_t. \tag{5.65}$$

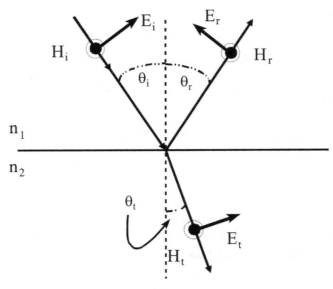

Figure 5.4. Boundary conditions for a TM or 'p-polarized' wave.

Using Maxwell's equations as before, the first equation of the above pair leads to

$$n_i E_i + n_i E_r = n_t E_t. \tag{5.66}$$

These equations give us the reflection and transmission coefficients

$$r_p = \left(\frac{E_r}{E_i}\right)_p = \frac{n_2 \cos \theta_i - n_1 \cos \theta_t}{n_1 \cos \theta_t + n_2 \cos \theta_i} = \frac{1 - \beta_p}{1 + \beta_p} \tag{5.67}$$

and

$$t_p = \left(\frac{E_t}{E_i}\right)_p = \frac{2\dfrac{n_1}{n_2}}{1 + \beta_p}, \tag{5.68}$$

where

$$\beta_p = \frac{n_1 \cos \theta_t}{n_2 \cos \theta_i}. \tag{5.69}$$

Again, as with TE waves, we can write the reflection and transmission coefficients in an alternative form:

$$r_p = \frac{\tan (\theta_i - \theta_t)}{\tan (\theta_i + \theta_t)} \tag{5.70}$$

$$t_p = \frac{2 \sin \theta_t \cos \theta_i}{\sin (\theta_i + \theta_t) \cos (\theta_i - \theta_t)}. \tag{5.71}$$

Physical interpretation of reflection coefficients

Now that we have the results for the reflection and transmission coefficients for both s- and p-polarized light, we should stop and take stock of the physical meaning. Recall that we used the boundary conditions for $\vec{E}$ and $\vec{H}$, as taken from figures 5.3 and 5.4, to arrive at equations (5.58), (5.61) and (5.67), (5.68). However, in the sketches, the choice of field directions is not unique. Let us look at the special case of normal incidence, $\theta_i = 0$ (and thus $\theta_t = 0$). For the TE wave this leads to $r_s = (1 - n_2/n_1)/(1 + n_2/n_1)$. If the wave is incident from air to another medium with a higher index of refraction, $r_s < 0$. Since we are looking at a *field* reflection coefficient this (i.e. the 'negative') simply means that the phase of the field reflected is different from that of the incident wave by 180°. With respect to our sketch, figure 5.3, we must draw the vector for $\vec{E}_r$ as coming out of the page (and of course redraw $\vec{H}_r$ accordingly, so that $\vec{E} \times \vec{H} = \vec{S}$, the direction of propagation). That is, the sign resulting from our application of the boundary conditions is relative to our initial choice of vector directions. Physically we see that $\vec{E}$ flips by 180° upon reflection from an interface with higher index.

Next we look at the case of a TM wave. Here our sketch (figure 5.4) shows that the component of the E-vector parallel to the interface is already reversed. At normal incidence, from equation (5.67), $r_p = (n_2/n_1 - 1)/(n_2/n_1 + 1) > 0$, if $n_2 > n_1$. Thus it essentially confirms our choice of drawing the vectors in figure 5.4. Again, physically we see that at normal incidence there is a 180° phase shift upon reflection from an interface where $n_2 > n_1$.

Note that at normal incidence there can be no distinction between TE and TM waves; since the normal to the interface and the incident ray coincide, there is no defined plane of incidence.

Reflectance and transmittance

All of the definitions discussed up to this point need to be related to measurable quantities, that is to the *power reflectances* and *transmittance*. For example, to obtain the reflected power for light incident at an angle θ_i on a surface of area A, one must determine $\int \langle \vec{S}_r \rangle \cdot \hat{n} \, dA$, where $\hat{n}$ is the normal to the surface and $\vec{S}_r$ is the reflected Poynting vector. We are interested in the ratio of the reflected power to the incident power, given by $\int \langle \vec{S}_0 \rangle \cdot \hat{n} \, dA$. These quantities are shown in figure 5.5. The power is calculated by integrating over the area A', corresponding to the cross-section of the incident beam. Relating the area of the incident and reflected beams to the area A on the surface, we find the following:

$$R = \frac{\int \langle \vec{S}_r \rangle \cdot \hat{n} \, dA}{\int \langle \vec{S}_0 \rangle \cdot \hat{n} \, dA}. \tag{5.72}$$

The dot product gives a factor of $\cos \theta_i$ ($\cos \theta_r$ for the reflected beam). We relate the areas of the beams to that of the surface area A; the projection gives a factor of $\cos \theta$ as well, but since $\theta_i = \theta_r$ everything reduces to

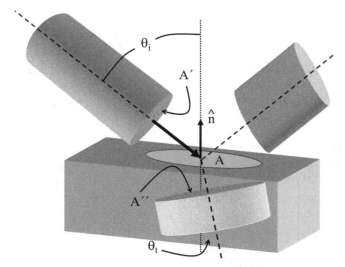

Figure 5.5. Illustration of the various quantities involved in calculating the transmittance and reflectance. The incident beam of light has area A', whereas the area of the beam as projected on the interface is A.

$$R = \frac{\frac{n_1 |E_r|^2}{2\mu_0 c} \cos \theta_r A}{n_1 \frac{|E_i|^2}{2\mu_0 c} \cos \theta_i A} = \frac{|E_r|^2}{|E_i|^2}, \tag{5.73}$$

where we have also taken into account the speed of the electromagnetic wave in the incident (and reflected) medium. Thus, the power reflectances for s- and p-polarizations are defined as $R_s = |r_s|^2$ and $R_p = |r_p|^2$.

Similarly, one can define the power transmittance through the interface as

$$T = \frac{\int \langle \vec{S}_t \rangle \cdot \hat{n} \, dA}{\int \langle \vec{S}_0 \rangle \cdot \hat{n} \, dA} = \frac{\frac{n_2 |E_t|^2}{2\mu_0 c} \cos \theta_t A}{n_1 \frac{|E_i|^2}{2\mu_0 c} \cos \theta_i A} = \frac{|E_t|^2}{|E_i|^2} \frac{n_2 \cos \theta_t}{n_1 \cos \theta_i}. \tag{5.74}$$

This can be rewritten for s- and p-polarizations as

$$T_s = \beta_s \frac{|E_t|^2}{|E_i|^2} = \beta_s |t_s|^2 \tag{5.75}$$

and

$$T_p = \frac{n_2^2}{n_1^2} \beta_p |t_p|^2 = \frac{n_2 \cos \theta_t}{n_1 \cos \theta_i} |t_p|^2. \tag{5.76}$$

It is in terms of these reflectances that one can speak of the conservation of energy, and thus arrive at a relation for the sum of reflectance and transmittance being unity (for an assumed lossless medium):

$$R + T = 1,$$

where R and T refer to either s-polarized or p-polarized incident fields.

Example: reflection at an air–glass interface
We can calculate the reflection and transmission coefficients for *s*- and *p*-polarized light incident from air onto a piece of glass. We can assume that the index of refraction for glass is $n_2 = n_g = 1.50$ and that $n_1 = n_{\text{air}} = 1.00$. From equations (5.58) and (5.67) we can simply substitute in the appropriate values for n_1 and n_2. At normal incidence the coefficients for *s*- and *p*-polarizations are identical, $r = \frac{n_1 - n_2}{n_1 + n_2}$. Substituting in the given values for the indices of refraction we find $r = -0.2$. Thus the reflectance is $R = 0.2^2 = 0.04$; this gives rise to the useful rule-of-thumb that about 4% of incident light is reflected from a glass surface. Note as well that reflection from a glass–air interface gives the same result for the reflectance. As a function of incidence angle, the results for both the reflection and transmission coefficients are shown in figure 5.6. In figure 5.7 we show both the power reflectance and transmittance for the same parameters as in figure 5.6.

5.9.3 Jones matrices for reflection

We can write out Jones matrices for reflection and transmission using the relations we have derived above. There is one subtlety which arises due to the essentially random choice of phases for the fields when we set up the boundary conditions to derive r_s and r_p. As this is discussed in general above, we will simply write down the result for the Jones matrices here, and accept by example that they give the proper physical answers.

For reflection we have

$$M_r = \begin{pmatrix} r_p & 0 \\ 0 & -r_s \end{pmatrix} = \begin{pmatrix} \dfrac{1 - \beta_p}{1 + \beta_p} & 0 \\ 0 & -\dfrac{1 - \beta_s}{1 + \beta_s} \end{pmatrix}.$$

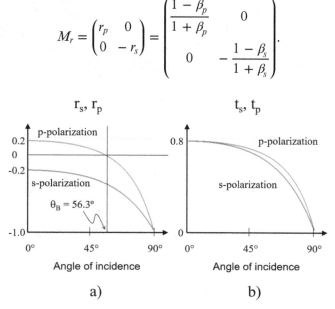

Figure 5.6. (a) Plot of the reflection coefficients r_s and r_p for light incident from air onto glass with $n_g = 1.50$. The curves for *s*- and *p*-polarization are labeled, as is the Brewster angle, θ_B, for which $r_p = 0$, i.e. for which the reflected light is purely *s*-polarized. (b) Similar plot for the transmission coefficients for both *s*- and *p*-polarized light.

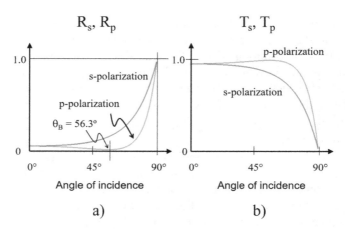

Figure 5.7. (a) Plot of the power reflectance R_s and R_p for light incident from air onto glass with $n_g = 1.50$. The curves for s- and p-polarization are labeled. Note that at the Brewster angle, θ_B, $R_p = 0$, i.e. the reflected light is purely s-polarized. (b) Similar plot for the power transmittance for both s- and p-polarized light.

The Jones matrix for transmission is given by

$$M_t = \begin{pmatrix} t_p & 0 \\ 0 & t_s \end{pmatrix} = \begin{pmatrix} \dfrac{2n_1/n_2}{1 + \beta_p} & 0 \\ 0 & \dfrac{2}{1 + \beta_s} \end{pmatrix}.$$

Example. As an example of the use of the reflection Jones matrices, consider the action of reflection on a right-circularly polarized incident beam of light. We use normal incidence for the sake of simplicity and assume $n_2 = 1.5$:

$$\begin{pmatrix} E_{in,\,x} \\ E_{in,\,y} \end{pmatrix} = \frac{1}{\sqrt{2}} \begin{pmatrix} 1 \\ i \end{pmatrix}$$

$$\begin{pmatrix} E_{out,\,x} \\ E_{out,\,y} \end{pmatrix} = \frac{1}{\sqrt{2}} \begin{pmatrix} \dfrac{1 - \beta_p}{1 + \beta_p} & 0 \\ 0 & -\dfrac{1 - \beta_s}{1 + \beta_s} \end{pmatrix} \begin{pmatrix} 1 \\ i \end{pmatrix}$$

$$= \frac{1}{\sqrt{2}} \begin{pmatrix} \dfrac{1 - \frac{2}{3}}{1 + \frac{2}{3}} & 0 \\ 0 & -\dfrac{1 - \frac{3}{2}}{1 + \frac{3}{2}} \end{pmatrix} \begin{pmatrix} 1 \\ i \end{pmatrix}$$

$$= \frac{1}{\sqrt{2}} (0.2) \begin{pmatrix} 1 \\ i \end{pmatrix}.$$

Thus we find that RCP light reflected at normal incidence from a dielectric material remains RCP and has its amplitude reduced.

5.9.4 Stokes relations

Consider a ray incident on an interface as we have discussed up until now. If we consider the wave equation, the time-reversed version of the incident wave splitting into refracted and reflected waves must also be valid. That is, the reverse situation must also be true, for which the angle of incidence is now the angle of refraction for one wave and the angle of reflection for another (see figures 5.8(a) and 5.8(b)). Consider for simplicity TE waves for now. An analysis of this reversed situation gives

$$r_s' = -\frac{1 - \beta_s}{1 + \beta_s} = -r_s \qquad (5.77)$$

$$t_s' = \frac{2\beta_s}{1 + \beta_s} = \beta_s t_s, \qquad (5.78)$$

where the primes denote the time-reversed rays. To arrive at this result, we consider the general case corresponding to figure 5.8(c). We see that in general, each incident ray gives rise to a reflected and a refracted ray at the interface. However, the argument about the validity of the time-reversed picture leads us to conclude that the situations in figures 5.8(b) and 5.8(c) must be physically equivalent. This in turn can only be true if

$$(r_s' t_s + t_s r_s) E_i = 0$$
$$(r_s^2 + t_s t_s') E_i = E_i.$$

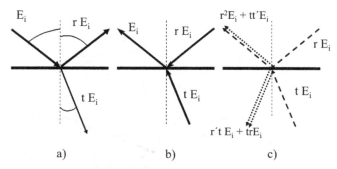

a)   b)   c)

Figure 5.8. Illustration of the Stokes relations. In (a) the usual configuration is shown. The time-reversed equivalent is shown in (b); each ray, traveling 'backward in time' must satisfy the wave equation. If we consider rays incident on the surface, one of amplitude rE_i and the other of amplitude tE_i, they will both be reflected and transmitted, with coefficients r' and t', respectively. In (c) the original rays are shown as dashed lines and the rays that 'should' be present (and that are used to calculate the Stokes relations) are shown as dotted lines.

Thus we have that $r_s = -r_s'$ and $t_s\, t_s' = 1 - r_s^2$. These latter two are known as the 'Stokes relations'. Using equation (5.59) we can arrive at equations (5.77) and (5.78). Also, it can be shown that

$$t_s t_s' - r_s r_s' = 1. \qquad (5.79)$$

5.9.5 Specific angles of reflection

There are some special cases which give interesting physical results for specific angles of incidence. The specific angles of incidence are the critical angle and the polarization angle (also called Brewster's angle). First we will consider the critical angle and work with s-polarization only.

The critical angle
When β is pure imaginary, the power reflectance is unity, since, from equation (5.60), the modulus of $\frac{1-\beta_s}{1+\beta_s}$ is unity (figure 5.9).

Now the expression for β_s becomes

$$\beta_s = \frac{n_2 \cos \theta_t}{n_1 \cos \theta_i} = \frac{[n_2^2 - n_1^2 \sin^2 \theta_i]^{\frac{1}{2}}}{n_1 \cos \theta_i}. \qquad (5.80)$$

β_s will be pure imaginary if $n_2 < n_1 \sin \theta_i$. This expression gives us the condition for the special angle of incidence, called the critical angle θ_c, the angle at which total reflection occurs because it is the angle for which the power reflectance is unity:

$$\theta_c = \sin^{-1} (n_2/n_1). \qquad (5.81)$$

Let us ask about the nature of the wave in the second medium. If $\theta_i = \theta_c$, then β_s is zero and we must have $\theta_t = \pi/2$. One can show (left for a homework problem) that the *evanescent wave*, i.e. the magnitude of the field in the second medium, can be written as

$$E = \mathcal{E}_t e^{-\alpha y} e^{i(k_2 x - \omega t)}, \qquad (5.82)$$

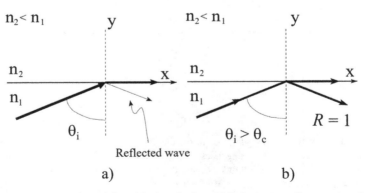

Figure 5.9. Illustration of the critical angle, for which the power reflectance is unity.

that is, that the direction of travel of the evanescent wave is parallel to the surface and that, in the direction normal to the surface, there is a wave which rapidly attenuates. It is left for a homework problem to show that no power is carried away by the evanescent wave. To determine the constants α and k_2 in equation (5.82) we consider the following arguments. The field incident on the interface and the transmitted field are given by

$$\vec{E}_i = \mathcal{E}_i \, \hat{z} \, e^{i(\vec{k}_i \cdot \vec{r} - \omega t)}$$
$$\vec{E}_t = \mathcal{E}_t \, \hat{z} \, e^{i(\vec{k}_t \cdot \vec{r} - \omega t)}.$$

For a coordinate system in which the x-direction is along the surface and the normal to the surface is the y-direction, we can rewrite the incident field as

$$\vec{E}_i = E_0 \, \hat{z} \, e^{i(k_i(x \sin \theta_i + y \cos \theta_i) - \omega t)}.$$

Using Snell's law we have $k_i \sin \theta_i = k_t \sin \theta_t$ and $\cos \theta_i = \sqrt{1 - \sin^2 \theta_i}$ and equating in the exponents for the transmitted and incident waves

$$k_{tx} \, x = \frac{\omega n_2}{c} \sin \theta_t \, x = k_i \, x \sin \theta_i = \frac{\omega n_1}{c} x \sin \theta_i = k_2 \, x$$
$$k_{ty} y = \frac{\omega n_2}{c} y \cos \theta_t = \frac{\omega n_2}{c} y \sqrt{1 - \sin^2 \theta_t} = \frac{\omega n_2}{c} y \sqrt{1 - (n_1^2/n_2^2) \sin^2 \theta_i}.$$

As noted earlier, for angles $\theta_i > \theta_c$ the square root becomes a pure imaginary number. In this case we can write

$$k_{ty} = \frac{i \omega n_2}{c} \sqrt{(n_1^2/n_2^2) \sin^2 \theta_i - 1} = i\alpha.$$

With this identification of variables we thus arrive at equation (5.82).

Although the total reflectance for incident angles greater than θ_c is unity, we should take a closer look at the reflected fields to see if there is any phase information that is of interest. Using our previous expressions for r_s and r_p we can write expressions for the case in which $\theta_i > \theta_c$,

$$r_s = \frac{a - ib}{a + ib}, \tag{5.83}$$

where $a = \frac{n_1}{n_2} \cos \theta_i$ and $b = \sqrt{\left(\frac{n_1}{n_2}\right)^2 \sin^2 \theta_i - 1}$. The reflection coefficient phase shift can be written as (the derivations are left for the homework problems)

$$\tan \frac{\delta_s}{2} = -\frac{\sqrt{\sin^2 \theta_i - \left(\frac{n_2}{n_1}\right)^2}}{\cos \theta_i}. \tag{5.84}$$

For a p-polarized field the corresponding phase shift is

$$\tan \frac{\delta_p}{2} = -\frac{\sqrt{\sin^2 \theta_i - \left(\frac{n_2}{n_1}\right)^2}}{\left(\frac{n_2}{n_1}\right)^2 \cos \theta_i}.$$ (5.85)

The numerator here is the same as the attenuation depth found for the wavevector k_y. Thus it appears that the reflected field upon internal reflection picks up a phase shift that is proportional to the distance to which the evanescent wave penetrates into the lower index medium. It is as if the reflected wave, although the reflected power is unity, has sampled the second medium out to the depth α and thus been delayed or phase shifted by that amount.

In figure 5.10 the reflection and transmission coefficients for both s- and p-polarized light are shown as a function of incident angle. For the case shown we have taken $n_1 = 1.50$ and $n_2 = 1.00$, representing internal incidence at a glass–air interface. Figure 5.11 shows the power reflectance and transmittance for the same interface, again for both s- and p-polarizations.

Brewster's angle
At an angle, θ_B, called Brewster's angle, the radiation reflected from a dielectric will be completely s-polarized. Thus at θ_B the reflectance for p-polarization is zero. We can determine this angle by looking at the condition for which r_p will be zero; we must have

$$1 - \beta_p = 0 \quad \Longrightarrow \quad \beta_p = 1$$

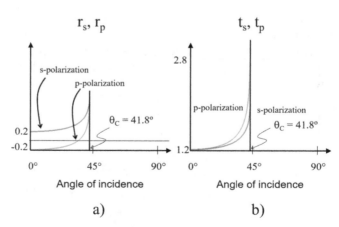

Figure 5.10. (a) Plot of the reflection coefficients r_s and r_p for light internally incident from glass onto air with $n_{glass} = 1.50$. The curves for s- and p-polarization are labeled. (b) Similar plot for the transmission coefficients for both s- and p-polarized light.

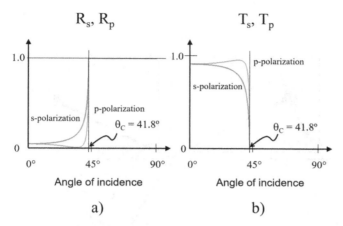

Figure 5.11. (a) Plot of the power reflectance R_s and R_p for light internally incident from glass onto air with $n_g = 1.50$. The curves for s- and p-polarization are labeled. Note that at the critical angle, θ_C, $R_p = R_s = 1$. (b) Similar plot for the power transmittance for both s- and p-polarized light.

and therefore

$$\beta_p = 1 = \frac{n_1 \cos \theta_t}{n_2 \cos \theta_B}.$$

Thus we have, using Snell's Law to eliminate θ_t,

$$n_2 \cos \theta_B = \frac{n_1}{n_2} \sqrt{n_2^2 - n_1^2 \sin^2 \theta_B}.$$

We can find from the above that

$$\tan \theta_B = \frac{n_2}{n_1}. \tag{5.86}$$

The angle θ_B given by equation (5.86) is called the polarization angle or Brewster's angle; figure 5.12 illustrates the results shown above.

We can consider the Brewster angle from another point of view to perhaps obtain a better physical feeling for why the p-wave is not reflected. If we look at equation (5.70), we see that $r_p \to 0$ for $\theta_i + \theta_t = \pi/2$, since $\tan(\pi/2)$ is infinite. Looking at figure 5.12 we recognize that this implies that the reflected wave and the transmitted wave are perpendicular to one another. Since the reflected wave must arise due to re-radiation of dipoles excited in the surface by the p-polarized incident wave, and thus aligned with that incident field, we see immediately that the direction of the reflected ray lies along the axis of the radiating dipoles. Thus, as we saw in chapter 3, we should expect no p-radiation along the direction given by θ_B ($=\theta_i$). Similarly, we can use this understanding to conclude that there is no direction for s-polarized radiation at which the dipole would be aligned with the wave propagation direction and thus lead to zero reflected s-radiation.

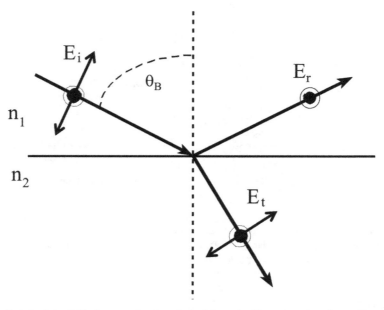

Figure 5.12. Sketch of the fields for a randomly polarized beam incident on an interface at Brewster's angle. Notice that the reflected beam is completely polarized, whereas the transmitted beam still has both of the orthogonal polarization components.

5.10 Thin-film anti-reflection (AR) coating

We now look in some detail at a more involved example in which light reflects off two surfaces in sequence. A common application of this configuration is that of a thin film deposited on a surface such as glass to create an 'anti-reflection' coating. A common example is with eyeglasses for which we often request (and pay for) such a coating. Our starting point is to consider incoherent light reflecting off the air–film interface and also off the film–substrate interface.

We have already seen that an external reflection (low n to higher n) leads to a phase shift of π in the reflected field. If we consider a film with index of refraction intermediate to the incident medium (e.g. air) and the substrate (e.g. glass), at each surface there will be a reflection with a phase shift. Our goal is to set the thickness of the film such that the light traversing the film picks up an additional propagation phase shift of π so that the net phase shift between the wave reflected off the film and that reflected off the substrate and re-emerging from the film is equal to π. A propagation phase shift of π is the equivalent of a half-wavelength difference in distance, therefore a film thickness $\lambda/4$, traversed twice, is needed, realizing that the λ we mean here is the wavelength in the film, i.e. $\lambda = \lambda_0/n_{\text{film}}$. Such a film coating is referred to as a 'quarter-wave' coating. This configuration is shown in figure 5.13, where the fields are shown at non-normal incidence for clarity of illustration; we will consider here only normally incident light but the result will be valid for a range of relatively small incidence angles.

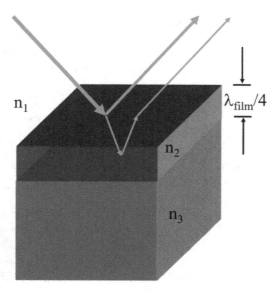

Figure 5.13. Sketch of the incident medium (n_1), the thin 'quarter-wave' film (n_2), and the substrate (n_3), with the incident, reflected, and transmitted rays.

Now we have the problem set up: our substrate will have a coating of thickness $\lambda/4$, but we need to know what index of refraction will work to fulfill the destructive interference condition, which can be found as follows. The total reflected field, assuming contributions only from one reflection at the film and one at the substrate, is given by

$$\frac{(E_r)_{\text{tot}}}{E_i} = r_{12} + e^{i\pi}t_{12}r_{23}t_{21} \tag{5.87}$$

$$= \frac{n_1 - n_2}{n_1 + n_2} - \left(\frac{2n_1}{n_1 + n_2}\right)\left(\frac{n_2 - n_3}{n_2 + n_3}\right)\left(\frac{2n_2}{n_1 + n_2}\right). \tag{5.88}$$

We must find a solution to this equation that yields zero reflected total field. The solution will involve a cubic equation in n_2; if we assume that $n_1 = 1.00$, the equation is given by

$$n_2^3 + (n_3 + 4)n_2^2 - (4n_3 + 1)n_2 - n_3 = 0.$$

With the assumption that $n_3 = 1.50$, typical for glass, and solving the equation numerically the result is $n_2 = 1.22$. Unfortunately, it is not always possible to find a dielectric material that has just the correct index of refraction; in practice, simple single-layer coatings are often made of MgO, with an index of refraction of $n = 1.38$, which can reduce the reflected intensity to less than 1% of the incident intensity.

Interestingly enough, solving this problem more exactly by assuming multiple reflections in the thin film, much as we did when discussing the field in a resonator in chapter 1, while being more difficult initially, allows one to find an analytic solution for the index of refraction in a simple form. It turns out that the ideal film index for anti-reflection coating, given the proper thickness, is given by $n_2 = \sqrt{n_1 n_3}$, yielding essentially the same numerical result we just found. We now turn to this more exact treatment, and will come back to this problem again later in the chapter when we consider metallic interfaces.

Our goal is to treat the incoming light as coherent, i.e. we will add up all the multiple reflections from various surfaces. In addition, we wish to consider the following two questions: if we have found an ideal, or nearly ideal quarter-wave anti-reflection film for normally incident light for a given wavelength, (a) how sensitive is that anti-reflection coating to changes in the angle of incidence and (b) how well does the anti-reflection coating work at wavelengths other than the design wavelength?

Again, we refer to chapter 1 where we looked at the example of light building up in an optical resonator and found that we could add fields at the output of the resonator, taking into account the field reflection and transmission coefficients along with a phase shift corresponding to the propagation distance through the resonator. Looking at the reflected field, we find that, upon successive reflections from the film–air and film–substrate interface, we can write

$$\mathcal{E}_r = \mathcal{E}_i(r_{12} + t_{12}r_{23}t_{21}e^{2i\delta} + t_{12}r_{23}r_{21}r_{23}t_{21}e^{4i\delta} + \cdots), \tag{5.89}$$

where r_{12} refers to the reflection coefficient for a field incident from the first medium onto the second medium, t_{21} refers to transmission from the second medium into the first medium, etc. The net phase shift between one reflected ray and the next is given by δ. If we rewrite the above and make use of the Stokes relations, $r_{21} = -r_{12}$ and $t_{12}t_{21} = T_{12} = 1 - |r_{12}|^2$, we find an infinite geometric series that can be summed to yield

$$\mathcal{E}_r = \mathcal{E}_i\left(r_{12} + \frac{\left(1 - r_{12}^2\right)r_{23}e^{2i\delta}}{1 - r_{23}r_{21}e^{2i\delta}}\right). \tag{5.90}$$

This may now be simplified to give

$$r \equiv \frac{\mathcal{E}_r}{\mathcal{E}_i} = \frac{r_{12} + r_{23}e^{2i\delta}}{1 + r_{23}r_{12}e^{2i\delta}}. \tag{5.91}$$

Equation (5.91) is our main result for this example. Since the reflection coefficients at the individual interfaces differ for s- and p-polarized light, the overall reflection coefficient will as well.

Before using this result to answer the questions we posed above, we must find an explicit expression for the phase shift δ. The derivation we leave for the problems, but with a little bit of geometry we can show that

$$\delta = \frac{2\pi n_2 d}{\lambda_0}\cos\theta_2 = \frac{2\pi}{\lambda_0}\sqrt{n_2^2 - n_1^2\sin^2\theta_i}. \tag{5.92}$$

5-32

Our first task is to find the index of refraction and the thickness for the second medium (the film) needed to make the reflected field vanish. From our expression for the reflection coefficient we see that a vanishing r implies

$$r_{12} + r_{23}e^{2i\delta} = 0. \tag{5.93}$$

This sets a condition on the phase, namely that $e^{2i\delta} = -1$ or (considering normal incidence for now) $2\pi d n_2/\lambda_0 = q\pi$, where q is an integer. Considering only $q = 1$, we find $d = \lambda/4$, with $\lambda = \lambda_0/n_2$, just the result we reached by qualitative arguments above.

Next we look for the index of refraction needed for the quarter-wave film by using $r_{12} - r_{23} = 0$, or

$$\frac{n_1 - n_2}{n_1 + n_2} = \frac{n_2 - n_3}{n_2 + n_3},$$

which can easily be solved to find the desired result, $n_2 = \sqrt{n_1 n_3}$.

To find the angle and wavelength sensitivity for a quarter-wave coating, we resort to numerical calculations using the expressions we have derived above. In figure 5.14 the reflectance, $R = r^2$, is shown for both s- and p-polarizations, taking $n_1 = 1.0$, $n_2 \equiv n_{\text{film}} = 1.38$, and $n_3 \equiv n_{\text{substrate}} = 1.50$. Recalling that the reflectance from the glass surface without a coating would be about 4%, we see from figure 5.14 that over a wide range of incidence angles there is a significantly reduced reflectance, even with this non-ideal (i.e. $n_{\text{film}} \neq 1.22$) thin film.

Figure 5.15 shows the normal-incidence reflectance (recall that the distinction between s- and p-polarizations disappears at normal incidence) as a function of the wavelength of the incident light. As expected, the reflectance has a minimum for the chosen wavelength (here $\lambda_0 = 500$ nm), but the deviation from that minimum is a rather weak function of wavelength. We see that throughout the visible range of the

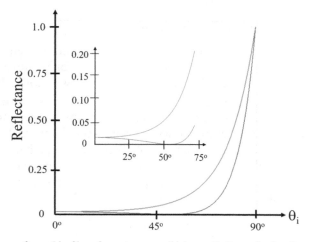

Figure 5.14. Reflectance for a thin film of quarter-wave thickness. Indices of refraction are: $n_1 = 1.0$ for the incident medium, $n_2 = 1.38$ for the thin 'quarter-wave' film, and $n_3 = 1.50$ for the substrate. The inset shows an expanded view for incidence angles up to 75°.

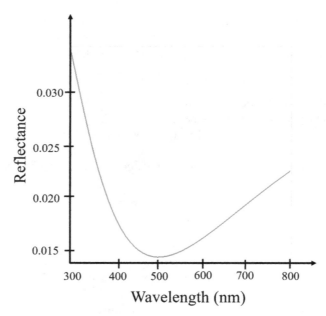

Figure 5.15. Reflectance for a thin film of quarter-wave thickness at the design wavelength of $\lambda_0 = 500$ nm. Indices of refraction are: $n_1 = 1.0$ for the incident medium, $n_2 = 1.38$ for the thin 'quarter-wave' film, and $n_3 = 1.50$ for the substrate.

spectrum, the reflectance remains below about 2.5%, significantly less than without the thin-film coating. However, the light that is reflected is mainly at the blue and red ends of the visible spectrum; when viewing a camera lens or eyeglasses that have been AR-coated, one typically sees a magenta-colored tint that is the result of a mixing in the eye of the blue and red reflected light.

5.11 Waves at a conducting interface

Now we wish to start over again to some extent, and present the details for calculation of reflection and transmission coefficients for conducting media. The calculations that follow are based on [6]. In figure 5.16 we show a layered medium with an incident wave designated by the wavevector $\vec{K}_0 = \vec{k}_0' + i\vec{k}_0''$. The incident medium is real and the subsequent media may be complex. The interfaces between the media are parallel. The figure shows the wavevectors in the three media. Our objective with this illustration will be to obtain relationships between the angle of incidence and the angles of refraction in the various media in terms of the parameters of the incident wave and the optical parameters of the media.

First surface. Consider the boundary conditions at the first interface:

$$\vec{k}_0' \cdot \hat{u}_t = \vec{k}_r' \cdot \hat{u}_t = \vec{k}_1' \cdot \hat{u}_t$$
$$\vec{k}_0'' \cdot \hat{u}_t = \vec{k}_r'' \cdot \hat{u}_t = \vec{k}_1'' \cdot \hat{u}_t,$$

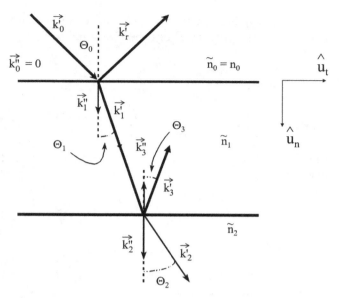

Figure 5.16. Layered medium where the first medium is non-conducting but the second two media are conductors. The angles of incidence and refraction are given by θ_0 and θ_1 at the first interface and θ_1 and θ_2 at the second interface, respectively.

where $\hat{u}_t$ is a unit vector along the surface, as indicated in figure 5.16. Now assume that the incident medium is non-attenuating and that the incident wave is homogeneous, that is, that $\vec{k}_0'' = 0$. From the above equations this leads us to conclude that $\vec{k}_r''$ and $\vec{k}_1''$ are either perpendicular to $\hat{u}_t$ or zero. In the incident medium, $\vec{k}_r''$ also must be either zero or perpendicular to $\vec{k}_r'$ because the incident medium is non-attenuating (see section 5.8). We will now argue that in fact only the former will be possible. If the medium is non-attenuating

$$\vec{k}_0'^2 = \frac{\omega^2}{c^2} n_0^2 = \vec{k}_r'^2 - \vec{k}_r''^2$$

giving the result that $\vec{k}_r'^2 \geqslant \vec{k}_0'^2$. However, if $\vec{k}_r''$ is not zero, then it is perpendicular to $\hat{u}_t$, again because of the fact that $\vec{k}_r'' \cdot \hat{u}_t = 0$. Furthermore, $\vec{k}_r'$ is also perpendicular to $\vec{k}_r''$ which then means that $\vec{k}_r'$ is parallel to $\hat{u}_t$. Now we also know that $\vec{k}_0' \cdot \hat{u}_t = \vec{k}_r' \cdot \hat{u}_t$. Thus if $\vec{k}_r'$ is parallel to $\hat{u}_t$, then $|\vec{k}_r'| < |\vec{k}_0'|$. But this result is not possible and leads us to conclude that $|\vec{k}_r''| = 0$ and $|\vec{k}_0'| = |\vec{k}_r'|$ and from the boundary condition equations this means that the angle of incidence equals the angle of reflection. Also $\vec{k}_0' \cdot \hat{u}_t = \vec{k}_1' \cdot \hat{u}_t$ gives the appropriate relationship (Snell's law) for the angles of refraction,

$$k_0' \sin \theta_0 = k_1' \sin \theta_1.$$

Second interface. Next apply the boundary conditions at the second interface. The incident wave is now inhomogeneous. From the first interface,

$$\vec{k}_1'' \cdot \hat{u}_t = 0,$$

therefore

$$\underbrace{\vec{k}_1'' \cdot \hat{u}_t}_{\text{incident}} = \underbrace{\vec{k}_2'' \cdot \hat{u}_t}_{\text{transmitted}} = \underbrace{\vec{k}_3'' \cdot \hat{u}_t}_{\text{reflected}} = 0,$$

which means that $\vec{k}_2''$ and $\vec{k}_3''$ are normal to the boundary. From the boundary conditions we also have

$$\vec{k}_1' \cdot \hat{u}_t = \vec{k}_2' \cdot \hat{u}_t = \vec{k}_3' \cdot \hat{u}_t$$

and therefore

$$k_1' \sin \theta_1 = k_2' \sin \Theta_2 = k_3' \sin \theta_3 = k_0' \sin \theta_0. \tag{5.94}$$

Equation (5.94) tells us that Snell's law applies for rays (wavevectors) that are normal to the surfaces of constant phase.

Consider next the reflected wave designated by $\vec{K}_3$:

$$\vec{K}_3 = (k_3' \sin \theta_3)\, \hat{u}_t + (k_3' \cos \theta_3 + ik_3'')(-\hat{u}_n)$$
$$\equiv \frac{\omega}{c}\tilde{n}_1[\, \hat{u}_t \sin \tilde{\theta}_3 - \hat{u}_n \cos \tilde{\theta}_3],$$

where $\hat{u}_n$ is a unit vector normal to the interface and in which the complex angle is defined by

$$\frac{\omega}{c}\tilde{n}_1 \sin \tilde{\theta}_3 \equiv k_3' \sin \theta_3$$
$$\frac{\omega}{c}\tilde{n}_1 \cos \tilde{\theta}_3 \equiv k_3' \cos \theta_3 + ik_3''.$$

Note that the complex angle $\tilde{\theta}$ is just a mathematical quantity which is introduced for algebraic convenience. Also, $\tilde{\theta}$ obeys the same trigonometric relations as a real angle, for example $\sin^2 \tilde{\theta} + \cos^2 \tilde{\theta} = 1$, so that from the above expressions we obtain

$$\frac{\omega}{c}\tilde{n}_1 \sin \tilde{\theta}_3 = k_3' \sin \theta_3 = k_0' \sin \theta_0 = k_1' \sin \theta_1$$
$$= \frac{\omega}{c}\tilde{n}_1 \sin \tilde{\theta}_1$$

and therefore $\tilde{\theta}_3 = \tilde{\theta}_1$. Thus the law of reflection also holds for the complex angles. Using this result we can write

$$\frac{\omega}{c}\tilde{n}_1 \cos \tilde{\theta}_3 = \frac{\omega}{c}\tilde{n}_1 \cos \tilde{\theta}_1$$
$$= k_1' \cos \theta_1 + ik_1''$$
$$= k_3' \cos \theta_3 + ik_3''.$$

Also, by using the fact that $\sin^2 \tilde{\theta} + \cos^2 \tilde{\theta} = 1$ and the definition of $\sin \theta$ above we find that

$$\frac{\omega}{c}\tilde{n}_1 \cos \tilde{\theta}_3 = \frac{\omega}{c}[\tilde{n}_1^2 - n_0^2 \sin^2 \theta_0]^{1/2}.$$

This is an important result in that it relates the parameters of the incident medium to those of the transmitting medium. Finally, from the above expressions we see that

$$k_3'' = k_1'', \quad k_1' = k_3', \quad \text{and} \quad \Theta_1 = \theta_3.$$

Now we consider the transmitted wave at the second interface,

$$\vec{K}_2 = (k_2' \sin \theta_2)\,\hat{u}_t + (k_2' \cos \theta_2 + ik_2'')\,\hat{u}_n$$
$$= \frac{\omega}{c}\tilde{n}_2[\,\hat{u}_t \sin \tilde{\theta}_2 + \hat{u}_n \cos \tilde{\theta}_2].$$

By equating vector components we find

$$\frac{\omega}{c}\tilde{n}_2 \sin \tilde{\theta}_2 = k_2' \sin \theta_2$$
$$\frac{\omega}{c}\tilde{n}_2 \cos \tilde{\theta}_2 = k_2' \cos \theta_2 + ik_2''$$

and using $\sin^2 \tilde{\theta} + \cos^2 \tilde{\theta} = 1$ we have

$$\frac{\omega}{c}\tilde{n}_2 \cos \tilde{\theta}_2 = \frac{\omega}{c}\left[\tilde{n}_2^2 - \frac{c^2}{\omega^2}k_2'^2 \sin^2 \theta_2\right]^{1/2}$$
$$= \frac{\omega}{c}[\tilde{n}_2^2 - n_0^2 \sin^2 \theta_0]^{1/2}.$$

Summary. For this parallel layered medium with a homogeneous source wave incident upon the system from a medium with real index, the waves in all layers behave according to

$$\vec{K} = (k' \sin \theta)\,\hat{u}_t + (k' \cos \theta + ik'')\,(\pm\,\hat{u}_n),$$

therefore

$$\vec{K} = \frac{\omega}{c}\tilde{n}[\,\hat{u}_t \sin \tilde{\theta} \pm \hat{u}_n \cos \tilde{\theta}], \tag{5.95}$$

where $\tilde{\theta}$ is the angle that the wavevector makes with the normal to the interface. Finally, all of the results of this illustration can be neatly summarized by two equations,

$$\tilde{n}\sin \tilde{\theta} = \frac{c}{\omega}k' \sin \theta = n_0 \sin \theta_0$$
$$\tilde{n}\cos \tilde{\theta} = \frac{c}{\omega}(k' \cos \theta + ik'') = [\tilde{n}^2 - n_0^2 \sin^2 \theta_0]^{1/2}.$$

5.11.1 Reflection and transmission for a conducting medium

We wish now to generalize our previous discussion of reflection and transmission coefficients to the case of conducting media. Consider only cases where $\vec{k}''$ is perpendicular to the surface. When attenuation is small, energy transport is mainly along $\vec{k}'$ (see the problems below). Thus in many cases it is reasonable to calculate the attenuation as though it produced an exponentially decreasing intensity along the direction of energy transport. The directions of the $\vec{k}'$ vectors may be found by assuming zero attenuation and then adding attenuation as though it does not alter the direction of propagation.

TE waves
Consider again the case of TE polarized waves incident on an interface:

$$\vec{K}_i \cdot \hat{u}_t = \vec{K}_r \cdot \hat{u}_t = \vec{K}_t \cdot \hat{u}_t.$$

For *s*-polarization

$$\vec{E}_i + \vec{E}_r = \vec{E}_t$$

but the phases are equal, therefore

$$\mathcal{E}_i + \mathcal{E}_r = \mathcal{E}_t. \tag{5.96}$$

Also $\vec{H}_i \cdot \hat{u}_t + \vec{H}_r \cdot \hat{u}_t = \vec{H}_t \cdot \hat{u}_t$. Again the phases are equal so that

$$\hat{u}_t \cdot (\mathcal{H}_i + \mathcal{H}_r) = \hat{u}_t \cdot \mathcal{H}_t. \tag{5.97}$$

Now from $\vec{K} \times \vec{\mathcal{E}} = \mu_0 \omega \mathcal{H}$ ($\mu_r = 1$) and $\vec{K} = \frac{\omega}{c}\tilde{n}(\hat{u}_t \sin\tilde{\theta} \pm \hat{u}_n \cos\tilde{\theta})$ (from equation (5.95)), we have

$$
\begin{aligned}
\hat{u}_t \cdot \mathcal{H} &= (\mu_0\omega)^{-1}\hat{u}_t \cdot \left(\vec{K} \times \vec{\mathcal{E}}\right) \\
&= (\mu_0\omega)^{-1}\left(\hat{u}_t \times \vec{K}\right) \cdot \vec{\mathcal{E}} \\
&= \frac{\tilde{n}}{\mu_0 c}[\hat{u}_t \times \hat{u}_t \sin\tilde{\theta} \pm \hat{u}_t \times \hat{u}_n \cos\tilde{\theta}] \cdot \vec{\mathcal{E}} \\
&= \frac{\tilde{n}}{\mu_0 c}[\hat{u}_s \cdot \vec{\mathcal{E}}(\pm\cos\tilde{\theta})],
\end{aligned}
\tag{5.98}
$$

where the minus sign refers to the reflected wave. Using equations (5.97) and (5.98) we obtain

$$\mathcal{E}_i - \mathcal{E}_r = \tilde{\beta}_s \mathcal{E}_t,$$

where

$$\tilde{\beta}_s = \frac{\tilde{n}_2 \cos\tilde{\theta}_t}{\tilde{n}_1 \cos\tilde{\theta}_i}. \tag{5.99}$$

Next we define amplitude reflection coefficient, $\tilde{r}_s$, as

$$\tilde{r}_s = \frac{\mathcal{E}_r}{\mathcal{E}_i}.$$

Equations (5.96) and (5.99) give us

$$\tilde{r}_s = \frac{1 - \tilde{\beta}_s}{1 + \tilde{\beta}_s}. \qquad (5.100)$$

Similarly defining the amplitude transmission coefficient, $\tilde{t}_s$, as

$$\tilde{t}_s = \frac{\mathcal{E}_t}{\mathcal{E}_i}$$

we find that

$$\tilde{t}_s = \frac{2}{1 + \tilde{\beta}_s}.$$

These expressions have the exact same form as those found in section 2.6, with the only difference being the introduction of the complex quantities $\tilde{\theta}$ and $\tilde{\beta}$.

Example.   Consider an s-polarized wave incident at $45°$ from air ($n_1 = 1$) on a medium whose index is $\tilde{n} = 2 + i$. Determine the amplitude reflection coefficient.
   From equation (5.99)

$$\tilde{\beta}_s = \frac{\tilde{n}_1 \cos \tilde{\theta}_t}{\tilde{n}_0 \cos \tilde{\theta}_i} = \frac{[\tilde{n}_1^2 - n_0^2 \sin^2 \theta_i]^{1/2}}{n_0 \cos \theta_i}$$

$$= [\tilde{n}_1^2/(n_0^2 \cos \theta_i) - \tan^2 \theta_i]^{1/2}$$

$$= [2(2 + i)^2 - 1]^{1/2} = [5 + 8i]^{1/2}$$

$$= [9.43 \, e^{1.012i}]^{1/2} = 3.07 \, e^{0.506i}.$$

To find the amplitude reflection coefficient we use equation (5.100)

$$\tilde{r}_s = \frac{1 - \tilde{\beta}_s}{1 + \tilde{\beta}_s} = \frac{1 - 3.07 \, e^{0.506i}}{1 + 3.07 \, e^{0.506i}}$$

$$= -0.533 - 0.188i$$

$$= 0.565 \, e^{3.48i} \quad (3.48 \text{ radians} \approx 199°).$$

Here the phase angle has been set such that $\varphi = \frac{-0.188}{-0.533} + \pi$ to correspond to the fact that it lies in the third quadrant of the complex plane.

TM waves
Next we want the amplitude reflection and transmission coefficients for TM polarized waves. The same analysis as for s-polarization above gives

$$\tilde{r}_p = \frac{1 - \tilde{\beta}_p}{1 + \tilde{\beta}_p} = \frac{\tilde{n}_t \cos \tilde{\theta}_i - \tilde{n}_i \cos \tilde{\theta}_t}{\tilde{n}_i \cos \tilde{\theta}_t + \tilde{n}_t \cos \tilde{\theta}_i}$$

$$\tilde{t}_p = \frac{2(\tilde{n}_i / \tilde{n}_t)}{1 + \tilde{\beta}_p},$$

where $\tilde{\beta}_p = \tilde{n}_1 \cos \tilde{\theta}_t / \tilde{n}_2 \cos \tilde{\theta}_i$.

5.11.2 Characteristic angles

Plane-polarized light with s- and p-polarization components (in phase with one another) given by E_s and E_p has an azimuthal angle defined by

$$\tan \Psi \equiv \frac{|\mathcal{E}_i|_s}{|\mathcal{E}_i|_p}$$

(see figure 5.17). This radiation will become elliptically polarized upon reflection. For a special angle of incidence Θ_p and special azimuthal angle Ψ_p, the reflected radiation will be circularly polarized, i.e. the s-and p-polarized components will be equal but have a phase difference of $\pi/2$. These two angles, Θ_p and Ψ_p, are called *characteristic angles*.

For the relative phase between s- and p-polarized components of reflected light to be $\pi/2$, since $e^{i\pi/2} = i$, we conclude that $r_p/r_s \equiv \frac{(\mathcal{E}_r)_p}{(\mathcal{E}_i)_p} / \frac{(\mathcal{E}_r)_s}{(\mathcal{E}_i)_s}$ will be a pure imaginary number. To see this consider the following derivation. For the ratio of the reflected s- and p-components we can write

$$\frac{(\mathcal{E}_r)_p}{(\mathcal{E}_r)_s} = e^{i\delta_r} \frac{|\mathcal{E}_r|_p}{|\mathcal{E}_r|_s},$$

where $\delta_r = \delta_{p,r} - \delta_{s,r}$ is the phase difference between the s- and p-reflected components; for the specific case in which we are interested here, $\delta_r = \pi/2$. Thus, using the definition of the ratio of reflection coefficients and since the reflected light is

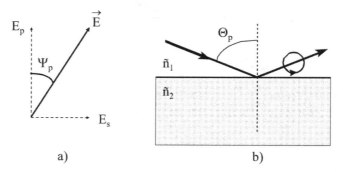

a)   b)

Figure 5.17. The characteristic angles, by means of which one can determine the optical properties of a conducting medium, are defined here. The angle Ψ_p gives the relative ellipticity and Θ_p is the principal (characteristic) angle of incidence.

circularly polarized (i.e. the amplitudes of the two components are equal and have a $\pi/2$ phase difference),

$$e^{i\pi/2} \equiv i = \frac{r_p(\mathcal{E}_i)_p}{r_s(\mathcal{E}_i)_s} = \frac{r_p}{r_s}\frac{|\mathcal{E}_i|_p}{|\mathcal{E}_i|_s}e^{i\delta_i},$$

where $\delta_i = \delta_{p,i} - \delta_{s,i}$ is the phase difference between incident s- and p-components. However, we begin by assuming that the incident radiation is plane polarized and therefore that $\delta_i = 0$, which means that we must have

$$\frac{r_p}{r_s} = i\frac{|\mathcal{E}_i|_s}{|\mathcal{E}_i|_p} = i\tan\Psi_p. \tag{5.101}$$

Now from previous expressions for r_p and r_s, the ratio of the two reflection coefficients can be written

$$\frac{r_p}{r_s} = \frac{\tilde{n}_2\cos\tilde{\Theta}_2\cos\tilde{\Theta}_1 + \tilde{n}_1\sin^2\tilde{\Theta}_1}{\tilde{n}_2\cos\tilde{\Theta}_2\cos\tilde{\Theta}_1 - \tilde{n}_1\sin^2\tilde{\Theta}_1}. \tag{5.102}$$

Consider the special case when $\tilde{n}_1 = 1$ (light incident from vacuum). From the two above equations, the following expressions can be derived:

$$\left(\tilde{n}_2^2 - \sin^2\Theta_p\right)^{1/2}\left(\sin\tilde{\Theta}_p\tan\tilde{\Theta}_p\right)^{-1} = \frac{1 + i\tan\tilde{\Psi}_p}{1 - i\tan\Psi_p}$$
$$= \cos 2\Psi_p + i\sin 2\Psi_p$$

and

$$\tilde{n}_2 = \left[\left(\cos 4\Psi_p + i\sin 4\Psi_p\right)\left(\sin\tilde{\Theta}_p\tan\tilde{\Theta}_p\right)^2 + \sin^2\tilde{\Theta}_p\right]^{1/2}. \tag{5.103}$$

The key point of this section is found in the above relation. From it we see that determination of the principal angles $\tilde{\Theta}_p$ and Ψ_p will determine $\tilde{n}_2$ completely. The spectroscopic technique of ellipsometry consists therefore of determining the relative amplitudes of incident polarizations, as well as the incidence angle, such that reflected light is circularly polarized.

For many practical situations the complex index of refraction will be so large that $|\tilde{n}_2| \gg \sin\Theta_p$, which then leads to the results we are seeking in this section,

$$\tilde{n}_2 = \left(\cos 2\Psi_p + i\sin 2\Psi_p\right)\sin\tilde{\Theta}_p\tan\tilde{\Theta}_p$$
$$n_2 = \sin\tilde{\Theta}_p\tan\tilde{\Theta}_p\cos 2\Psi_p$$
$$\kappa_2 = \sin\tilde{\Theta}_p\tan\tilde{\Theta}_p\sin 2\Psi_p.$$

Again, the main point is that by measuring the characteristic angles, the optical constants of a material can be determined.

5.12 Problems

1. Derive the boundary condition for the normal components of $\vec{B}$ and $\vec{D}$ from Maxwell's equations by using Gauss's divergence theorem.

2. Derive the boundary conditions for the tangential components of $\vec{E}$ and $\vec{H}$ from Maxwell's equations by using Stoke's theorem.

3. Within a dielectric sphere of radius R the polarization vector $\vec{P}$ is radially outward and its magnitude is proportional to the distance from the center of the sphere, i.e. $\vec{P} = P_0\vec{r}$. Find $\vec{D}$ and $\vec{E}$.

4. Show that the relation $\vec{j} = \sigma\vec{E}$ is equivalent to the usual statement of Ohm's law, i.e. $V = IR$.

5. An ac generator is connected to a parallel-plate capacitor. The plates are circular disks of area A. The charge on the plates is $q = q_0 \sin \omega t$. Neglecting edge effects:
   (a) Calculate the conduction and displacement currents and discuss how they compare.
   (b) Calculate the magnitude and direction of the magnetic field inside the capacitor.

6. A coaxial, cylindrical capacitor has inner and outer radii of 0.4 and 0.6 cm, respectively, and a length of 60 cm. The material between the cylinders has a relative dielectric constant of 6.5. The cylinders are kept at a potential difference $V = 250 \sin(350t)$ volts. Determine the displacement current I_D and the conduction current I.

7. The space between two concentric spherical conducting shells of inner and outer radii a and b is filled with a dielectric for which $\epsilon_1 = 6.0$. Given an applied voltage $V = 150 \sin(250t)$ volts, obtain the conduction and displacement currents.

8. A certain material has a conductivity of $\sigma_0 = 10^{-2}$ and a relative dielectric constant of $\epsilon_1 = 2.5$. If an applied electric field is given by $E_0 \sin \omega t$, where $E_0 = 6 \times 10^{-6}$ V m^{-1} and $\omega = 0.9 \times 10^9$ Hz, determine the displacement and conduction current density in the material.

9. An infinitely long cylindrical capacitor of radii a and b, where $b > a$, carries a free charge λ_f per unit length. The region between the plates is filled with a nonmagnetic dielectric of conductivity σ_0.
   (a) Show that at every point in the dielectric, the conduction current is exactly compensated by the displacement current so that no magnetic field is produced in the interior.
   (b) Find the rate of energy dissipation per unit volume at a point a distance R_0 from the axis.
   (c) Find the total rate of energy dissipation for a length l of the dielectric and show that this is equal to the rate of decrease of the electrostatic energy of the capacitor.

10. A parallel-plate capacitor is constructed from circular plates of area A separated by a material of uniform relative dielectric constant ϵ_1, permeability μ, and conductivity σ_0. It is initially charged to q on the two plates

and then it discharges slowly through the material between the plates. Find the electric and magnetic fields as a function of position and time. Neglect the fringing effects.

11. A parallel-plate capacitor has circular plates of area 10^{-2} m^2. The plate separation is 10^{-2} m, and the gap region is filled with a dielectric $\epsilon_1 = 3$. Connected in series with the condenser are an open switch, a 5 Ω resistor, and a 500 Volt battery. At $t = 0$ the switch is closed. Neglecting edge effects, compute:
    (a) The electric and magnetic field vectors as a function of time.
    (b) The Poynting vector and the total energy stored at any time t.
    (c) The energy stored in the polarization of the dielectric as a function of time.
    (d) The dielectric displacement current.

12. Given that

$$\langle \vec{S} \rangle = \frac{1}{2} \Re\left(\vec{E} \times \vec{H}^* \right)$$

and

$$\vec{E} = \vec{\mathcal{E}}\, e^{i(\vec{K}\cdot\vec{r} - \omega t)} = \left(\vec{E}' + i\vec{E}'' \right) e^{i(\vec{K}\cdot\vec{r} - \omega t)}$$

$$\vec{H} = \vec{\mathcal{H}}\, e^{i(\vec{K}\cdot\vec{r} - \omega t)} = \left(\vec{H}' + i\vec{H}'' \right) e^{i(\vec{K}\cdot\vec{r} - \omega t)},$$

where $\vec{K} = \vec{k}' + i\vec{k}''$:
    (a) For TE modes, determine the direction of power flow relative to the $\vec{k}'$ and $\vec{k}''$ directions, i.e. find an expression for $\langle \vec{S} \rangle$ in terms of $\vec{k}'$, $\vec{k}''$, and $\vec{E}'$.
    (b) Determine the direction of power flow for TM modes in terms of $\vec{k}'$, $\vec{k}''$, and $\vec{H}'$.

13. Sketch the vector relationship between $\vec{H}'$, $\vec{H}''$, $\vec{E}'$, $\vec{E}''$, $\vec{S}$, $\vec{k}'$, $\vec{k}''$ for both TE and TM waves.

14. Consider the plane-polarized electromagnetic wave

$$\vec{E} = \hat{y}\, E_0 \sin\left(kz - \omega t \right).$$

Find $\vec{D}$, $\vec{B}$, and $\vec{H}$.

15. Consider a monochromatic homogeneous plane wave in a medium with complex refractive index. Assume the wave is linearly polarized. Show that at any point the E and H fields oscillate with a difference of phase and find an expression for this difference in terms of n and κ.

16. At a boundary between air and a certain kind of glass, 5.25% of normally incident light is reflected. Find:
    (a) The index of refraction of the glass.
    (b) The amplitude ratios of the reflected and transmitted electric fields to the incident electric field.

17. A beam of circularly polarized light

$$\vec{E} = E_0(\hat{x} + i\hat{y})e^{i\left(\vec{k}'\cdot\vec{r}-\omega t\right)}$$

is incident from air on a glass ($n = 1.5$, $\kappa = 0$) surface at an angle of incidence $\theta_i = 45°$. Describe the state of polarization of the reflected wave.

18. At what angles will light, externally and internally reflected from a diamond–air interface, be completely linearly polarized? For diamond $n = 2.42$.

19.

   (a) For a perfect dielectric ($\sigma = 0$), show that all time varying fields are transverse.

   (b) For the case of $\sigma \neq 0$ show that, (i) the only longitudinal magnetic field is a static one and (ii) the longitudinal electric field is uniform in space, but may have a time variation.

   Hint: Assume $\vec{E} = \vec{E}_\perp + \vec{E}_\|$, $\vec{H} = \vec{H}_\perp + \vec{H}_\|$ and use Maxwell's equations. Also, for simplicity, assume the fields vary in only one spatial variable.

20. Plot the reflectance of water ($n = 1.33$) for both TE and TM polarizations for angles of incidence from 0 to 90 degrees.

21. A beam of light is incident at an angle θ on a dielectric (real) surface. Show that the sum of the energies of the reflected and refracted beams is equal to the energy in the incident beam.

22. The index of refraction for x-polarized light in an RTA (RbTiOAsO$_4$) crystal is given by the Sellmeier equation,

$$n^2(\lambda) = 2.226\,81 + \frac{0.996\,16\lambda^2}{\lambda^2 - (0.214\,23)^2} - 0.013\,69\lambda^2, \qquad (5.104)$$

with λ in microns and with $0.35\ \mu m < \lambda < 5.8\ \mu m$. Assume light incident from a vacuum.

   (a) Plot the reflectance as a function of incident angle for both s- and p-polarizations at $\lambda = 1.064\ \mu m$. Calculate Brewster's angle for this wavelength and see if the plot agrees.

   (b) Over what angular range is $R_p < 10^{-4}$?

23. For the same index of refraction as in the previous problem, equation (5.104):

   (a) Plot the reflectance as a function of wavelength for Brewster's angle at $\lambda = 1.064\ \mu m$. The Sellmeier equation is valid for $0.35\ \mu m < \lambda < 5.8\ \mu m$.

   (b) Determine the wavelength range over which the reflectance for p-polarized light is less than 10^{-5}.

24. The critical angle for total internal reflection at a certain air–glass interface is exactly 45 degrees. What is the Brewster angle for (a) external reflection and (b) internal reflection?

25. Consider a wave in an absorbing medium in which the angle between $\vec{k}'$ and $\vec{k}''$ is θ. If the optical parameters of the material are n and κ, find the magnitudes, k' and k'', of the wavevectors in terms of n, κ, and θ.

26. Show that equations (5.62) and (5.63) and equations (5.70) and (5.71) are true.

27. Derive the expression for the phase difference between the reflected and incident waves for total internal reflection, as a function of incidence angle and relative index of refraction (equations (5.84) and (5.85)).

28. Plot the phase difference between the reflected wave and the incident wave for the case of internal reflection at all angles of incidence for an interface between a dielectric medium with an index of refraction of $n = 1.6$ and air. Do the plot for both s- and p-polarizations.

29. A plane-polarized electromagnetic wave is totally reflected at the interface between two non-attenuating media and the E vector of the incident wave has components parallel and normal to the plane of incidence. Show that the reflected wave is elliptically polarized and that the component whose E vector is parallel to the plane of incidence leads the other component by

$$2 \tan^{-1}\left[\frac{\sin^2 \theta_i}{\cos \theta_i [\sin^2 \theta_i - (n_2/n_1)^2]^{1/2}} \right].$$

30. Consider total internal reflection at a dielectric interface. Consider the inhomogeneous plane wave in the second medium. What are the surfaces of constant amplitude? What are the surfaces of constant phase?

31. Show by direct calculation of $\vec{E} \times \vec{H}$ that, for total internal reflection, the normal component of the Poynting vector has zero time average in the second (real) medium.

32. Consider total internal reflection where the wave is incident at an angle θ and the plane of incidence is the x–y plane (n_1 is the index of the incident medium and n_2 is the transmitting medium). Show that the evanescent wave can be written as

$$E = E_0 e^{-\alpha y} e^{i(k_2 x - \omega t)}$$

and find α and k_2 in terms of the parameters given in the problem.

33. The reflectance of a metal is 80% at normal incidence and the extinction coefficient is $\kappa = 4$. Find the real part, n, of the index of refraction.

34. Show that the phase change that takes place on reflection at normal incidence is equal to

$$\tan^{-1}\left[\frac{2\kappa}{n^2 + \kappa^2 - 1} \right],$$

where the incident medium has an index of unity. (Note: In using the above result it is necessary to take the correct branch of the arctan function. This is done by requiring that as $\kappa \to 0$ the phase change is π, for $n > 1$ and 0 for $n < 1$.)

35. Show that the relative phase between plane waves reflected from the front and back surfaces of a thin film of thickness d and index of refraction n_2 is given by

$$\delta = \frac{2\pi n_2 d}{\lambda_0} \cos \theta_2,$$

where λ_0 is the vacuum wavelength of the incident light and θ_2 is the refraction angle in the film

36. For light with wavelength $\lambda = 500$ nm incident on aluminum, $n = 0.77$ and $\kappa = 6.1$, find the normal reflectance and phase change on reflection.

37. Verify the statement that, in the case of small attenuation, it is valid to treat attenuation as an exponentially decreasing intensity along the direction of energy transport.

38. We found an expression for the power reflectances in terms of the reflection coefficient. Find the power transmittance, T, in terms of the transmission coefficients, t. The power transmittance is defined in the same manner as the power reflectance, that is, it is the ratio of the transmitted power incident flux to power flux. Hint: only the normal component of the Poynting vector contributes to the transmittance. Why?

39. The complex index of refraction of a material is $4 + 2i$. Find the complex relative dielectric constant. What is the phase relationship between the electric and magnetic fields? Assume that the material is nonmagnetic.

References

[1] Wikipedia: Willebrord Snellius https://en.wikipedia.org/wiki/Willebrord_Snellius (Accessed: 11 Jan. 2021)

[2] Wikipedia: René Descartes https://en.wikipedia.org/wiki/Ren%C3%A9_Descartes (Accessed: 11 Jan. 2021)

[3] Wikipedia: Isaac Newton https://en.wikipedia.org/wiki/Isaac_Newton (Accessed: 11 Jan. 2021)

[4] Wikipedia: David Brewster https://en.wikipedia.org/wiki/David_Brewster (Accessed: 11 Jan. 2021)

[5] Jackson J D 1998 *Classical Electrodynamics* 3rd edn (Wiley: New York)

[6] Bell E E 1967 Optical constants and their measurement *Handbuch der Physik: Light and Matter* (Springer: Berlin)

IOP Publishing

Optical Radiation and Matter

Robert J Brecha and J Michael O'Hare

Chapter 6

Optical properties of simple systems

In the previous several chapters we have built up some basic concepts and tools for understanding the interaction of radiation and matter. In chapter 5 we looked at some macroscopic properties of matter, characterized mainly by the complex index of refraction, and how these properties can be used to describe reflection and absorption. In this chapter we begin to take a more detailed look at some of the properties of solids in general.

In the classical approach to the interaction of light with matter, we have seen that the optical properties of matter are determined by the motion of an electron responding to an external electric field, and that absorption of radiation by free atoms and molecules can often occur in the visible and ultraviolet (UV) portions of the spectrum. If we would have considered the motion of the nuclei as well, we would have seen additional optical responses. The vibration of a diatomic molecule is an example of such motion. These molecular motions, however, are much slower than electronic motions because of the mass of the atoms, with the result that the molecular motions contribute to the infrared (IR) properties of the spectrum. This model can be extended somewhat to treat dielectrics and metals as well; we will see that it is mainly infrared and visible radiation that produce characteristic responses in these materials.

6.1 Normal modes of motion

In this section we will briefly examine the motions of electrons and molecules in response to oscillating electromagnetic fields and how these motions determine the optical properties of a medium consisting of these particles.

In the mechanics of particle motion, the position coordinates of the particle are referred to as *degrees of freedom*. For every particle that makes up a particular system of matter there are three degrees of freedom (three position coordinates) that are associated with the particle; thus an N-particle system would have $3N$ degrees of freedom. For many applications in optics we will only be concerned with the valence

doi:10.1088/978-0-7503-2624-7ch6

electron associated with each atom or molecule, which means that each atom of an N-atom system would contribute two particles to the motion which determines the optical properties of the system. Thus each atom would have six degree of freedom, three for the optically active electron and three for the parent atom. Therefore, for an N-atom system there would be $6N$ degrees of freedom. The position coordinates associated with these degrees of freedom can be designated as x_α ($\alpha = 1, 2, \ldots, 6N$). These particles can participate in three types of motion, translational, vibrational, and rotational. In a purely classical picture we regard the electron as a point charge which is bound to the atom or molecule by springs (except in the case of a metal where the valence electron is considered as free from the parent atom). For small departures from equilibrium, the vibration of electrons and intermolecular vibrations can be treated as simple harmonic oscillators as was done previously in treating propagation in a dilute gas.

We begin by considering an extremely simple model, consisting of two coupled harmonic oscillators. In general, the equation of motion of a molecule with its electron can be a complicated set of equations, depending upon the coordinate system chosen. However, there is a coordinate system in which the six degrees of freedom are decoupled. That is, the equations of motion for each degree of freedom are independent of each other. These decoupled motions are called *normal modes* and they are described mathematically by so-called *normal coordinates*. This can probably be best described by an example [1]. The purpose of this example is not to explore the mathematical details of determining normal coordinates, but just to give an example of how the equations of motion decouple into normal coordinates for a relatively simple system. Consider the system in figure 6.1 which consists of two pendulums connected by a spring. If we let x_1 and x_2 be the position coordinates of the two masses, ζ the spring constant, and l the length of the string in each case, then, if there is no damping, the equations of motion for small displacements are

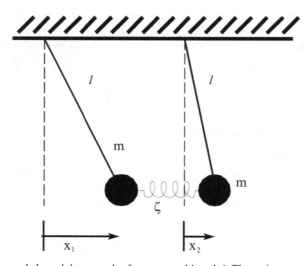

Figure 6.1. Two coupled pendulums, each of mass m and length l. The spring constant is given by ζ.

$$m\ddot{x}_1 = -\frac{mg}{l}x_1 - \zeta(x_1 - x_2) \tag{6.1}$$

$$m\ddot{x}_2 = -\frac{mg}{l}x_2 + \zeta(x_1 - x_2). \tag{6.2}$$

The procedure is to try solutions of the type

$$x_1 = Ae^{i\omega t} \quad \text{and} \quad x_2 = Be^{i\omega t} \tag{6.3}$$

in which the two coordinates are both contributing to an oscillatory motion with a common frequency ω; A and B are the constant amplitudes of oscillation. The next step is to substitute these expressions for x_1 and x_2 into the equations of motion and solve for the allowed frequencies of oscillation and for x_1 and x_2. There will be two allowed frequencies, ω_1 and ω_2, for these equations and for each allowed frequency we can determine the ratio between A and B. In this example it turns out to be $A = B$ for one of the frequencies and $A = -B$ for the other frequency. In the final result, x_1 and x_2 will be functions of both.

To see how this works out formally, we first substitute the trial solutions into equation (6.2) and rewrite the resulting equations in the form of a matrix relation:

$$\begin{pmatrix} \left(\frac{g}{l} + \frac{\zeta}{m}\right) & -\frac{\zeta}{m} \\ -\frac{\zeta}{m} & \left(\frac{g}{l} + \frac{\zeta}{m}\right) \end{pmatrix} \begin{pmatrix} A \\ B \end{pmatrix} = \begin{pmatrix} \omega^2 & 0 \\ 0 & \omega^2 \end{pmatrix} \begin{pmatrix} A \\ B \end{pmatrix}$$

$$\begin{pmatrix} \left(\frac{g}{l} + \frac{\zeta}{m}\right) - \omega^2 & -\frac{\zeta}{m} \\ -\frac{\zeta}{m} & \left(\frac{g}{l} + \frac{\zeta}{m}\right) - \omega^2 \end{pmatrix} \begin{pmatrix} A \\ B \end{pmatrix} = \begin{pmatrix} 0 \\ 0 \end{pmatrix}.$$

Thus we have set up an eigenvalue problem, familiar from linear algebra. Our goal is to find the eigenvalues or natural frequencies of the system. In this case that becomes a simple problem of solving a quadratic equation, since the solution here is given by setting the *determinant* of the above matrix equal to zero, where the determinant is given by

$$\begin{vmatrix} p & q \\ r & s \end{vmatrix} \equiv ps - rq.$$

For our oscillator problem this gives

$$\left[\left(\frac{g}{l} + \frac{\zeta}{m}\right) - \omega^2\right]^2 - \left(\frac{\zeta}{m}\right)^2 = 0$$

$$\omega^4 + \left(\frac{g}{l} + \frac{\zeta}{m}\right)^2 - 2\omega^2\left(\frac{g}{l} + \frac{\zeta}{m}\right) - \left(\frac{\zeta}{m}\right)^2 = 0,$$

which has solutions for ω^2 given by

$$\omega_+^2 = \frac{g}{l} + \frac{2\varsigma}{m}$$

$$\omega_-^2 = \frac{g}{l}.$$

We find the eigenvectors associated with each eigenvalue by substituting the eigenvalues one at a time into the original matrix equation. For the eigenvalue ω_+ this leads to

$$\begin{pmatrix} \left(\frac{g}{l} + \frac{\varsigma}{m}\right) - \omega_+^2 & -\frac{\varsigma}{m} \\ -\frac{\varsigma}{m} & \left(\frac{g}{l} + \frac{\varsigma}{m}\right) - \omega_+^2 \end{pmatrix} \begin{pmatrix} A \\ B \end{pmatrix} = \begin{pmatrix} 0 \\ 0 \end{pmatrix}$$

$$\begin{pmatrix} -\frac{\varsigma}{m} & -\frac{\varsigma}{m} \\ -\frac{\varsigma}{m} & -\frac{\varsigma}{m} \end{pmatrix} \begin{pmatrix} A \\ B \end{pmatrix} = \begin{pmatrix} 0 \\ 0 \end{pmatrix},$$

which has the solution $A = -B$. A similar procedure for the eigenvalue ω_- leads to the solution for the amplitudes $A = B$. Physically these correspond to pulling the pendulums equal amounts in opposite directions initially ($\omega = \omega_+$) or pulling them equal amounts in the same direction ($\omega = \omega_-$) and then letting them go. The fact that these are eigenvectors associated with a given eigenfrequency, or characteristic frequency, means that once started in the oscillation mode given by $A = -B$ or $A = B$ the system will continue to oscillate in that mode (neglecting friction). Figure 6.2 shows a freeze-frame look at how these modes oscillate once started properly.

There is another approach to solving the above problem, but it relies more on the fact that the system is so simple (only two masses, and thus two simultaneous equations). The general formalism given above is useful to know, and in particular with mathematical manipulation programs such as MathCad, Matlab, Maple, or Mathematica, solutions can be readily found.

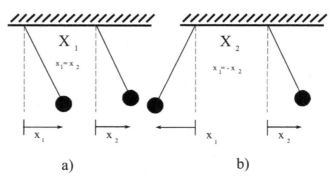

Figure 6.2. Normal modes for coupled oscillators.

To find the simple solution, we note by inspection that it is possible to form linear combinations of x_1 and x_2 such that each combination will be a function of ω_1 or ω_2 only. If we designate these linear combinations as X_1 and X_2, we find that X_1 and X_2 (which we now regard as the new coordinates of the system) each vary harmonically with their own frequency. The quantities X_1 and X_2 are called the *normal coordinates* of the system. Except when special boundary conditions exist, both are excited simultaneously. However, if the boundary conditions are such that only one normal coordinate is excited and the other is zero initially, the latter will remain zero. Energy will not pass from one normal coordinate to the other. The normal coordinates for this problem are

$$X_1 = \frac{x_1 + x_2}{2} \quad \text{and} \quad X_2 = \frac{x_1 - x_2}{2}.$$

The equations of motion in this coordinate system are

$$m\ddot{X_1} = -\frac{mg}{l}X_1 = -\omega_1^2 X_1$$

$$m\ddot{X_2} = -\left(\frac{mg}{l} + 2\zeta\right)X_2 = -\omega_2^2 X_2. \tag{6.4}$$

The frequencies $\omega_{1,2}$ are called the *normal modes* of vibration and are just the same as the eigenfrequencies $\omega_{-,+}$ found formally above. Again, figure 6.2 shows how these normal modes would oscillate.

Generalizing from the above example, the equations of motion of the normal coordinates are as follows. Designating the normal coordinates as x_α, the force component along the αth normal coordinate is

$$F_\alpha = -\frac{\partial V}{\partial X_\alpha}, \tag{6.5}$$

where V is the potential energy. From our example, for vibrational modes of motion the equations of the normal coordinates are

$$\ddot{X_\alpha} + \omega_\alpha^2 X_\alpha = 0. \tag{6.6}$$

Solutions to the above equation describe a simple harmonic oscillator with solution

$$X_\alpha = Ae^{i\omega_\alpha t} + Be^{-i\omega_\alpha t}, \tag{6.7}$$

where the constants A and B are determined from the initial conditions on X_α and $\dot{X_\alpha}$.

We wish to show from this example a common property of many materials, namely the presence of 'allowed' or preferred modes of motion. In the case of the coupled oscillators, we can clearly see that it is possible to excite certain types of motion that are characteristic of the system (out-of-phase and in-phase oscillations of the pendulums, for example). All other combinations of initial conditions lead to more complicated motions, but these can in principle always be thought of as a mixture of the two basic modes.

Once we start considering more realistic material systems in the following sections, it will not be as easy to pick out these modes, but the model serves as a starting point for further development. We will now use the idea of coupled oscillators, a whole string of them, to begin building a model for the behavior of solids, both for the propagation of sound waves (useful for material to be covered in chapter 9) and for the interaction of light (mostly IR radiation) with solids of various types.

Finally, we will see later in this chapter that the idea of coupled oscillators can be useful in another context when we talk about the coupling between photons (light) and phonons (sound waves in a crystal). We will see there that the harmonic oscillator coupling leads to the creation of a new entity that is neither photon nor phonon, called a 'polariton'.

6.2 Local and collective modes

We are interested in two types of normal modes, 'local modes' and 'collective modes'. In a local mode, the molecule is the basic unit and the normal coordinate X_α describes the motions within that molecule. The motion associated with local modes occurs over distances which are small compared to a wavelength, whereas collective modes involve motions that are large compared to a wavelength.

6.2.1 Local modes

Local modes can be illustrated by looking at two different examples. First, we consider a single atom with an optically active electron. Here we have two particles, the electron and the parent atom, a total of six degrees of freedom. The normal modes of motion turn out to be three translational degrees of freedom for the center of mass of the atom and electron and three vibrational degrees of motion associated with the bound electron. The vibrational electronic modes of motion can be modeled by springs along the normal coordinate axes. In this situation, neither the atom or the electron has any preferred orientation, so any three orthogonal translational axes and any three vibrational axes can be used. In this case the axes are said to be 'degenerate'. This is effectively the model we have already seen, in which we further simplified by considering only one of the equivalent directions of oscillation.

Consider next a diatomic molecule with an optically active electron. Now we have three particles, the two atoms which make up the molecule and one optically active electron. In this case, we would have nine degrees of freedom; three degrees for each atom and three degrees of freedom for the optical electron. The degrees of freedom break down into normal modes as follows: (i) three degenerate translations of the center of mass; (ii) two degenerate molecular rotations; (iii) one molecular vibration; and (iv) three electronic vibrations.

We wish to consider collective modes in more detail. A collective mode is a motion involving the entire medium. A sound wave would be an example of a collective excitation or mode. Since the sound wave involves forces acting between all of the molecules of the medium. We will now build up a first model of a solid material, in which such collective modes can appear.

6.2.2 The linear monatomic lattice

The simplest model to consider is a one-dimensional linear chain of oscillators. The displacement of the atoms from equilibrium is designated by x_α as shown in figure 6.3(a).

If ζ is the spring constant, the force on the αth atom due to nearest neighbor atoms only is [2]

$$F_\alpha = \zeta(x_{\alpha+1} - x_\alpha) - \zeta(x_\alpha - x_{\alpha-1})$$

$$m\ddot{x}_\alpha = \zeta(x_{\alpha+1} + x_{\alpha-1} - 2x_\alpha). \tag{6.8}$$

The usual procedure is to attempt solutions of the form

$$x_\alpha = Ae^{i(k_\beta \alpha a - \omega_\beta t)}, \tag{6.9}$$

where αa is the equilibrium position of the αth atom (a is the interatomic spacing) and β is a label to identify the βth mode. A solution of this form is possible because of the nature of the linear chain, i.e. equal masses at regular intervals. If the masses had random unequal values or were spaced at unequal intervals, the solution would be an attenuated wave. The above solution in which all masses of the system oscillate at the same frequency is a *normal mode*. If the above equations are solved for the frequency ω_β, we would find

$$\omega_\beta = \left[\frac{4\zeta}{m}\right]^{1/2} \sin\left(\frac{k_\beta a}{2}\right). \tag{6.10}$$

The relationship between ω_β and k_β is called a *dispersion relationship*. There are generally boundary conditions which determine the allowed values that k_β can assume. In this particular case we can assume periodic boundary conditions which means that the first and last atoms have exactly the same oscillation, i.e. $x_1 = x_N$, where N is the total number of atoms. Carrying this through we find that

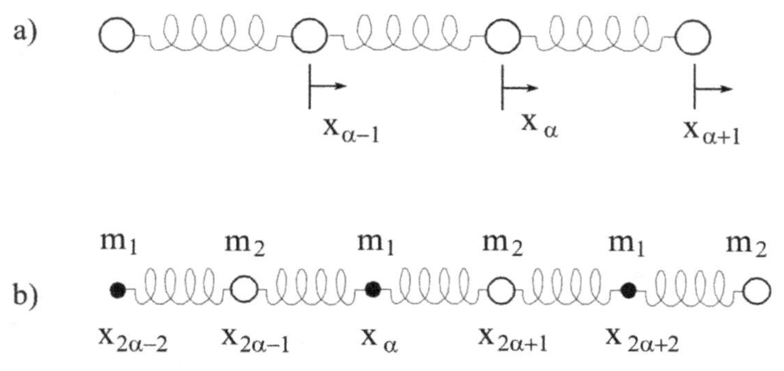

a)

$$x_{\alpha-1} \qquad x_\alpha \qquad x_{\alpha+1}$$

$$m_1 \qquad m_2 \qquad m_1 \qquad m_2 \qquad m_1 \qquad m_2$$

b)

$$x_{2\alpha-2} \qquad x_{2\alpha-1} \qquad x_\alpha \qquad x_{2\alpha+1} \qquad x_{2\alpha+2}$$

Figure 6.3. (a) The monatomic linear chain of coupled oscillators, which forms the model for a one-dimensional solid. (b) The diatomic chain of oscillators is used as a model for an ionic solid.

$$k_\beta = \beta\frac{2\pi}{Na}, \qquad \text{where } \beta = 0, \pm1, \pm2, \ldots\pm N/2. \tag{6.11}$$

Thus, not counting the trivial solution, $\beta = 0$, there are N modes. This is exactly what you would expect since, for a linear chain there would be one degree of freedom for each atom. Figure 6.4 shows the dispersion relation ω_β versus k_β for the linear chain. This represents the allowed normal modes of wave propagation for such a physical configuration. In solid state physics these are called the acoustic modes; in a quantum mechanical picture they are called acoustic phonons. Of course, in a real three-dimensional picture there would be additional degrees of freedom, however, this simple one-dimensional picture gives a good representation of the properties of the acoustic modes.

6.2.3 One-dimensional diatomic lattice

Consider now a one-dimensional diatomic lattice. In addition to having the properties of the monatomic lattice, the diatomic lattice also exhibits important features of its own. Figure 6.3(b) shows a diatomic lattice composed of two atoms of masses m_1 and m_2. Again, the distance between the atoms is a.

The motion of this lattice can be treated in the same manner as the motion of the monatomic lattice. However, since there are two different types of atoms, there will be two distinct equations of motion. In analogy with the above treatment, the equations of motion of the two atoms are (again considering only nearest neighbor interactions)

$$\begin{aligned}
m_2\ddot{x}_{2\alpha+1} &= \zeta(x_{2\alpha+2} + x_{2\alpha} - 2x_{2\alpha+1}) \\
m_1\ddot{x}_{2\alpha+2} &= \zeta(x_{2\alpha+3} + x_{2\alpha+1} - 2x_{2\alpha+2}).
\end{aligned} \tag{6.12}$$

The subscripts are such that all atoms with mass m_2 are labeled with odd integers while those with mass m_1 are labeled with even integers. Attempting solutions of the form

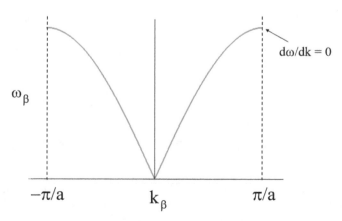

Figure 6.4. Dispersion curve (i.e. plot of ω_β versus k_β) for the monatomic linear chain.

$$x_{2\alpha+1} = A_2 e^{i[k_\beta(2\alpha+1)a-\omega_\beta t]}$$
$$x_{2\alpha+2} = A_1 e^{i[k_\beta(2\alpha+2)a-\omega_\beta t]}$$

(6.13)

we find the following result for the allowed frequencies:

$$\omega_\beta^2 = \zeta\left[\frac{1}{m_1} + \frac{1}{m_2}\right] \pm \zeta\sqrt{\left[\frac{1}{m_1} + \frac{1}{m_2}\right]^2 - 4\frac{\sin^2(k_\beta a)}{m_1 m_2}}.$$

(6.14)

This is the dispersion relation for the linear diatomic lattice. As was done for the case of the linear chain, boundary conditions can be applied to find the allowed values for k_β. Figure 6.5 shows the dispersion relation for the equation above and illustrates a new property that appears for the diatomic lattice. Namely, because of the additional degrees of freedom, each allowed value of k_β has two possible modes, an *acoustic mode* and an *optical mode*. The acoustic mode corresponds to oscillations in which the displacements of atoms m_1 and m_2 are in the same direction, whereas the optical mode corresponds to oscillations in which the displacement of the masses m_1 and m_2 are in opposite directions.

We would like to see a more concrete illustration of what these normal modes and dispersion relations mean, at least in an artificial system. With that basis, it should then be possible to extrapolate to a real solid. Consider a system of six masses connected by seven springs, to each other and to a rigid wall on each end. It is possible to realize this using an air track and six carts, for example [3]. Now we can write down the equations of motion for each mass using equations (6.12). It is convenient to write the equations in matrix form,

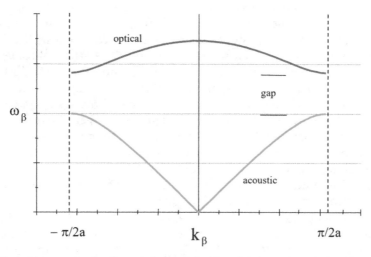

Figure 6.5. Dispersion curve for the diatomic linear chain. Note the change in scale for k_β compared to the monatomic lattice, figure 6.4.

$$\begin{pmatrix} \ddot{x}_1 \\ \ddot{x}_2 \\ \ddot{x}_3 \\ \ddot{x}_4 \\ \ddot{x}_5 \\ \ddot{x}_6 \end{pmatrix} - \begin{bmatrix} 2\dfrac{\zeta}{m_2} & -\dfrac{\zeta}{m_2} & 0 & 0 & 0 & 0 \\ -\dfrac{\zeta}{m_1} & 2\dfrac{\zeta}{m_1} & -\dfrac{\zeta}{m_1} & 0 & 0 & 0 \\ 0 & -\dfrac{\zeta}{m_2} & 2\dfrac{\zeta}{m_2} & -\dfrac{\zeta}{m_2} & 0 & 0 \\ 0 & 0 & -\dfrac{\zeta}{m_1} & 2\dfrac{\zeta}{m_1} & -\dfrac{\zeta}{m_1} & 0 \\ 0 & 0 & 0 & -\dfrac{\zeta}{m_2} & 2\dfrac{\zeta}{m_2} & -\dfrac{\zeta}{m_2} \\ 0 & 0 & 0 & 0 & -\dfrac{\zeta}{m_1} & 2\dfrac{\zeta}{m_1} \end{bmatrix} \begin{pmatrix} x_1 \\ x_2 \\ x_3 \\ x_4 \\ x_5 \\ x_6 \end{pmatrix} = 0.$$

Once we have written down the equations in the above form, our problem is reduced to that of finding the eigenvalues of the matrix, which we could do according to the techniques shown for the coupled pendulums. Alternatively, we can use one of the mathematical manipulation packages. If we solve for the eigenvalues, or normal-mode frequencies for the chain of oscillators, we find six values. The corresponding eigenvectors represent a mode of oscillatory motion for the six masses. For example, choosing equal masses $m_1 = m_2 = 0.2$ kg and a coupling constant $\zeta = 2$ N m^{-1} (the units are actually irrelevant here), we find frequencies of $\omega_\beta = \sqrt{1.98}, \sqrt{7.53}, \sqrt{15.55}, \sqrt{24.45}, \sqrt{32.47}, \sqrt{38.02}$ s^{-1}. The physical meaning of these frequencies can be understood as follows. If the masses are stationary initially, a driving force oscillating at the normal-mode frequency will resonate, i.e. it will start a collective motion of the masses. Other frequencies will cause the masses to move, but the motion will be a seemingly random pattern of individual motions.

To clarify the collective mode concept, we can return to the eigenvalue and eigenvector results obtained above. For each frequency ω_i there is an associated vector v_i that represents the positions of the masses; we can consider this to be the initial displacements of the masses that will result in the collective mode of oscillation at frequency ω_i. We can also solve equation (6.10) for k_β for each of the values of ω_β. The k_β corresponding to the frequencies found above are, respectively, $k_\beta = 0.449, 0.898, 1.35, 1.80, 2.24,$ and 2.69. Using these values we can make a plot of ω_β versus k_β, as shown in figure 6.6, where we have plotted both the individual points and the full curve that would be found by going to the limit of an infinite number of coupled masses. We can see that even for a small number of masses, the dispersion curve is starting to 'fill in'.

It is useful to take this example one step further. We can determine the wavelength corresponding to each of the eigenfrequencies by using $k_\beta = 2\pi/\lambda$. The results for our example are: $\lambda_\beta = 14a, 7a, 4.667a, 3.50a, 2.80a,$ and $2.33a$. Since the total length of our chain of masses is $L = 7a$, we see that the wavelengths correspond to waves of wavelength $\lambda = 2L, 2L/2, 2L/3, \ldots, 2L/N$ for $N = 1$ to 6. This principle is illustrated in figure 6.7 for the modes with $\lambda = L = 7a$ and for $\lambda = 2L/6 = 2a$.

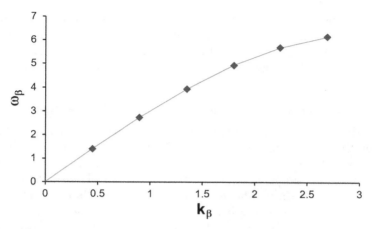

Figure 6.6. Dispersion curve for the monatomic linear chain, equation (6.10), along with the six data points for the chain of coupled oscillators.

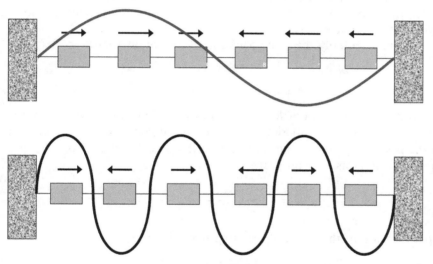

Figure 6.7. An illustration of the relative motions of six coupled masses in two different normal modes. In the upper frame the mode has a wavelength of $\lambda = L = 7a$, while in the lower frame the wavelength is $\lambda = 2L/6 = 2a$. These two modes correspond to the frequencies $\omega_\beta = 2.74$ s^{-1} and $\omega_\beta = 6.17$ s^{-1}, respectively.

6.3 Optical properties of simple classical systems

The dielectric constant of a material arises from a number of different physical mechanisms which couple the EM wave to molecules and their constituents. The first case we will consider in our classical picture will be that of the bound oscillator, for example the oscillations of a diatomic molecule. The approach here is essentially the same as in chapter 4, where we examined the properties of a bound electron to determine the optical properties of a dilute medium, except that in this case the masses of the oscillator are different and we let the driving field be the macroscopic electric field. This mechanism will be important in the infrared region for ionic crystals. The approach used here is essentially the same for all mechanisms involving bound charges which contribute to the dielectric constant and, except for minor details, it will typify the general frequency dependence of all these mechanisms.

6.3.1 Frequency dependence of the dielectric constant of bound charges

Consider a diatomic, ionic crystal (a crystal composed of positive and negative ions) under the driving influence of an EM wave. Also, just consider a small portion of the lattice, but a portion which is large enough to contain a large number of ion pairs and yet small compared to the wavelength of the driving EM field. The lattice of positive and negative ions will oscillate in simple harmonic motion. The parameters of the problem are the masses, m_+ (m_-) = mass of the positive (negative) ions, $m_r \equiv$ reduced mass $= \frac{m_+ m_-}{m_+ + m_-}$, and $\vec{x}_\beta$, the displacement of the center of mass of the positive ion from the center of mass of the negative ions along the βth normal coordinate. Introducing the reduced mass allows us to consider the problem in more generality; in chapter 4 we really looked at the motion of one particle only.

We obtain essentially the same equation of motion as we did in chapter 4,

$$m_r \ddot{x}_\beta + \underbrace{m_r \Gamma \dot{x}_\beta}_{\text{damping force}} + \underbrace{m_r \omega_0^2 x_\beta}_{\text{restoring force}} = -eE_\beta, \tag{6.15}$$

where E_β is the component of E along the βth vibrational normal coordinate. In this simple diatomic molecule model, the vibrational modes are degenerate, so we will drop the β subscript for the normal mode. If $\vec{E} \propto e^{-i\omega t}$, then $\vec{x} \propto e^{-i\omega t}$ and, as in previous calculations, we can write

$$\vec{x} = \frac{-e\vec{E}}{m_r[\omega_0^2 - \omega^2 - i\Gamma\omega]}. \tag{6.16}$$

The polarization of the material resulting from this motion is

$$\vec{P} = -N_v e\vec{x}, \tag{6.17}$$

where N_v is the number of ion pairs per unit volume. Therefore we have

$$\vec{P} = \frac{N_v e^2 \vec{E}}{m_r[\omega_0^2 - \omega^2 - i\Gamma\omega]}. \tag{6.18}$$

Now

$$\vec{D} = \varepsilon_0 \varepsilon_r \vec{E} = \varepsilon_0 \vec{E} + \vec{P}$$
$$= \varepsilon_0 \vec{E} + \frac{N_v e^2 \vec{E}}{m_r[\omega_0^2 - \omega^2 - i\Gamma\omega]} \tag{6.19}$$

so that

$$\tilde{\varepsilon}_r = 1 + \frac{N_v e^2}{\varepsilon_0 m_r [\omega_0^2 - \omega^2 - i\Gamma\omega]} \tag{6.20}$$

and therefore

$$\tilde{\varepsilon}_r = 1 + \frac{N_v e^2 (\omega_0^2 - \omega^2)}{m_r \varepsilon_0 \left[(\omega_0^2 - \omega^2)^2 + \Gamma^2 \omega^2 \right]} + i \frac{N_v e^2 \Gamma\omega}{m_r \varepsilon_0 \left[(\omega_0^2 - \omega^2)^2 + \Gamma^2 \omega^2 \right]} \tag{6.21}$$

$$\tilde{\varepsilon}_r = \varepsilon_1 + i\varepsilon_2.$$

We see, for equation (6.21), that the frequency dependence of the optical constants of this system is strongest in the neighborhood of the natural undamped resonant frequency ω_0.

6.3.2 Electronic contributions to the dielectric constant

The dielectric constant of a material will depend on all processes which allow interaction of the EM wave with the material. The classical oscillator model can also be used to describe resonant electronic oscillations of an electron bound to a parent atom. Again from chapter 4, the equation of motion is

$$m_e \ddot{\vec{x}} + m_e \Gamma_e \dot{\vec{x}} + m_e \omega_e^2 \vec{x} = -e\vec{E}, \tag{6.22}$$

where now m_e is the electron mass rather than the reduced mass of an oscillating ion-pair, ω_e is frequency of the electronic oscillator, and Γ_e is the damping factor for the bound electrons.

This leads to the very same form as before for $\tilde{\varepsilon}_r$, namely

$$\tilde{\varepsilon}_r = 1 + \frac{N_v^e e^2 (\omega_e^2 - \omega^2)}{m_e \varepsilon_0 \left[(\omega_e^2 - \omega^2)^2 + \Gamma_e^2 \omega^2 \right]} + i \frac{N_v^e e^2 \Gamma\omega}{m_e \varepsilon_0 \left[(\omega_e^2 - \omega^2)^2 + \Gamma_e^2 \omega^2 \right]}, \tag{6.23}$$

where N_v^e is now the electron density.

Equations (6.21) and (6.23) give the complex dielectric constant if only the ionic lattice contributes or if only bound electrons contribute, respectively. However, both mechanisms contribute to the dielectric constant, although in different parts of the frequency spectrum. In fact, there may be several molecular resonances and several electronic resonances throughout the spectrum. To take into account the fact that different charges may be bound differently, one may assume that a certain fraction f_1 of charges have a resonance frequency ω_1, a certain fraction f_2 have resonance frequency ω_2, etc. The resulting complex dielectric constant or complex index of refraction squared is

$$\tilde{\varepsilon}_r = \tilde{n}^2 = 1 + \frac{N_v e^2}{\varepsilon_0} \sum_j \left[\frac{f_j / m_j}{\omega_j^2 - \omega^2 - i\Gamma_j \omega} \right], \tag{6.24}$$

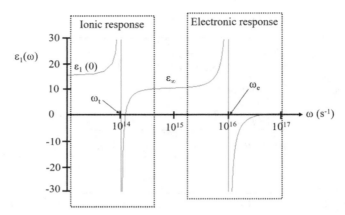

Figure 6.8. A plot of $\varepsilon_1(\omega)$ versus ω showing the ionic resonant frequency, ω_t, referred to as the transverse optical phonon frequency, and the electronic resonance, ω_e.

where Γ_j is the damping factor and ω_j is the resonant frequency for each type of oscillator. The f_j are called oscillator strengths and can be determined from quantum theory. The mass m_j can be either reduced mass or electron mass depending upon the 'oscillator'.

We can illustrate the spectrum for a single ionic and a single electronic resonance. In figure 6.8 we plot $\varepsilon_1(\omega)$ as a function of ω. At very low frequencies (below the ionic response resonance) the value of $\varepsilon_1(\omega)$ is that of the static dielectric constant, often called K or κ in undergraduate electricity and magnetism texts. The value of $\varepsilon_1(\omega)$ between the two resonances is what we will refer to as ε_∞ when we wish to concentrate only on the ionic resonance.

If the damping terms are sufficiently small, so that the terms $\Gamma_j\omega$ can be neglected in comparison with $\omega_j^2 - \omega^2$, the index is essentially real and its square can be given by

$$n^2 = 1 + \frac{Ne^2}{\varepsilon_0} \sum_j \left[\frac{f_j/m_j}{\omega_j^2 - \omega^2} \right]. \tag{6.25}$$

It is possible, by empirical curve fitting, to make this formula match the experimental data quite well for many transparent substances. The empirical formula is called Sellmeier's equation and is given below:

$$n^2 = A_0 + \sum_j \frac{A_j\lambda^2}{\lambda_j^2 - \lambda^2}.$$

Example. We will use the single classical oscillator model to find the *long* wavelength index of refraction of Ge if $\omega\varepsilon_2$ peaks at 0.295 μm. The atomic density of Ge is $N_v = 4.5 \times 10^{22}$ cm^{-3}; each atom has four valence electrons.

Solution. The peak in wavelength at 0.295 μm corresponds to $f = c/\lambda = 3 \times 10^8/0.295 \times 10^{-6} = 1.02 \times 10^{15}$ s^{-1}. From $\omega = 2\pi f$ we have $\omega_e = 6.4 \times 10^{15}$ s^{-1}.

Thus we can use equation (6.23) to find ε_1 and ε_2. In the long wavelength limit, which we are interested in here, $\omega \to 0$. Thus we have

$$\varepsilon_1 \simeq n^2 \simeq 1 + \frac{N_v e^2 \omega_e^2}{m_e \varepsilon_0 \omega_e^4} = 1 + \frac{N_v e^2}{m_e \varepsilon_0 \omega_e^2}.$$

The electron density is given by $4 \times (4.5 \times 10^{22}$ cm$^{-3}) = 18 \times 10^{28}$ m^{-3}. Thus

$$\varepsilon_1 = 1 + \frac{(18 \times 10^{28})(1.6 \times 10^{-19})^2}{(9.1 \times 10^{-31})(8.85 \times 10^{-12})(6.4 \times 10^{15})^2} = 15.$$

Finally, $n = \sqrt{\varepsilon_1} = 3.87$.

6.3.3 Reststrahlen bands

For components with significant ionic bonding there is always an intense absorption band in the infrared. This high absorption results in high reflectivity, often greater than 90% over a region known as the *Reststrahlen band*. For many materials the single classical oscillator of the proceeding section provides a good model for this phenomena.

From our previous model for the single ionic oscillator we have

$$\varepsilon_1(\omega) = \varepsilon_\infty + \frac{N_v e^2 (\omega_0^2 - \omega^2)}{m_r \varepsilon_0 [(\omega_0^2 - \omega^2)^2 + \Gamma^2 \omega^2]}. \tag{6.26}$$

Except for the term ε_∞, which we discussed in the previous section, equation (6.26) is just the real part of equation (6.21).

At this point it is interesting to relate the results expressed by equation (6.26) to the collective modes of oscillation as discussed earlier in this chapter. The collective modes in an ionic lattice that are stimulated by an electromagnetic wave may be either longitudinal or transverse optical phonons. A phonon is an excitation of the crystal lattice that correlates to a specific frequency, much as a photon is an excitation of the electromagnetic field corresponding to a wave of a given frequency. In the case of phonons, one can imagine the two different orthogonal possibilities for exciting a coordinated motion of the atoms (about their respective equilibrium positions). One possibility is to start the lattice vibrating in a direction parallel to a surface, for example, with an excitation force much like that used to create a wave on a rope. The second possibility is to cause a compressional motion, similar to what can be done with a 'Slinky'. In figure 6.9 the two different cases are shown for light incident at an oblique angle to the surface, and having two orthogonal polarization components.

Returning to equation (6.26), we will take the resonance frequency for our one-dimensional ionic dielectric to be $\omega_0 \equiv \omega_t$ the transverse optical phonon frequency. Again, ε_∞ is the contribution of the bound electrons to the dielectric constant. In other words, we are only considering the frequency dependence due to the ionic

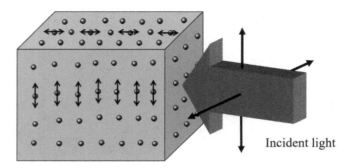

Incident light

Figure 6.9. An illustration of the relative motions for both longitudinal optical (LO) and transverse optical (TO) phonons. Light is incident on the surface of the material at some angle, and the electric field has both *s*- and *p*-components. The collective motion shown on the 'top' of the sample corresponds to an LO phonon, whereas that shown on the front face corresponds to a TO phonon.

lattice oscillations, with all other higher frequency contributions being included in the constant ε_∞ as indicated by equation (6.24).

Consider the case when $\Gamma = 0$, which gives $\tilde{\varepsilon}_r = \varepsilon_1$ and equation (6.26) gives

$$\varepsilon_1(\omega) = \varepsilon_\infty + \frac{N_v e^2}{m_r \varepsilon_0 [\omega_t^2 - \omega^2]}$$
$$= \varepsilon_\infty + \frac{N_v e^2}{m_r \varepsilon_0 \omega_t^2 [1 - \omega^2/\omega_t^2]}. \tag{6.27}$$

For $\omega \ll \omega_t$, both the first and second terms contribute to ε_1, resulting in the static dielectric constant

$$\varepsilon_1(0) = \varepsilon_\infty + \frac{N_v e^2}{m_r \varepsilon_0 \omega_t^2}. \tag{6.28}$$

Rewriting the above equation for $\varepsilon_1(\omega)$ we obtain

$$\varepsilon_1(\omega) = \varepsilon_\infty + \frac{\varepsilon_1(0) - \varepsilon_\infty}{1 - \omega^2/\omega_t^2}, \tag{6.29}$$

where the ionic contribution is now contained entirely in the second term. In this form the relative dielectric constant is expressed in terms of quantities which are directly measurable.

Figure 6.10 shows ε_1 as a function of frequency in the vicinity of the transverse optical phonon frequency. We see that $\varepsilon_1 < 0$, when $\omega_t < \omega < \omega_\ell$ (where ω_ℓ is the longitudinal optical phonon and is the frequency at which $\varepsilon_1 = 0$). From this we obtain the so-called Lyddane–Sachs–Teller relation:

$$\omega_\ell = \left[\frac{\varepsilon_1(0)}{\varepsilon_\infty} \right]^{1/2} \omega_t. \tag{6.30}$$

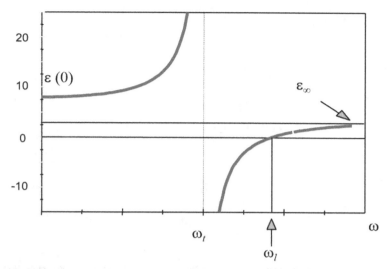

Figure 6.10. A zoomed-in view of figure 6.8, ε_1 versus ω for a bound oscillator, in the region around ω_t. The longitudinal optical phonon frequency ω_ℓ is where $\varepsilon_1(\omega) = 0$.

Since ε_1 is negative in the region $\omega_t < \omega < \omega_\ell$, from equations (5.33) and (5.34), we have $n = 0$ and $\kappa \neq 0$ in this region; this leads to a reflectivity of $R = 1$ at normal incidence (see problem 6.11). Therefore, in the frequency band $\omega_t < \omega < \omega_\ell$ we obtain a band or so-called forbidden frequency gap in which the wave does not propagate and is entirely reflected.

If we consider the more general case when $\Gamma \neq 0$, the classical oscillator model leads to results for $\varepsilon_1(\omega)$ and $\varepsilon_2(\omega)$ which are somewhat 'rounded off', i.e. the reflectivity does not go to unity and there is no divergence in $\varepsilon_1(\omega)$. This is illustrated in figure 6.11. In figure 6.12 the index of refraction and the absorption coefficient are shown on the same frequency scale and, finally, in figure 6.13 the reflectance is shown as a function of frequency for the same parameters.

We return briefly to examine the origin of the frequency ω_ℓ. From Maxwell's equations, in the absence of free charges, we have $\nabla \cdot \vec{D} = \nabla \cdot (\varepsilon_0 \vec{E} + \vec{P}) = 0$. Here $\vec{E}$ is the total field at the site of the ions, i.e. it is the applied field plus the field produced by the induced polarization, as discussed in chapter 5. For a transverse wave incident on the dielectric medium at normal incidence, only transverse modes, or phonons, can be excited, and therefore only a transverse polarization of the medium can be generated. But that implies that $\nabla \cdot \vec{P} = 0$ (recall, $\nabla \cdot \vec{P} \equiv \vec{k} \cdot \vec{P}$), which in turn leaves us with $\nabla \cdot \vec{E} = 0$.

For electromagnetic waves at oblique incidence, the p-polarized component can induce a polarization of the medium longitudinally (perpendicular to the surface) so $\nabla \cdot \vec{P}$ does not necessarily vanish and thus neither does $\nabla \cdot \vec{E}$. From Maxwell's equations, we still must have $\nabla \cdot \vec{D} = 0$, however. This we can write as $\varepsilon_1(\omega)\nabla \cdot \vec{E}$, using $\vec{D} = \varepsilon_1(\omega)\vec{E}$ (we still consider a lossless medium). To satisfy Maxwell's

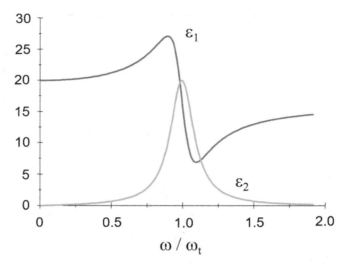

Figure 6.11. Real and imaginary parts of $\tilde{\epsilon}(\omega)$ as a function of normalized frequency, ω/ω_ℓ. Parameters are $\epsilon_1(0) = 20$, $\epsilon_\infty = 16$, and $\Gamma_e/\omega_e = 0.2$.

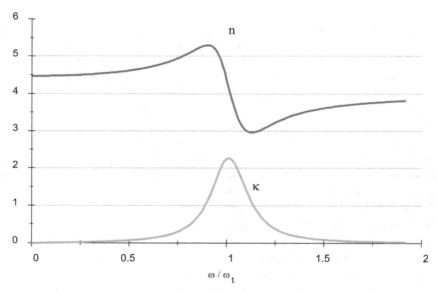

Figure 6.12. Index of refraction (n) and the absorption coefficient (κ) for the same parameters as in the previous figure.

equations, however, our only possibility now is that $\varepsilon_1(\omega) = 0$ to allow this longitudinal mode to propagate. The conclusions we reach here is that the frequency at which $\varepsilon_1(\omega) = 0$ is the resonant frequency for the longitudinal optical phonon mode, ω_ℓ.

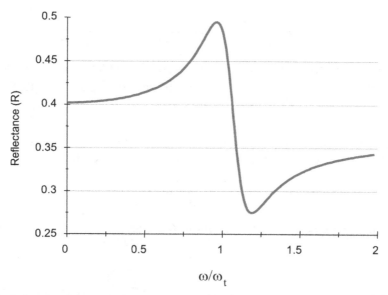

Figure 6.13. Reflectance (R) for the classical oscillator model with parameters $\epsilon_1(0) = 20$, $\epsilon_\infty = 16$, and $\Gamma_e/\omega_e = 0.2$.

6.4 Drude theory of metals

Next we will consider the classical theory of EM propagation in a metal using the Drude theory. The so-called Drude theory of metals consists of a model which treats the metal as a 'gas' of free charge carriers in a viscous material. This provides a simple model for conductivity. Consider the equation of motion of a normal mode of a charge in a viscous medium under the action of a sinusoidal electric field. First, the interaction force is $F_{\text{int}} = -eE$, where e is the electronic charge. The equation of motion is then

$$m\ddot{x}_\beta + m\Gamma_\beta \dot{x}_\beta = -eE_\beta, \tag{6.31}$$

where E_β is the component of E along the βth normal coordinate. Since the particles are free and $\vec{E}$ does not couple normal modes, each particle will have three degenerate translational modes, thus we can drop the β subscript for the normal modes. In this case the equation of motion could just as well have been written in terms of the velocity $\vec{v}$,

$$m\frac{d\vec{v}}{dt} + m\Gamma\vec{v} = -e\vec{E}. \tag{6.32}$$

As we have done previously, $E \propto e^{-i\omega t}$ so that $v \propto e^{-i\omega t}$ and the equation of motion gives

$$m(-i\omega)\vec{v} + m\Gamma\vec{v} = -e\vec{E}$$

$$\vec{v} = -\frac{e\vec{E}}{m(\Gamma - i\omega)} = -\frac{e\vec{E}\tau}{m(1 - i\omega\tau)}, \tag{6.33}$$

where τ (=$1/\Gamma$) the relaxation time, which physically is a measure of the mean time between electron scattering events. The current density $\vec{j}$ can be written

$$\vec{j} = -eN_v\vec{v}, \tag{6.34}$$

where N_v is the number of charges per unit volume. From Ohm's law, $\vec{j} = \sigma\vec{E}$, we see that σ is complex, which we designate as $\tilde{\sigma}$, and is given by

$$\tilde{\sigma} = \frac{-eN_v v}{E} = \frac{N_v e^2 \tau}{m(1 - i\omega\tau)}. \tag{6.35}$$

Letting σ_0 be the dc conductivity, equation (6.35) gives

$$\sigma_0 = \frac{N_v e^2 \tau}{m} \tag{6.36}$$

and the complex conductivity is written

$$\tilde{\sigma} = \frac{\sigma_0}{(1 - i\omega\tau)}. \tag{6.37}$$

The complex index of refraction, $\tilde{n} = n + i\kappa$, is readily found to be

$$\tilde{n}^2 = \varepsilon_r + \frac{i\tilde{\sigma}}{\varepsilon_0\omega} = \varepsilon_r + \frac{i\sigma_0}{\varepsilon_0\omega(1 - i\omega\tau)}. \tag{6.38}$$

What happens to the above expression at very high frequencies? Now we are talking about frequencies so high that even a light particle like the electron will be too sluggish to respond. At such frequencies, for example in the x-ray region of the spectrum, the index of refraction will be unity. This argument leads to

$$n^2 = \varepsilon_r = 1.$$

Therefore we write the complex index of refraction as

$$\tilde{n}^2 = 1 + \frac{i\sigma_0}{\varepsilon_0\omega}\left[\frac{1 + i\omega\tau}{1 + \omega^2\tau^2}\right]. \tag{6.39}$$

As in chapter 5, we can define a complex dielectric constant

$$\tilde{n}^2 = \tilde{\varepsilon}_r = \varepsilon_1 + i\varepsilon_2 = 1 + \frac{i\sigma_0}{\varepsilon_0\omega}\left[\frac{1 + i\omega\tau}{1 + \omega^2\tau^2}\right] \tag{6.40}$$

so that the real and imaginary parts are given by

$$\varepsilon_1 = 1 - \frac{\sigma_0\tau}{\varepsilon_0(1 + \omega^2\tau^2)} \tag{6.41}$$

$$\epsilon_2 = \frac{\sigma_0}{\epsilon_0 \omega (1 + \omega^2 \tau^2)}. \tag{6.42}$$

Also from chapter 5,

$$\tilde{n} = n + i\kappa = \sqrt{\tilde{\epsilon}_r} \tag{6.43}$$

and

$$n^2 = \frac{1}{2}\left[(\epsilon_1^2 + \epsilon_2^2)^{\frac{1}{2}} + \epsilon_1 \right] \tag{6.44}$$

$$\kappa^2 = \frac{1}{2}\left[(\epsilon_1^2 + \epsilon_2^2)^{\frac{1}{2}} - \epsilon_1 \right]. \tag{6.45}$$

Next we will define a quantity called the plasma frequency, which is

$$\omega_p^2 \equiv \frac{N_v e^2}{\epsilon_0 m^*} = \frac{\sigma_0}{\epsilon_0 \tau}, \tag{6.46}$$

where m^* is the called the 'effective mass' of the electrons. With this definition, the complex index can be written in the following form:

$$\tilde{n}^2 = \tilde{\epsilon}_r = \left[1 + \frac{i\sigma_0}{\epsilon_0 \omega (1 - i\omega\tau)} \right]$$
$$= \left[1 + \frac{i\omega_p^2 \tau}{\omega (1 - i\omega\tau)} \right] \tag{6.47}$$

$$= \left[1 - \frac{\omega_p^2 \tau^2}{1 + \omega^2 \tau^2} \right] + i \frac{\omega_p^2 \tau}{\omega (1 + \omega^2 \tau^2)}. \tag{6.48}$$

We will now investigate the behavior in different frequency limits of the dielectric constant and other optical properties using the above general result as a starting point.

6.4.1 Low-frequency region ($\omega\tau \ll 1$): the Hagen–Rubens relation

In the low-frequency region, defined by $\omega\tau \ll 1$, the above expressions readily reduce to

$$\epsilon_1 = (1 - \omega_p^2 \tau^2) = 1 - \frac{\sigma_0 \tau}{\epsilon_0} \tag{6.49}$$

$$\epsilon_2 = \frac{\omega_p^2 \tau}{\omega} = \frac{\sigma_0}{\epsilon_0 \omega} \tag{6.50}$$

with the following results for n and κ,

$$n^2 = \frac{1}{2}\left[\left[\left(1 - \omega_p^2\tau^2\right)^2 + \left(\frac{\omega_p^2\tau}{\omega}\right)^2\right]^{\frac{1}{2}} + \left(1 - \omega_p^2\tau^2\right)\right] \qquad (6.51)$$

$$\kappa^2 = \frac{1}{2}\left[\left[\left(1 - \omega_p^2\tau^2\right)^2 + \left(\frac{\omega_p^2\tau}{\omega}\right)^2\right]^{\frac{1}{2}} - \left(1 - \omega_p^2\tau^2\right)\right]. \qquad (6.52)$$

Since we are considering the region in which $\omega\tau \ll 1$, we see that as ω decreases, the term $(\omega_p\tau/\omega)$ in the above expressions for n and κ will dominate and we have

$$n \simeq \kappa \simeq \left[\frac{\omega_p^2\tau}{2\omega}\right]^{\frac{1}{2}}. \qquad (6.53)$$

Here both n and κ increase with decreasing frequency.

Example. What is the reflectivity at an air–metal interface in the infrared region at normal incidence?

Solution. We begin with the definition of the plasma frequency,

$$\omega_p^2 = \frac{N_v e^2}{\varepsilon_0 m_e} = \frac{\sigma_0}{\varepsilon_0 \tau}.$$

From the definition of the reflection coefficient

$$r_s = \frac{1 - \beta_s}{1 + \beta_s},$$

where $\tilde{n}_2 = n + i\kappa$ and $\tilde{n}_1 = 1$, the reflectance at normal incidence becomes

$$R = |r_s|^2 = \frac{(1 - n)^2 + \kappa^2}{(1 + n)^2 + \kappa^2}. \qquad (6.54)$$

Now if we assume the low-frequency condition ($\omega\tau \ll 1$),

$$n \simeq \kappa \simeq \left[\frac{\omega_p^2\tau}{2\omega}\right]^{\frac{1}{2}} = \left[\frac{\sigma_0}{2\varepsilon_0\omega}\right]^{\frac{1}{2}}$$

and we obtain

$$R = \frac{1 - 2n + 2n^2}{1 + 2n + 2n^2} = 1 - \frac{4n}{1 + 2n + 2n^2}. \qquad (6.55)$$

In addition, from equation (6.53) we have that for long wavelengths $n^2 \gg n \gg 1$ which, along with the above expression gives us the so-called Hagen–Rubens relation, which is the long wavelength expression for reflectance:

$$R = 1 - \frac{2}{n} = 1 - \left[\frac{8\epsilon_0 \omega}{\sigma_0}\right]^{\frac{1}{2}}. \tag{6.56}$$

$$\underbrace{\qquad\qquad\qquad\qquad\qquad\qquad}_{\text{Hagen–Rubens relation}}$$

6.4.2 The high-frequency region ($\omega\tau \gg 1$)

The next limiting case is the high-frequency region, specified by the condition that $\omega\tau \gg 1$. Under this condition the components of the dielectric constant and the index become

$$\varepsilon_1 = \left[1 - \frac{\omega_p^2}{\omega^2}\right], \qquad \varepsilon_2 \simeq \frac{\omega_p^2}{\omega^3 \tau}$$

and

$$n = \left[\frac{1}{2}\left(\varepsilon_1^2 + \varepsilon_2^2\right)^{\frac{1}{2}} + \frac{1}{2}\varepsilon_1\right]^{\frac{1}{2}} \simeq \sqrt{\varepsilon_1} \quad (\text{when } \varepsilon_2 = 0)$$

$$\simeq \left[1 - \frac{\omega_p^2}{\omega^2}\right]^{\frac{1}{2}} \tag{6.57}$$

$$\simeq \left[1 - \frac{\omega_p^2}{2\omega^2}\right].$$

Therefore, as $\omega \to \infty$, $n \to 1$. What this says is that at high frequencies, $\omega \gg \omega_p$, the real part of the dielectric function ε_1 approaches $\simeq 1$ as $\varepsilon_2 \to 0$. Under these conditions n becomes unity and $\kappa \to 0$, and the metal or semiconductor becomes transparent (this generally occurs in the UV part of the spectrum).

6.4.3 Plasma reflection

The last case of interest concerns the phenomena of plasma reflection. Assuming a pure metal (electron gas) again, with $\epsilon_\infty = 1$, from equation (6.48),

$$\varepsilon_1 = \left[1 - \frac{\omega_p^2 \tau^2}{1 + \omega^2\tau^2}\right]. \tag{6.58}$$

Now, for the sake of algebraic simplicity, consider the case when there is no damping, i.e. $\tau \to \infty$. This leads to

$$\varepsilon_1 = \left[1 - \frac{\omega_p^2}{\omega^2}\right]. \tag{6.59}$$

In this case, when $\omega < \omega_p$, ϵ_1 becomes negative. Also in this case, n will increase with decreasing frequency, see equation (6.53) above and the example given below. This means that a relatively large reflected wave amplitude occurs (Why?). In the absence of damping, total reflection occurs even at normal incidence. When damping is not zero there is some penetration of the wave into the metal or semiconductor, but the small transmitted wave is rapidly attenuated. Losses cause the near IR reflectivity of metals to fall a little below the value of 1.0. In the very far infrared the damping due to conduction electrons cannot be neglected and the reflectivity is given by the Hagen–Rubens relation.

Example. Consider a metal for which we can take $\varepsilon_\infty = 1$. Suppose that $\omega_p\tau = 100$. Find n and κ as a function of ω. Calculate the reflectance as a function of frequency as well.

Solution. We begin with equation (6.48) and rewrite it in a form which is useful for the information we are given:

$$\varepsilon_1 = 1 - \left[\frac{1}{\omega_p^2 \tau^2} + \frac{\omega^2}{\omega_p^2} \right]^{-1}$$

$$\varepsilon_2 = \left[\frac{\omega}{\omega_p}(\omega_p\tau)\left(\frac{1}{\omega_p^2\tau^2} + \frac{\omega^2}{\omega_p^2} \right) \right]^{-1}.$$

We have thus put everything in terms of $\omega_p\tau$ and the relative frequency ω/ω_p. Figure 6.14 shows the values of ε_1 and ε_2 as a function of the normalized frequency. Similarly we can use equations (6.44) and (6.45) to find n and κ. These are shown in figure 6.15. Finally we can use equation (6.54) to find the reflectance R, which is then illustrated in figure 6.16. In the latter figure, note the sharp plasma reflection edge.

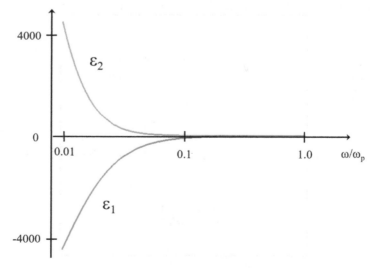

Figure 6.14. Plot of ε_1 and ε_2 versus the normalized frequency ω/ω_p for the parameters $\omega_p\tau = 100$ and $\varepsilon_\infty = 1$.

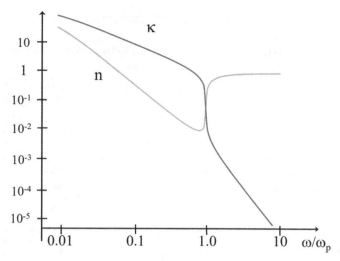

Figure 6.15. Plot of n and κ versus the normalized frequency ω/ω_p for the parameters $\omega_p\tau = 100$ and $\epsilon_\infty = 1$.

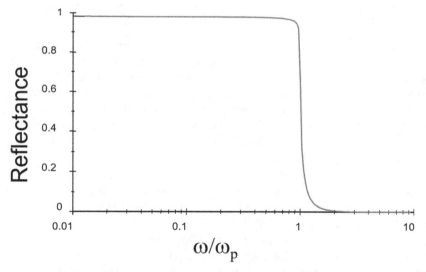

Figure 6.16. Plot of reflectance R versus the normalized frequency ω/ω_p for the parameters $\omega_p\tau = 100$ and $\epsilon_\infty = 1$.

6.5 Semiconductors—the example of InSb

We will look at the semiconductor material InSb as a way of pulling together many of the concepts discussed in this chapter. Some of the specific information used here is taken from [4]. Semiconductors, as the name implies, lie between insulators (dielectrics) and conductors in their electrical conductivity properties. For our purposes, semiconductors also lie somewhere between dielectrics and conductors in there optical properties, i.e. they have contributions from both free electrons and from polarizable ionic cores, as well as electronic transitions in the visible part of the electromagnetic spectrum.

In a semiconductor, when two different carriers are present, the plasma frequency is given by

$$\omega_p^2 = \frac{e^2}{\epsilon_0}\left[\frac{N_e}{m_e^*} + \frac{N_h}{m_h^*}\right], \tag{6.60}$$

where N_e, N_h, m_e^*, and m_h^* are the electron carrier density, the hole carrier density, the electron effective mass, and the hole effective mass, respectively. For InSb we can find the effective masses in the literature to be $m_e^* = 0.017\,4m_e$ and $m_h^* = 0.18m_e$. We will wish to concentrate for now on the lower frequency resonance and therefore use

$$\tilde{\varepsilon}_r = \left[\varepsilon_\infty + (\varepsilon_1(0) - \varepsilon_\infty)\frac{1 - \omega^2/\omega_t^2}{(1 - \omega^2/\omega_t^2)^2 + \Gamma^2\omega^2/\omega_t^4} - \frac{\omega_p^2\tau^2}{1 + \omega^2\tau^2}\right]$$
$$+ i\left[(\varepsilon_1(0) - \varepsilon_\infty)\frac{\omega\Gamma/\omega_t^2}{(1 - \omega^2/\omega_t^2)^2 + \Gamma^2\omega^2/\omega_t^4} + \frac{\omega_p^2\tau}{\omega(1 + \omega^2\tau^2)}\right]. \tag{6.61}$$

From [4] we find the values for the low-frequency and intermediate frequency dielectric constants to be $\varepsilon(0) = 17.72$ and $\varepsilon_\infty = 15.68$. These values are somewhat dependent on the temperature at which experiments are carried out, but we will assume here room temperature, $T \simeq 300$ K. Also from the same reference we find values for the transverse optical phonon frequency, $\omega_t = 3.38 \times 10^{13}$ s^{-1} and for the plasma frequency of the free electrons, $\omega_p = 6.04 \times 10^{13}$ s^{-1}. Two additional important parameters are the damping rate for the ionic resonance, $\Gamma = 5.4 \times 10^{11}$ s^{-1} and the collision time, $\tau = 5.0 \times 10^{-13}$ s.

Our final goal is to calculate the reflectance as a function of frequency; the aim of [4] was to theoretically fit to measured data for certain samples of InSb. In figure 6.17 we show the reflectance over a limited range of frequencies near the ionic and plasma resonances.

There is one additional important frequency range at which we should take a closer look. The term ε_∞ in the complex dielectric function plays the role of taking into account contributions from (relatively) high-frequency resonances due to electronic transitions in a material. In the case of a semiconductor, the transition in question is that of absorption of an optical or infrared photon to promote an electron from the valence band to the conduction band [2, 5]. The so-called band gap for InSb is at an energy of 0.17 eV at $T = 300$ K, which corresponds to a wavelength of 7.3 μm or a frequency of $\omega_0 = 2.58 \times 10^{14}$ s^{-1}. For radiation near this frequency, we must consider the resonance behavior here as well. On the other hand, for frequencies significantly above the resonance, the material will no longer interact with the external radiation and the index of refraction goes to unity.

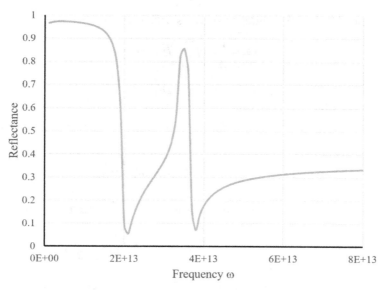

Figure 6.17. Plot of the reflectance R versus frequency ω for the parameters used in the text. Note the resonance ω_t (transverse optical phonon) and that for low frequencies the reflectance goes toward unity. ω_p, the plasma frequency does not have a visible effect on the reflectance.

6.6 Kramers–Kronig relations

We have found in this chapter that the dielectric constant is a strong function of the frequency near lattice and electron resonant frequencies. The lattice resonances in the infrared frequency region ($\sim 10^{13}$ Hz) and the electron resonances in the ultraviolet frequency region (10^{15} Hz) divide the spectrum into three general regions in which $\varepsilon_1(\omega)$ is nearly constant. In the region below the infrared, that is the radio frequency range, $\varepsilon_1(\omega) \approx \varepsilon_1(0)$; in the optical range $\varepsilon_1(\omega) \approx \varepsilon_\infty = n^2$; and in the far ultraviolet $\varepsilon_1(\omega) = 1$.

From any of the dispersion relations developed in this chapter, for example equations (6.21) and (6.23), it can be seen that the two optical constants, $\varepsilon_1(\omega)$ and $\varepsilon_2(\omega)$, or n and κ, are not completely independent. In equations (6.21) and (6.23), both the real and imaginary parts of the dielectric constant at a given frequency are determined by the same parameters, N_v, ω_0, and Γ. It turns out that a functional relation exists between ε_1 and ε_2 and if one is known for all frequencies, then the other can be found at any frequency. Actually, one can in a completely general way, using causality arguments and the theory of complex variables, show that the real and imaginary parts of the dielectric constant must be inter-related. This was first done by Kramers and Kronig working independently and thus the inter-relation is called the Kramers–Kronig relation.

Suppose we write the connection between the displacement $\vec{D}$ and the electric field $\vec{E}$ at the frequency ω as

$$\vec{D}(\omega) = \varepsilon_0 \tilde{\varepsilon}(\omega) \vec{E}(\omega). \qquad (6.62)$$

Let us write the time dependence of $\vec{D}$ as a Fourier superposition according to

$$\vec{D}(t) = \frac{1}{\sqrt{2\pi}} \int_{-\infty}^{\infty} \vec{D}(\omega)e^{-i\omega t}d\omega$$

$$\vec{D}(\omega) = \frac{1}{\sqrt{2\pi}} \int_{-\infty}^{\infty} \vec{D}(t')e^{i\omega t'}dt'.$$

Substituting with equation (6.62) above, the Fourier superposition of $\vec{D}$ can be written,

$$\vec{D}(t) = \frac{\varepsilon_0}{\sqrt{2\pi}} \int_{-\infty}^{\infty} \tilde{\varepsilon}(\omega)\vec{E}(\omega)e^{-i\omega t}d\omega,$$

and if we substitute a Fourier superposition for $\vec{E}$ into the above equation, we have

$$\vec{D}(t) = \frac{\varepsilon_0}{2\pi} \int_{-\infty}^{\infty} d\omega \, \tilde{\varepsilon}(\omega) \, e^{-i\omega t} \int_{-\infty}^{\infty} dt' \vec{E}(t')e^{i\omega t'}.$$

Next we make the substitution $\tilde{\varepsilon}(\omega) = [1 + (\tilde{\varepsilon}(\omega) - 1)]$ so that the above expression becomes

$$\begin{aligned}\vec{D}(t) &= \frac{\varepsilon_0}{2\pi} \int_{-\infty}^{\infty} d\omega \, e^{-i\omega t} \int_{-\infty}^{\infty} dt' \vec{E}(t')e^{i\omega t'} \\ &+ \frac{\varepsilon_0}{2\pi} \int_{-\infty}^{\infty} d\omega \, [\tilde{\varepsilon}(\omega) - 1] \, e^{-i\omega t} \int_{-\infty}^{\infty} dt' \vec{E}(t')e^{i\omega t'}.\end{aligned} \qquad (6.63)$$

Using the fact that

$$\int_{-\infty}^{\infty} d\omega \, e^{-i\omega t} = 2\pi \, \delta(t),$$

equation (6.63) becomes

$$\vec{D}(t) = \varepsilon_0\vec{E}(t) + \frac{\varepsilon_0}{2\pi} \int_{-\infty}^{\infty} d\omega \, [\tilde{\varepsilon}(\omega) - 1] \, e^{-i\omega t} \int_{-\infty}^{\infty} dt' \vec{E}(t')e^{i\omega t'}.$$

Changing the order of integration yields

$$\begin{aligned}\vec{D}(t) &= \varepsilon_0\vec{E}(t) + \int_{-\infty}^{\infty} dt' \vec{E}(t')\left[\frac{\varepsilon_0}{2\pi} \int_{-\infty}^{\infty} d\omega \, [\tilde{\varepsilon}(\omega) - 1] \, e^{-i\omega(t-t')}\right] \\ &= \varepsilon_0\vec{E}(t) - \int_{-\infty}^{\infty} g(\tau)\vec{E}(t - \tau)d\tau,\end{aligned} \qquad (6.64)$$

where

$$g(\tau) = \frac{\varepsilon_0}{2\pi} \int_{-\infty}^{\infty} d\omega \, [\tilde{\varepsilon}(\omega) - 1] \, e^{-i\omega\tau}.$$

Finally, taking the inverse Fourier transform of the above equation, we obtain

$$\int_{-\infty}^{\infty} g(\tau)e^{i\omega'\tau}d\tau = \frac{\varepsilon_0}{2\pi} \int_{-\infty}^{\infty} d\omega \, [\tilde{\varepsilon}(\omega) - 1] \int_{-\infty}^{\infty} e^{-i(\omega-\omega')\tau}d\tau$$

$$= \frac{\varepsilon_0}{2\pi} \int_{-\infty}^{\infty} d\omega \, [\tilde{\varepsilon}(\omega) - 1][2\pi\delta(\omega - \omega')]$$

$$= \varepsilon_0[\tilde{\varepsilon}(\omega') - 1].$$

Therefore the dielectric constant can be written in terms of $g(\tau)$ according to

$$\tilde{\varepsilon}(\omega) = 1 + \frac{1}{\varepsilon_0} \int_{-\infty}^{\infty} g(\tau)e^{i\omega\tau}d\tau.$$

Equation (6.64) shows that $\vec{D}(t)$ at time t depends on the electric field at times other than t. Now we will invoke the concept of causality in dealing with equation (6.64). In determining $\vec{D}(t)$ at time t, only values of the electric field prior to t should have any effect on the value of $\vec{D}(t)$. Therefore equation (6.64) can be written

$$\vec{D}(t) = \varepsilon_0\vec{E}(t) - \int_0^{\infty} g(\tau)\vec{E}(t - \tau)d\tau,$$

which then allows us to write

$$\tilde{\varepsilon}(\omega) = 1 + \frac{1}{\varepsilon_0} \int_0^{\infty} g(\tau)e^{i\omega\tau}d\tau. \tag{6.65}$$

Now there is no reason why we cannot mathematically treat frequency as a complex quantity of the form $\tilde{\omega} = \omega' + i\omega''$. From equation (6.65), $\tilde{\varepsilon}$ is analytic in the upper half of the complex plane, a direct consequence of the fact that the integral goes from 0 to ∞. Since equation (6.65) tells us that $\tilde{\varepsilon}$ is analytic, we can make use of the fact that the integral of an analytic function over a contour that encloses no singularities is zero. Pick some frequency ω on the real axis and integrate the function $(\tilde{\varepsilon}-1)/(\omega' - \omega)$ around the contour as shown in figure 6.18.

The imaginary part of ω, i.e. ω'', introduces a damping factor $\exp(-\omega''t)$ so that the integral over the large contour contributes nothing. Counterclockwise integration around the point at ω gives $i\pi(\tilde{\varepsilon}-1)$ and we have the principal part of an integral from $-\infty$ to ∞ over ω' along the real axis

$$P \int_{-\infty}^{\infty} \frac{\tilde{\varepsilon}-1}{\omega' - \omega}d\omega' + i\pi(\tilde{\varepsilon}-1) = 0.$$

Writing $\tilde{\varepsilon} = \varepsilon_1 + i\varepsilon_2$ and equating real and imaginary parts, we have

$$\varepsilon_1(\omega) = 1 + \frac{1}{\pi}P \int_{-\infty}^{\infty} \frac{\varepsilon_2(\omega')}{\omega' - \omega}d\omega' \tag{6.66}$$

and

$$\varepsilon_2(\omega) = -\frac{1}{\pi}P \int_{-\infty}^{\infty} \frac{\varepsilon_1(\omega') - 1}{\omega' - \omega}d\omega' . \tag{6.67}$$

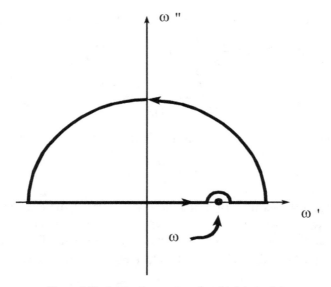

Figure 6.18. Integration contour for obtaining $\epsilon_2(\omega)$.

It can also be shown (we will not do it here) that $\varepsilon_1(\omega)$ is an even function and $\varepsilon_2(\omega)$ must be an odd function. Using this fact, we may write equation (6.66) as

$$\varepsilon_1(\omega) = 1 + \frac{1}{\pi}P\int_0^\infty \frac{\varepsilon_2(\omega')}{\omega' + \omega}d\omega' + \frac{1}{\pi}P\int_0^\infty \frac{\varepsilon_2(\omega')}{\omega' - \omega}d\omega',$$

which, with a little additional algebra, can be shown to be

$$\varepsilon_1(\omega) = \frac{2}{\pi}P\int_0^\infty \frac{\omega'\varepsilon_2(\omega')}{\omega'^2 - \omega^2}d\omega'. \tag{6.68}$$

A similar approach for $\epsilon_2(\omega)$ gives

$$\epsilon_2(\omega) = -\frac{2\omega}{\pi}P\int_0^\infty \frac{\varepsilon_1(\omega')}{\omega'^2 - \omega^2}d\omega'. \tag{6.69}$$

Equations (6.68) and (6.69) are known as the Kramers–Kronig relations and they can be used to calculate ϵ_2 if ε_1 is known for all frequencies and vice versa.

6.7 Problems

1. Two masses m_1 and m_2 are joined by a massless spring of force constant ζ and unstretched length a. Obtain the normal coordinates of the system for motion along the line joining the particles.
2. Show that the normal-mode frequency solution ω_- leads to the condition for the amplitudes $A = B$.
3. Consider InSb with 10^{16} negative carriers per cm^3, $\tau = 10^{-13}$ s, $m_e^* = 0.013\,9m_e$, and $\epsilon_\infty = 16$. Plot the power reflectance in the frequency

range from 10^9 s^{-1} to 10^{15} s^{-1} for normally incident light. What is the wavelength corresponding to the plasma frequency?

4. Show that if the imaginary part κ of the complex index of refraction is much smaller that the real part, n, then for the case of a single resonance frequency, ω_0, the following approximate equations are valid:

$$n \simeq \sqrt{\epsilon_\infty} + \frac{N_v e^2}{2m\epsilon_0\sqrt{\epsilon_\infty}}\left[\frac{1}{(\omega_0^2 - \omega^2)}\right]$$

$$\kappa \simeq \frac{N_v e^2}{2nm\epsilon_0}\left[\frac{\Gamma\omega}{(\omega_0^2 - \omega^2)^2}\right].$$

5. A hypothetical metal has a plasma frequency $\omega_p = 10^{15}$ Hz and a relaxation time of $\tau = 10^{-13}$ s. Plot the real and imaginary parts of the index of refraction for $\omega_2 \leqslant \omega \leqslant 2\omega_p$.

6. The static conductivity of silver is 6.21×10^7 mho m^{-1}. Assuming that the charge carriers are free electrons of density 5.85×10^{28} m^{-3}, find the following quantities: (a) the plasma frequency, (b) the relaxation time, (c) the real and imaginary parts of the index of refraction, (d) the reflectance at $\lambda = 1$ μm, and (e) the effective mass of the electrons; assume normal incidence.

7. Show that, when there is no damping, i.e. $\tau \to \infty$, and if $\omega < \omega_p$, the reflectance R, equals unity.

8. Given a certain material for which $\epsilon_\infty = 14$, $\epsilon_1(0) = 18$, and $\omega_0\Gamma = 20, 60, 200$ plot the Reststrahlen band, i.e. plot R versus ω/ω_0.

9. Consider the Drude theory of metals for the case for which there are no losses ($\tau \to \infty$).
   (a) Find n and κ when (i) $\omega > \omega_p$ and (ii) $\omega < \omega_p$.
   (b) Find the phase between the E and H vectors when $\omega > \omega_p$.
   (c) What is the reflectance in part (b) and how does the reflectance behave as the wavelength is decreased?
   (d) Find the phase between the E and H vectors when $\omega < \omega_p$. What is the Poynting vector? What does the Poynting vector tell us about the reflectance?

10. A particular dye has a deep red color. It appears red because it absorbs out the green component of the spectrum. Imagine that you have a thin walled prism filled with this dye. How will this prism disperse incident white light? Hint: How does the index of refraction of this substance vary as a function of wavelength?

11. Show that a negative value for the real part of the dielectric constant leads to a reflectance of unity.

12. Light, whose vacuum wavelength is 500 nm, is incident normally on a sample with index of refraction $n = 1.653$ and extinction coefficient $\kappa = 2.35 \times 10^{-2}$.

Table 6.1. Important parameters for several elemental metallic substances.

Element	σ	N_v	τ
	($\times 10^7$ Ω^{-1}m^{-1})	($\times 10^{28}$ m^{-3})	($\times 10^{-14}$ s)
Li	1.07	4.6	0.9
Na	2.11	2.50	3.1
K	1.39	1.30	4.3
Rb	0.80	1.10	2.75
Cu	5.88	8.45	2.7
Ag	6.21	5.85	4.1
Au	4.55	5.90	2.9

    (a) What is the speed of the wave in this sample?

    (b) What is the wavelength of the wave in the sample?

    (c) What is the reflectance of the sample?

13. A certain material has a reflectance of 0.250 for light with angular frequency of 2.56×10^{15} s^{-1} which is incident normally. In this material, the intensity of the light decreases by half in a distance of 5 mm. Find:

    (a) The extinction coefficient.

    (b) The index of refraction.

    (c) The real and imaginary parts of the relative dielectric constant.

14. A certain solid has an electronic resonance frequency of ω_0.

    (a) Show that the index of refraction n is less than unity but approaches unity in the limit as ω becomes large.

    (b) Show that the extinction coefficient and reflectance both approach zero in the same limit.

15. Use the data from table 6.1 for silver at room temperature to:

    (a) Plot the refractive index (n) and extinction coefficient (κ) of Ag versus ω on a log scale.

    (b) Plot the optical reflectance versus ω.

    (c) Repeat for rubidium (Rb) and compare to silver.

16. The crystal NaCl has a static dielectric constant $\varepsilon_1(0)$ of 5.62 and an index of refraction of $n = 1.47$ at frequencies in the visible range.

    (a) What is the physical difference between $\varepsilon_1(0)$ and n^2?

    (b) Calculate the percentage contribution of the ionic polarizability.

    (c) If the optical phonon frequency for NaCl is 3.1×10^{13} Hz and $\omega_\ell = 5.0 \times 10^{13}$ Hz, plot ε_1 versus frequency in the frequency range $0.1\omega_t$ to $10\omega_t$.

17. Why do both n and κ increase for decreasing frequency for frequencies below the plasma frequency?

18. Can you give a physical explanation of why $\omega_\ell > \omega_t$ for an ionic crystal?

19. Give a physical explanation of why the collective mode of oscillation in an ionic lattice that is stimulated by an EM wave is a transverse optical phonon.

20. Why do we designate the frequency at which $\varepsilon_1(\omega) = 0$ as the *longitudinal optical phonon*?

21. The skin depth of a metal is defined as the distance into the metal in which the amplitude of an electromagnetic wave falls to $1/e$ of its initial value due to attenuation. Obtain an expression for the skin depth in terms of the material parameters of the metal, ω_p and τ.

22. Why do we replace the lead term of unity in the expression for $\tilde{n}^2$ of a metal to ε_∞ when writing $\tilde{n}^2$ for a semiconductor?

23. Assume that aluminum has one free electron per atom and a static conductivity of $3.54 \times 10^7 \Omega^{-1}\,\mathrm{m}^{-1}$. Determine:
    (a) The damping coefficient Γ.
    (b) The plasma frequency.
    (c) The index of refraction and the extinction coefficient at $\lambda = 500$ nm.

24. Suppose you are given a pure sample with a good polished surface of an unknown opaque material with a high reflectance at a certain wavelength and at normal incidence. How could you tell whether the large value of R was due to a large value of n or a large value of κ?

25. Consider that in the metallic reflection region ε_1 is negative and ε_2 is very small. Show that to first order in ε_2, the reflectance R at normal incidence can be written

$$1 - R = \frac{2\varepsilon_2/\kappa}{1 + \kappa^2}.$$

26. The index of refraction of the crystal $LiNbO_3$ is given by the Sellmeier equation

$$n^2(\lambda) = 4.913\,00 + \frac{0.118\,717}{\lambda^2 - 0.045\,932} - 0.027\,8\lambda^2,$$

    where λ is in microns (μm). The relation is valid for the wavelength range $\lambda = 0.4$–5.5 μm.
    (a) Find n at $\lambda = 1.064$ μm.
    (b) For an incident plane wave from a vacuum, plot the reflectance (power reflection coefficient) as a function of incident angle from 0 to 90° for both s- and p-polarizations. Again, use $\lambda = 1.064$ μm.
    (c) Calculate Brewster's angle for this wavelength and check your plot.
    (d) Plot the phase of the reflected waves as a function of angle.

27. Using the Sellmeier equation from the previous problems and assuming that light is incident from vacuum onto a crystal of $LiNbO_3$, calculate Brewster's angle for $\lambda = 2.0$ μm. Plot the reflectance as a function wavelength for $\lambda = 1$–4 μm at the angle found initially.

28. Use the Kramers–Kronig relation to calculate $\varepsilon_1(\omega)$ given that:
    (a) $\varepsilon_2(\omega) = A[H(\omega - \omega_1) - H(\omega - \omega_2)]$, $\omega_2 > \omega_1 > 0$ where A is a constant and $H(\omega)$ is the Heaviside step function.
    (b)

$$\varepsilon_2(\omega) = \frac{A\Gamma\omega}{(\omega_0^2 - \omega)^2 + \Gamma\omega^2}.$$

References

[1] Symon K 1971 *Mechanics* 3rd edn (Reading, MA: Addison-Wesley)
[2] Kittel C 1970 *Introduction to Solid State Physics* 5th edn (New York: Wiley)
[3] Runk R B, Stull J L and Anderson O L 1963 A laboratory linear analog for lattice dynamics *Am. J. Phys.* **31** 915–21
[4] Sanderson R B 1965 Far infrared optical properties of indium antimonide *J. Phys. Chem. Solids* **26** 803–10
[5] Omar M A 1975 *Elementary Solid State Physics* (Reading, MA: Addison-Wesley)

IOP Publishing

Optical Radiation and Matter

Robert J Brecha and J Michael O'Hare

Chapter 7

Crystal optics

When electromagnetic radiation propagates in a crystalline medium, the polarization of the light becomes an important consideration. If the medium is isotropic, which has been true for all the cases we have considered up to this point, then the polarization state of the light does not affect how the light propagates (this is because the medium appears the same to the light beam in all directions). We then say that the propagation modes (polarization states) are 'degenerate'. In an anisotropic medium, however, it is a different story and the propagation modes of the light will depend on both the state of polarization of the light and the direction of propagation in the crystal. In this chapter, we will study birefringence, or double refraction, in some detail. We will take a look as well at two further examples of anisotropic media, optical activity and magnetic rotation. In each of these the birefringence in question is not based on linear, but rather on circular polarization modes.

7.1 Historical introduction

Soon after the discovery of the law of refraction by Snell, the Danish scientist Rasmus Bartholin (1625–1698) [1] noticed the curious fact that a transparent crystal of Iceland spar transmits two images of an object placed beneath the crystal. He also measured the angles of refraction for the two transmitted rays and found that one obeyed Snell's law as expected (the so-called ordinary ray), whereas the other transmitted ray did not. Huygens was able to determine the exact relation between the incident and refracted 'extraordinary' ray.

While carrying out his investigations, Huygens also found that the two rays transmitted by the Iceland spar have the properties of what we would now call polarized light. It was based on this work and further experimentation that Fresnel was able to surmise the transverse nature of light waves. In the meantime, and again within the space of only a few years, Jean Baptiste Biot (1774–1862) [2] and David Brewster found the additional properties of doubly refracting (birefringent) crystals which we now call 'positive' and 'negative', depending on the relative sizes of the ordinary and extraordinary indices of refraction. Brewster found that some crystals do not appear to be described as simply as the common uniaxial crystals (two indices of refraction) in that they seem to have two optic axes and three different indices of refraction, depending on the polarization and propagation directions. The key to the observation of the characteristics of these 'biaxial' crystals was that no ray obeyed the ordinary law of refraction for all crystal orientations.

Fundamentally, birefringence is the result of differing indices of refraction and, therefore, differing phase velocities, for various propagation directions and polarizations. It was also Biot who made the first careful measurements of the rotation of the plane of polarization of light as it propagates through certain crystals and other substances. Fresnel was able to explain this effect by postulating that left- and right-circularly polarized light could 'see' different polarizations; since linearly polarized light can be considered a superposition of circularly polarized rays of opposite helicity, the predicted result was a rotation of the polarization orientation, as Faraday had observed. This effect we now call optical activity.

Because he was so convinced of the necessity of an interaction between light and electric and magnetic fields, Faraday carried out several experiments to search for such an effect. He did observe that certain glasses, when placed in a strong magnetic field, caused a rotation of the plane of polarization of incident light (now called the Faraday effect). The explanation for the Faraday effect is essentially a magnetically induced change in index of refraction between left- and right-circularly polarized fields, again leading to a net rotation upon traversing the medium.

7.2 Polarizers

Before actually starting to study propagation in crystals, a brief discussion about how polarized light is produced is in order. The primary mechanisms for producing polarized light are (i) reflection (e.g. reflection at Brewster's angle), (ii) scattering, (iii) double refraction or birefringence, and (iv) polarization by selective absorption (dichroism). The first two mechanisms have been discussed in previous chapters.

A polarizer is a device which ideally transmits only the component of the light wave E-vector along a particular transverse axis. For example, in figure 7.1 below an initially unpolarized beam of light becomes polarized by passing through polarizer 1. The resulting beam of plane polarized light is incident on polarizer 2 such that the direction of vibration of the E-vector makes an angle of Θ to the transmission axis of the polarizer. Thus the light wave that is transmitted by the second polarizer has an E-vector given by $E = E_0 \cos \Theta$, where E_0 is the incident E-vector.

Since the irradiance is proportional to the square of the E-vector, the transmitted irradiance will be given by the expression below, called *Malus' law*,

$$I(\Theta) = \frac{|E_0|^2}{2\mu_0 c} \cos^2 \Theta = I_0 \cos^2 \Theta. \tag{7.1}$$

In the above expression for Malus' law, I_0 is the incident irradiance defined from chapter 1 as

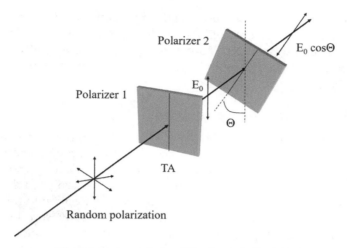

Figure 7.1. Transmission of light through a polarizer. The first polarizer is used to define a polarization direction for the initially unpolarized light.

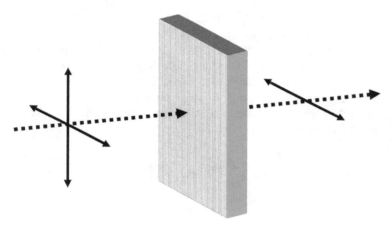

Figure 7.2. Wire-grid polarizer. Polaroid polarizers work on the same principle.

$$I_0 = \frac{|E_0|^2}{2\mu_0 c}. \tag{7.2}$$

A *dichroic* polarizer is one that selectively absorbs light with electric-field vibrations along a particular axis in the dichroic material. Thus the dichroic material will transmit light with its *E*-vector perpendicular to the axis of selective absorption. One example of a dichroic polarizer is the wire-grid polarizer shown in figure 7.2, which is a dichroic polarizer because it selectively absorbs light along the wire grids and transmits light orthogonal to the wire grids. We explain this by saying that the *y*-component is absorbed and re-radiated forward and backward. The forward radiation is canceled by the (non-absorbed) incident radiation, whereas the backward radiation just appears as reflected radiation. Other types of polarizers

7-3

include dichroic crystals, which absorb light differently for different polarizations and different directions of propagation, and polaroids, which are the molecular analog of the wire-grid polarizer. Wire-grid polarizers are typically used for long-wavelength radiation (middle- to far-infrared) whereas polaroid polarizers are used for visible light.

7.3 Birefringence (double refraction)

Crystals are generally optically anisotropic, that is their physical properties are not the same in every direction. As a consequence, the speed of propagation is a function of direction and of polarization of the light wave. Considering transverse waves only, the above statement means that there are in general two possible values of the phase velocity for a given direction of propagation; an exception to this general rule is found in crystals with cubic symmetry, which are optically isotropic. All crystals other than cubic crystals display a polarization-dependent phase velocity, or index of refraction, which we call *birefringence* or *double refraction*.

Figure 7.3, which shows the spring model of an atom in a crystal with anisotropic spring constants (due to different bonding strengths along the three axes), indicates why the propagation will depend on both direction of propagation and the polarization of the EM wave. From our previous study of the induced oscillator model, the spring constant determines the resonant frequency. Thus the different spring constants along the different axes would yield different indices of refraction for the different directions of propagation and polarization.

We will now use the macroscopic Maxwell equations to take a quantitative look at wave propagation in a birefringent (anisotropic) material. Recall the procedure for obtaining the wave equation:

$$\nabla \times (\nabla \times \vec{E}) = -\mu_0 \nabla \times \left(\frac{\partial \vec{H}}{\partial t} \right)$$
$$= -\mu_0 \frac{\partial \nabla \times \vec{H}}{\partial t}. \qquad (7.3)$$

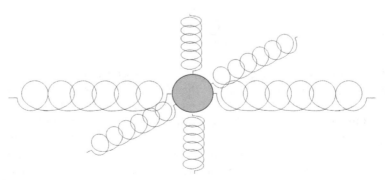

Figure 7.3. Spring model of crystal bonding. The 'bond lengths' can be different between neighboring atoms depending on the direction, as can the coupling strengths.

Using Maxwell's equation $\nabla \times \vec{H} = \frac{\partial \vec{D}}{\partial t}$,

$$\nabla \times (\nabla \times \vec{E}) = -\mu_0 \frac{\partial^2 \vec{D}}{\partial t^2}$$

$$= -\mu_0 \varepsilon_0 \frac{\partial^2 (\varepsilon_r \vec{E})}{\partial t^2}$$

$$= -\mu_0 \varepsilon_0 [1 + \chi] \frac{\partial^2 \vec{E}}{\partial t^2} \qquad (7.4)$$

$$= -\frac{1}{c^2} \frac{\partial^2 \vec{E}}{\partial t^2} - \frac{1}{c^2} \frac{\partial^2 [\chi \vec{E}]}{\partial t^2}.$$

However, in an anisotropic medium, χ is no longer a scalar quantity but a second rank tensor, designated as $\overset{\leftrightarrow}{\chi}$. The above equation should therefore be written as

$$\nabla \times [\nabla \times \vec{E}] = -\frac{1}{c^2} \frac{\partial^2 \vec{E}}{\partial t^2} - \frac{1}{c^2} \frac{\partial^2 [\overset{\leftrightarrow}{\chi} \vec{E}]}{\partial t^2}, \qquad (7.5)$$

where $\overset{\leftrightarrow}{\chi}$ is defined by the equation

$$\vec{P} = \epsilon_0 \overset{\leftrightarrow}{\chi} \vec{E}, \qquad (7.6)$$

which in component form can be written as a matrix equation,

$$\begin{pmatrix} P_x \\ P_y \\ P_z \end{pmatrix} = \varepsilon_0 \begin{pmatrix} \chi_{11} & \chi_{12} & \chi_{13} \\ \chi_{21} & \chi_{22} & \chi_{23} \\ \chi_{31} & \chi_{32} & \chi_{33} \end{pmatrix} \begin{pmatrix} E_x \\ E_y \\ E_z \end{pmatrix} \qquad (7.7)$$

or in terms of the field quantities $\vec{D}$ and $\vec{E}$ we would write,

$$\vec{D} = \varepsilon_0 [1 + \overset{\leftrightarrow}{\chi}] \vec{E} = \varepsilon_0 \overset{\leftrightarrow}{\varepsilon} \vec{E}. \qquad (7.8)$$

In the above expressions $\overset{\leftrightarrow}{\chi}$ is called the electric susceptibility tensor and $\overset{\leftrightarrow}{\varepsilon}$ is called the relative dielectric tensor. The subscripts on χ_{ij} stand for x, y, and z; i.e. $1 \rightarrow x$, $2 \rightarrow y$, and $3 \rightarrow z$. For 'ordinary' nonabsorbing crystals the χ-tensor is symmetric and a set of coordinate axes exist, called 'principal axes', for which $\bar{\chi}$ assumes a diagonal form:

$$\overset{\leftrightarrow}{\chi} = \begin{pmatrix} \chi'_{11} & 0 & 0 \\ 0 & \chi'_{22} & 0 \\ 0 & 0 & \chi'_{33} \end{pmatrix}. \qquad (7.9)$$

These χs are called the principal susceptibilities. The 'principal axes' are just the normal coordinates associated with the normal modes of motion that we discussed in chapter 5. Alternatively, we can use what are called the 'principal dielectric constants' defined by

$$\epsilon_{11} = 1 + \chi'_{11} \tag{7.10}$$

$$\epsilon_{22} = 1 + \chi'_{22} \tag{7.11}$$

$$\epsilon_{33} = 1 + \chi'_{33}. \tag{7.12}$$

For monochromatic plane waves recall that we have

$$\nabla \times \rightarrow i\vec{k}\times \tag{7.13}$$

so that the wave equation can be written

$$\vec{k} \times [\vec{k} \times \vec{E}] = -\frac{\omega^2}{c^2}\vec{E} - \frac{\omega^2}{c^2}\overset{\leftrightarrow}{\chi}\vec{E} \tag{7.14}$$

$$\vec{k}(\vec{k} \cdot \vec{E}) - k^2\vec{E} = -\frac{\omega^2}{c^2}\vec{E} - \frac{\omega^2}{c^2}\overset{\leftrightarrow}{\chi}\vec{E}. \tag{7.15}$$

Note that $\vec{k} \cdot \vec{E}$ is no longer necessarily zero (Why not?). It is somewhat instructive to see the above tensor equation in component form. The development below closely follows that in [3]:

$$\left[\frac{\omega^2}{c^2} - k_y^2 - k_z^2\right]E_x + k_xk_yE_y + k_xk_zE_z = -\frac{\omega^2}{c^2}\chi'_{11}E_x \tag{7.16}$$

$$k_yk_xE_x + \left[\frac{\omega^2}{c^2} - k_x^2 - k_z^2\right]E_y + k_yk_zE_z = -\frac{\omega^2}{c^2}\chi'_{22}E_y \tag{7.17}$$

$$k_zk_xE_x + k_zk_yE_y + \left[\frac{\omega^2}{c^2} - k_x^2 - k_y^2\right]E_z = -\frac{\omega^2}{c^2}\chi'_{33}E_z. \tag{7.18}$$

To interpret the physical meaning of the above component set of equations we can look at a specific case, letting

$$k_x = k$$
$$k_y = k_z = 0.$$

For the x-component equation this gives

$$\frac{\omega^2}{c^2}E_x = -\frac{\omega^2}{c^2}\chi'_{11}E_x,$$

which means that $E_x = 0$. The y- and z-component equations are

$$\left[\frac{\omega^2}{c^2} - k^2\right]E_y = -\frac{\omega^2}{c^2}\chi'_{22}E_y$$

$$\left[\frac{\omega^2}{c^2} - k^2\right]E_z = -\frac{\omega^2}{c^2}\chi'_{33}E_z.$$

Now, if $E_y \neq 0$, we find that

$$k = \frac{\omega}{c}\sqrt{1 + \chi_{22}'} = \frac{\omega}{c}\sqrt{\epsilon_{22}} = \frac{\omega}{c}n_2 \qquad (7.19)$$

and, if $E_z \neq 0$, we find

$$k = \frac{\omega}{c}\sqrt{1 + \chi_{33}'} = \frac{\omega}{c}\sqrt{\epsilon_{33}} = \frac{\omega}{c}n_3. \qquad (7.20)$$

What we have just done was to assume a direction of propagation ($\vec{k} = \hat{x}k_x$) and shown that there are two modes of propagation; one for the beam polarized along y and the other for the beam polarized along z. These are called modes or degrees of freedom for propagation of the electromagnetic wave. More generally, we can say that for any direction of propagation indicated by the vector, there are two possible values for the magnitude of k, and hence two values for the phase velocity. These are determined by the principle indices of refraction:

$$n_1 = \sqrt{1 + \chi_{11}'} = \sqrt{\epsilon_{11}}$$
$$n_2 = \sqrt{1 + \chi_{22}'} = \sqrt{\epsilon_{22}} \qquad (7.21)$$

$$n_3 = \sqrt{1 + \chi_{33}'} = \sqrt{\epsilon_{33}}. \qquad (7.22)$$

Remember that we are assuming lossless materials, i.e. $\kappa = 0$. In equation (7.21) the meaning of n_1 is that it is the index of refraction seen by a propagating wave polarized along the x-axis. Similarly, n_2 would be the index seen by a wave polarized along the y-axis and n_3 would be the index seen by a wave polarized along the z-axis. We are assuming the x, y, and z are the principal axes of the crystal.

In a more general treatment, for nontrivial solutions to equations (7.16)–(7.18) to exist, the determinant of coefficients of E_x, E_y, and E_z must vanish: Therefore

$$\begin{vmatrix} \left[\dfrac{n_1\omega}{c}\right]^2 - k_y^2 - k_z^2 & k_xk_y & k_xk_z \\[2ex] k_yk_x & \left[\dfrac{n_2\omega}{c}\right]^2 - k_x^2 - k_z^2 & k_yk_z \\[2ex] k_zk_x & k_zk_y & \left[\dfrac{n_3\omega}{c}\right]^2 - k_x^2 - k_y^2 \end{vmatrix} = 0. \qquad (7.23)$$

The above determinant defines a wavevector surface in k-space. For example, consider the plane $k_z = 0$. The above determinant then gives the resulting expression on the $k_z = 0$ plane,

$$\left[\frac{n_3^2\omega^2}{c^2} - k_x^2 - k_y^2\right]\left[\left(\frac{n_1^2\omega^2}{c^2} - k_y^2\right)\left(\frac{n_2^2\omega^2}{c^2} - k_x^2\right) - k_x^2 k_y^2\right] = 0.$$

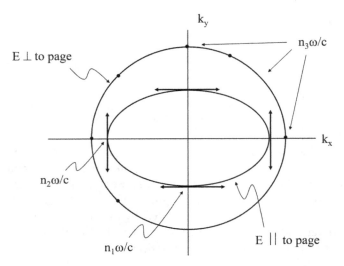

Figure 7.4. Wavevector surface for the $k_z = 0$ plane and $n_3 > n_2 > n_1$. The dots represent the polarization of the E-field pointing into and out of the page.

Either or both factors in the above equation must vanish, thus yielding

$$k_x^2 + k_y^2 = \left[\frac{n_3\omega}{c}\right]^2 \qquad \text{(equation of a circle)} \tag{7.24}$$

$$\frac{k_x^2}{\left(\frac{n_2^2\omega^2}{c^2}\right)} + \frac{k_y^2}{\left(\frac{n_1^2\omega^2}{c^2}\right)} = 1 \qquad \text{(equation of an ellipse).} \tag{7.25}$$

These two equations are sketched in figure 7.4 below which show the wavevector surface on the plane $k_z = 0$ and for the case of $n_3 > n_2 > n_1$.

7.3.1 Uniaxial crystals

A *uniaxial crystal* is defined as a crystal in which two of the principal indices are equivalent. Let us now look at the wavevector surface in three dimensions for a uniaxial crystal ($n_1 = n_2 \neq n_3$). In figure 7.5 we have drawn the case for which $n_3 > n_1, n_2$. In that figure, the k_z-axis is the *optic axis*, defined as the symmetry axis of the crystal. The optic axis is the propagation axis in the crystal for which the phase velocity is independent of the polarization of the EM wave; thus the k-vectors have the same magnitude (they are degenerate) for propagation along the optic axis.

From the above figure we see some interesting features regarding the direction of propagation and polarization of the wave. First, for light propagating along the optic axis the phase velocity is the same for both polarizations. In addition, for any wave which is polarized perpendicular to the optic axis it will see an index n_o for which $n_o = n_1 = n_2$. This is called the *o-ray* or ordinary ray. In a uniaxial crystal, the

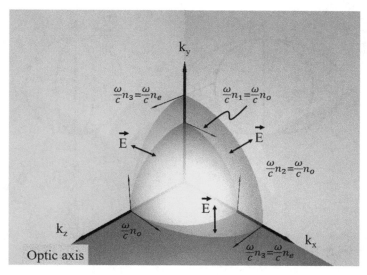

Figure 7.5. Wavevector surface in three dimensions for a uniaxial crystal.

o-ray is the ray that is always polarized perpendicular to the optic axis. The ray with the other polarization state—the ray that is not polarized perpendicular to the optic axis—is called the *e-ray* or extraordinary ray. The e-ray sees an index of refraction $n_e = n_3$ when it is propagating perpendicular to the optic axis. When the e-ray is propagating in a direction other than perpendicular to the optic axis it will see an index of refraction between n_o and n_e. In this case $(n_o < n_e)$, we can write $n_o \leqslant n(\theta) \leqslant n_e$ for the index of refraction, $n(\theta)$, of the e-ray. The angle θ is the angle that the k-vector of the e-ray makes with the optic axis. Later, we will see how to calculate $n(\theta)$. From this discussion, we see that, in general, unpolarized light propagating through a crystal can be considered as two independent orthogonal waves with different phase velocities. These phase velocities in a uniaxial crystal are determined by the values of the two indices of refraction, n_o and $n_e(\theta)$, of the o-ray and e-ray, respectively.

If the three principal axes are all unique, that is $n_1 \neq n_2 \neq n_3$, the crystal is called a biaxial crystal. A biaxial crystal will have two optic axes. In this course we will be concerned only with uniaxial crystals.

Uniaxial crystals are classified as either uniaxial positive $(n_e > n_o)$ or uniaxial negative $(n_o > n_e)$. These two cases are sketched in figure 7.6 below which shows the k-vector surfaces in a plane containing the optic axis.

In uniaxial crystals the index corresponding to the equal elements, $\chi_{11} = \chi_{22}$, is the ordinary ray, n_o. The index for χ_{33} is the extraordinary ray, n_e. Note: It is worthwhile to again remind ourselves that n_e is the index for the e-ray only when it is propagating perpendicular to optic axis. In general, $n_e \equiv n_e(\theta)$, that is, n_e is a function of the angle between the propagation direction of the extraordinary ray and the optic axis. Also $k_e = n_e\omega/c$ only when the e-ray is propagating perpendicular to the optic axis.

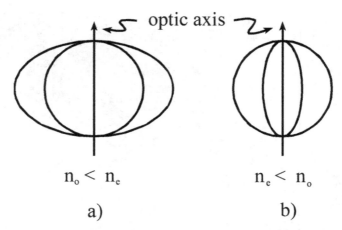

Figure 7.6. Comparison for wavevector surfaces for uniaxial positive and uniaxial negative crystals. (a) $n_o < n_e$; uniaxial positive. (b) $n_o > n_e$; uniaxial negative.

Table 7.1. Values for the ordinary and extraordinary indices of refraction for some common uniaxial materials for a wavelength of 589 nm.

Material	n_o	n_e
Ice (H_2O)	1.309	1.313
Quartz	1.54424	1.55335
Calcite ($CaO \cdot CO_2$)	1.658	1.486
Sapphire (Al_2O_3)	1.7681	1.7599
KDP	1.507	1.467
ADP	1.522	1.478
$LiNbO_3$	2.300	2.208

We would like to get a sense of how different the indices of refraction actually are for the uniaxial crystal directions. Table 7.1 gives the values of the indices of refraction for several common materials at 589 nm.

7.3.2 Ray direction and the Poynting vector

In an anisotropic medium $\vec{k}$ and $\vec{E}$ are not generally perpendicular (see equation (7.15) to see why this is true). However, $\vec{H}$ is still perpendicular to both $\vec{E}$ and $\vec{k}$ as can be seen from the Maxwell equation

$$\vec{k} \times \vec{E} = \mu_0 \omega \vec{H}.$$

From what we have already seen in this chapter we cannot assume that $\vec{D}$ and $\vec{E}$ are parallel. Finally, we can make use of the definition of the Poynting vector, $\vec{S} = \vec{E} \times \vec{H}$, to relate the propagation direction of the beam energy to the other vector quantities. In figure 7.7 $\vec{k}$ defines the normal to surfaces of constant phase, whereas $\vec{S}$ defines energy

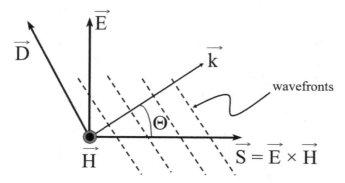

Figure 7.7. Orientation of electromagnetic field vectors for an anisotropic material.

propagation or ray direction, and $\vec{E}$ and $\vec{k}$ are coplanar but not necessarily perpendicular. The planes of constant phase can be thought of as moving along the ray $\vec{S}$, but are not perpendicular to $\vec{S}$.

Now we wish to see how these vector relations are manifested in the actual propagation of light through an anisotropic crystal.

7.3.3 Double refraction at the boundary of an anisotropic medium

We start by recalling that Snell's law comes from the boundary condition,

$$\vec{k}_i \cdot \vec{r} = \vec{k}_t \cdot \vec{r}.$$

The boundary condition must be valid for both o- and e-rays. In the above expression, $\vec{k}_i$ is the k-vector of the wave incident on the crystal and $\vec{k}_t$ is the wavevector of the wave transmitted into the crystal. For the o- and e-rays we have $\vec{k}_i \cdot \vec{r} = \vec{k}_o \cdot \vec{r} = \vec{k}_e \cdot \vec{r}$, and thus we can write

$$k_i \sin \theta_i = k_o \sin \theta_o \qquad (7.26)$$

$$k_i \sin \theta_i = k_e \sin \theta_e. \qquad (7.27)$$

The first equality gives us the usual Snell's law for refraction. This is valid for the ordinary ray. However, for the case of the e-ray, both k_e and θ_e are variable because k_e will depend on θ, i.e. it will depend on how the wave enters into the crystal. Figure 7.8 below depicts a wavevector k_o incident on a crystal and the refraction of the o- and e-rays. The projections of the wave vectors along the interface, corresponding to the boundary conditions, are shown in figure 7.8 as well. To summarize the point we are trying to make here, for the ordinary ray, we may apply Snell's law to find the angle of refraction in the medium, since the ordinary index of refraction is independent of propagation direction of that ray. For the extraordinary ray, on the other hand, we must write

$$n_i \sin \theta_i = n_e(\theta) \sin \theta_e.$$

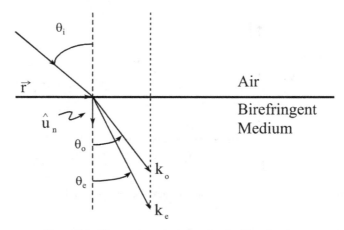

Figure 7.8. Wavevector orientation for double refraction.

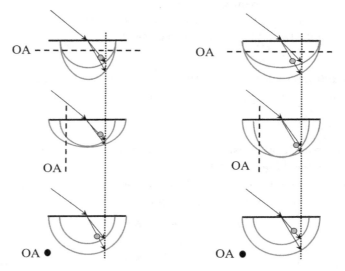

Figure 7.9. Wavevector surfaces for various double refraction configurations. The set of figures on the left corresponds to a uniaxial positive crystal, those on the right to a uniaxial negative crystal. The optical axis is indicated in each figure, by either a dashed line or a filled circle if perpendicular to the page. The filled circle on the refracted rays represent the field polarization perpendicular to the page; the polarization for the other ray lies in the plane of the page.

Here we cannot find the refraction angle directly because we do not know the index of refraction, which in turn depends on the angle of propagation in the medium, i.e. on the refraction angle.

The sketches in figure 7.9 show the refraction angles for positive and negative crystals, and for the optic axis oriented in different directions with respect to the surface of the anisotropic medium.

7.3.4 The optical indicatrix

The normal-mode solutions discussed in previous sections can be obtained by a geometrical construction called variously the *optical indicatrix, index ellipsoid*, or *ellipsoid of wave normals*. This concept is developed by considering the energy stored in a unit volume of the dielectric. In the principal axis system this is [4, 5]

$$U = \frac{1}{2}\vec{E} \cdot \vec{D} = \frac{1}{2\epsilon_0}\left(\frac{D_x^2}{\epsilon_{11}} + \frac{D_y^2}{\epsilon_{22}} + \frac{D_z^2}{\epsilon_{33}}\right). \tag{7.28}$$

What this equation says is that the constant energy surfaces in $\vec{D}$-space are ellipsoids given by the above equation. If we make a change of variables, $\vec{x} = \vec{D}/\sqrt{2U\epsilon_0}$, and define the principle indices of refraction by n_1, n_2, n_3, the above equation may be written

$$\frac{x^2}{n_1^2} + \frac{y^2}{n_2^2} + \frac{z^2}{n_3^2} = 1. \tag{7.29}$$

This is the equation of a general ellipsoid with major axes parallel to the x-, y-, and z-directions whose respective lengths are $2n_1$, $2n_2$, $2n_3$. The index ellipsoid is used to find the two indices of refraction and the two corresponding directions of $\vec{D}$ associated with the two independent plane waves that can propagate along an arbitrary direction in the crystal. This is done according to the following recipe:

    (a) Find the intersection ellipse between a plane through the origin normal to the direction of propagation $\vec{k}$ and the index ellipsoid.

    (b) The two axes of the intersection ellipse are equal in length to $2n_1'$ and $2n_2'$.

    (c) These axes are parallel, respectively, to the direction of the D_1, D_2 vectors of the two allowed solutions.

Now n_1' and n_2' in the above prescription correspond to the two indices of refraction for the propagation vector $\vec{k}$. $\vec{D}_{1,2}$ are the allowed solutions and the ellipse axes are parallel to $\vec{D}_{1,2}$.

The ellipse is specified by the two surfaces:

    (i) The index ellipsoid

$$\frac{x^2}{n_1^2} + \frac{y^2}{n_2^2} + \frac{z^2}{n_3^2} = 1. \tag{7.30}$$

    (ii) The normal to $\vec{k}$

$$\vec{r} \cdot \vec{k} = xk_x + yk_y + zk_z = 0.$$

For example, consider propagation in uniaxial crystals:

$$\frac{x^2}{n_1^2} + \frac{y^2}{n_1^2} + \frac{z^2}{n_3^2} = 1 \quad (n_1 = n_2). \tag{7.31}$$

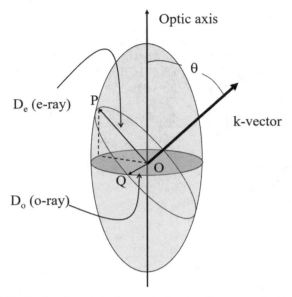

Figure 7.10. Finding the principal axes of refraction from the optical indicatrix.

The length of the semi-major axis OP is equal to $n_e(\theta)$ and the electric displacement $D_e(\theta)$ is parallel to OP. The ordinary ray is polarized along OQ. According to figure 7.10 the index of the e-ray is

$$\frac{1}{n_e^2(\theta)} = \frac{\cos^2 \theta}{n_0^2} + \frac{\sin^2 \theta}{n_e^2}. \tag{7.32}$$

To see how this was obtained, we consider the equation for the index ellipsoid,

$$\frac{x^2}{n_o^2} + \frac{y^2}{n_o^2} + \frac{z^2}{n_e^2} = 1.$$

Now for $n_e(\theta)$, $x = 0$, $y = -n_e(\theta) \cos \theta$, $z = n_e(\theta) \sin \theta$,

$$n_e^2(\theta)\left[\frac{\cos^2 \theta}{n_o^2} + \frac{\sin^2 \theta}{n_e^2} \right] = 1. \tag{7.33}$$

Although the correspondence between the indicatrix construction has been demonstrated for uniaxial crystals only, it can be shown that it also holds for the general ellipsoid.

The result we just derived gives us the additional tool we need to solve the Snell's law problem for an extraordinary wave. As we will see in the following example, Snell's law gives us one equation involving both the propagation direction and the index of refraction $n_e(\theta)$. The index ellipsoid equation provides the second relation needed so that we can determine both the refraction angle and the corresponding index of refraction.

Example. Let us examine one case that will illustrate the above discussion. Assume that an unpolarized beam of light is incident at $\theta_i = 45°$ on a uniaxial crystal, for example sapphire, Al_2O_3, which has its optic axis perpendicular to the surface of the plate. Looking at table 7.1 we find that the values of the indices of refraction are $n_o = 1.7681$ and $n_e = 1.7599$, i.e. this is a uniaxial negative crystal. The other piece of information we need is the result derived above, equation (7.32), and relates $n_e(\theta)$ to n_e, n_o, and θ. First we can calculate θ_o, the angle of refraction for the ordinary ray. From Snell's law,

$$n_i \sin \theta_i = n_o \sin \theta_o$$

$$\theta_o = \sin^{-1}\left[\frac{n_i}{n_o} \sin \theta_i\right]$$

$$= 23.57°.$$

Here we have assumed that $n_i = 1.00$. Now we need to find θ_e. From figure 7.9 we see that $\theta_e = \theta$, where θ is the angle between the optic axis and the k-vector direction. Using equation (7.32) we can eliminate θ with the help of Snell's law. This gives

$$\frac{1}{n_e^2(\theta)} = \frac{1}{n_o^2}\left(1 - \frac{n_i^2}{n_e^2(\theta)} \sin^2 \theta_i\right) + \frac{1}{n_e^2}\left(\frac{n_i^2}{n_e^2(\theta)} \sin^2 \theta_i\right).$$

This can be rearranged to give

$$n_e^2(\theta) = n_o^2 + n_i^2 \sin^2 \theta_i \left(1 - \frac{n_o^2}{n_e^2}\right).$$

Thus we find that $n_e(\theta) = 1.7668$. This can in turn be used to find θ_e from equation (7.27):

$$\theta_e = \sin^{-1}\left[\frac{n_i}{n_e(\theta)} \sin \theta_i\right]$$

$$= 23.59°.$$

We have two different rough checks we can make to see if our answers at least make sense. First, we see that $n_e(\theta)$ does indeed lie between n_o and n_e. Second, we expect that since $n_e(\theta)$ is smaller than n_o, θ_e should be larger than θ_o, as it is. The results we have found here are also consistent with figure 7.9.

There is one other important case which is not shown in figure 7.9, namely, when the optic axis is neither perpendicular nor parallel to the surface. This is illustrated in figure 7.11. Shown is the surface of a uniaxial crystal, along with the corresponding k-vector surfaces. For unpolarized light at normal incidence, from equations (7.26) and (7.27) we have that the two k-vectors, k_e and k_o for the refracted extraordinary and ordinary rays must be normal to the surface as well. Recall, however, that the direction of $\vec{E}$ is given by the tangent to the k-vector surface. Further, since $\vec{S} = \vec{E} \times \vec{H}$ is orthogonal to $\vec{E}$ we can see from the figure that $\vec{S}_e$ will not be in the same direction as k_e. Qualitatively we find that $\vec{S}_e$ is at some angle ϕ with respect to the normal while the

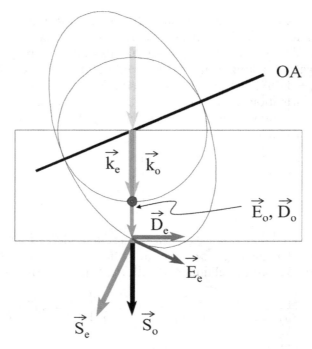

Figure 7.11. The various fields, k-vectors, and the Poynting vector directions for the case in which the optic axis is not parallel to or perpendicular to the surface of the crystal.

light propagates in the crystal. When the e-ray reaches the other end of the crystal, it is bent back to the normal and emerges parallel to the o-ray, but displaced from it. This is illustrated in figure 7.12.

The above discussion is actually very easy to test experimentally. It corresponds to the common demonstration of laser light incident on a piece of calcite or quartz. In that case the optic axis is at an angle to the surface. Laser light incident is split into two beams, one of which propagates straight through the crystal while the other is displaced. As the crystal is rotated about the axis along which the input light is propagating, the e-ray (the displaced beam) moves in a circle around the undisplaced o-ray. One can also use a sheet polarizer to check that the two output beams are orthogonally polarized.

7.3.5 Wave-velocity surfaces

We can also look at propagation in anisotropic crystals from a real space viewpoint, i.e. in terms of wave-velocity surfaces rather than the k-vector surface approach in the previous sections. In figure 7.13 the sketches show the propagation of phase velocity surfaces for both uniaxial positive and uniaxial negative crystals. Since the phase velocity goes as the inverse of the index of refraction, the pictures are just the inverse of the k-vector surfaces for the corresponding crystals.

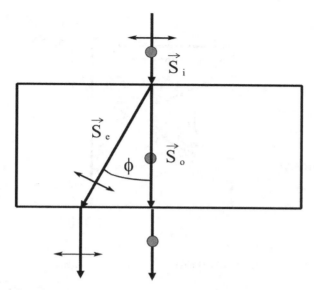

Figure 7.12. Illustration of the Poynting vectors for the o-ray and the e-ray for the case when the optic axis is at some angle to the surface.

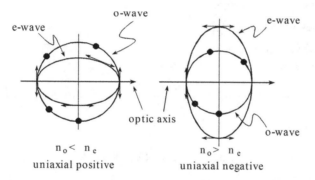

Figure 7.13. Wave-velocity surfaces for a doubly refracting crystal.

7.4 Retarders

The state of polarization of an incident wave can be changed by an optical device called a phase retarder, as we have already seen. Recall that we mentioned in chapter 2 that a retarder performs its role by inducing a phase change (lag) in one component of the field with respect to the other. Now we can understand in more detail the origin of the relative phase delays. As a wave propagates through an anisotropic crystal, the two orthogonal polarization states will acquire a phase difference or optical path difference due to the differing indices of refraction seen by the two polarization, as illustrated in figure 7.14. The optical path difference is defined by $\Lambda \equiv d|n_o - n_e|$ and the phase difference is $\Delta\phi = k_0\Lambda = \frac{2\pi}{\lambda_0}d|n_o - n_e|$.

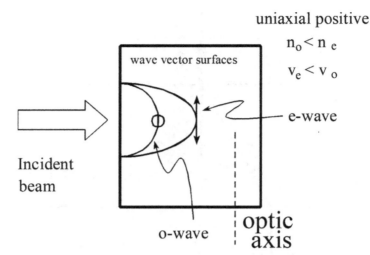

Figure 7.14. Phase retardation by a retarder.

Some examples of retarders are:
  (i) Full-wave plate: $\Delta\phi = 2\pi$.
    Optical effect: none.
    Jones matrix: $\begin{pmatrix} 1 & 0 \\ 0 & 1 \end{pmatrix}$.
  (ii) Half-wave plate: $\Delta\phi = \pi$.
    Optical effect: path difference given by

$$d|n_o - n_e| = \frac{(2m + 1)\lambda_0}{2} \quad m \equiv \text{integer}.$$

    Jones matrix: $\begin{pmatrix} 1 & 0 \\ 0 & -1 \end{pmatrix}$.
  (iii) Quarter-wave plate: $\Delta\phi = \pi/2$ or $(2m + 1)\pi/2$.
    Jones matrix: $\begin{pmatrix} 1 & 0 \\ 0 & \pm i \end{pmatrix}$.
    Optical effect: path difference given by

$$d(n_o - n_e) = \frac{(2m + 1)\lambda_0}{4},$$

where m is an integer.

In chapter 2 we worked through some examples using waveplates, and we will have the opportunity to see these devices again in chapter 8.

Another class of retarder is the compensator or variable retarder. A typical compensator is the Babinet compensator shown in figure 7.15. In this device the phase difference acquired by the two polarization states can be varied to achieve a

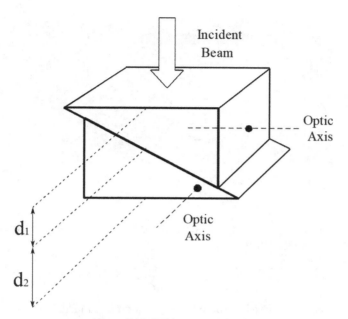

Figure 7.15. Babinet compensator.

particular phase difference for the particular wavelength. A retarder, on the other hand, induces a specified phase difference for a specific wavelength only. For example, a retarder of particular dimension will act as a quarter-waveplate for only one wavelength.

7.5 Optical activity

Many substances are found to rotate the plane of polarization as light passes through. Figure 7.16 illustrates this property, called *optical activity*. Optically active materials are designated by their *specific rotatory power*, which is defined to be the rotation per unit length of the plane of polarization. There are two types of optically active materials, right-handed (*dextrorotatory*) and left-handed (*levorotatory*). We will make this more quantitative by considering the susceptibility tensor for an optically active medium.

7.5.1 Susceptibility tensor of optically active medium

If $\overset{\leftrightarrow}{\chi}$ has conjugate imaginary off-diagonal elements such that [3]

$$\bar{\chi} = \begin{pmatrix} \chi_{11} & i\chi_{12} & 0 \\ -i\chi_{12} & \chi_{22} & 0 \\ 0 & 0 & \chi_{33} \end{pmatrix} \quad \text{uniaxial,} \tag{7.34}$$

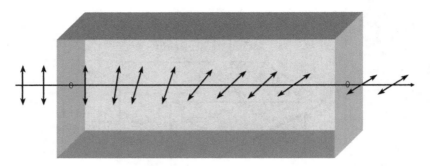

Figure 7.16. Rotation of the plane of polarization by an optically active medium.

where χ_{12} is real, then the medium will be optically active. This can be seen by looking at the wave equation

$$\vec{k}(\vec{k} \cdot \vec{E}) - k^2\vec{E} = -\frac{\omega^2}{c^2}\vec{E} - \frac{\omega^2}{c^2}\overleftrightarrow{\chi}\,\vec{E}.$$

Writing the above expression in component form and letting the wave propagate in the z-direction, $\vec{k} = k\hat{z}$, we have

$$-k^2E_x + \frac{\omega^2}{c^2}E_x = -\frac{\omega^2}{c^2}[\chi_{11}E_x + i\chi_{12}E_y] \tag{7.35}$$

$$-k^2E_y + \frac{\omega^2}{c^2}E_y = -\frac{\omega^2}{c^2}[-i\chi_{12}E_x + \chi_{11}E_y] \tag{7.36}$$

$$\frac{\omega^2}{c^2}E_z = -\frac{\omega^2}{c^2}\chi_{33}E_z. \tag{7.37}$$

The latter equation implies that $E_z = 0$.

As before, for nontrivial solutions to exist, the determinant of the coefficients of the E-components must vanish; therefore

$$\begin{vmatrix} -k^2 + \dfrac{\omega^2}{c^2}[1 + \chi_{11}] & i\dfrac{\omega^2}{c^2}\chi_{12} \\[2mm] -i\dfrac{\omega^2}{c^2}\chi_{12} & -k^2 + \dfrac{\omega^2}{c^2}[1 + \chi_{11}] \end{vmatrix} = 0.$$

The solution to this determinant equation will give two solutions for k, given by

$$k = \frac{\omega}{c}\sqrt{1 + \chi_{11} \pm \chi_{12}}. \tag{7.38}$$

Substituting these values for k into equation (7.35) or equation (7.36), we obtain

$$E_y = \pm iE_x = E_x e^{\pm i\pi/2}.$$

The two relations above imply that the eigenmodes or normal modes for this optically active system are left- and right-circularly polarized:

$$' + ' \quad \rightarrow E_y = +iE_x \quad \rightarrow \begin{pmatrix} 1 \\ i \end{pmatrix}.$$

Similarly,

$$' - ' \quad \rightarrow E_y = -iE_x \quad \rightarrow \begin{pmatrix} 1 \\ -i \end{pmatrix}.$$

The two values of k which give the two modes of propagation for the wave propagating along the optic axis in the optically active medium give different indices of refraction for right- and left-circularly polarized light:

$$n_L = \sqrt{1 + \chi_{11} + \chi_{12}} \quad ; \quad n_R = \sqrt{1 + \chi_{11} - \chi_{12}}. \tag{7.39}$$

The difference between these two indices is

$$n_L - n_R \simeq \frac{\chi_{12}}{\sqrt{1 + \chi_{11}}} = \frac{\chi_{12}}{n_0}, \tag{7.40}$$

where we have made the approximation that $\chi_{12} \ll \chi_{11}$. This difference in indices leads to a rotation, β, of the plane of polarization given by

$$\beta = \text{rotation angle} = \frac{[n_L - n_R]\pi d}{\lambda_0}, \tag{7.41}$$

where d is the distance traveled through the medium. From our definition of the specific rotatory power,

$$\text{Specific rotatory power:} \quad \frac{\beta}{d} \equiv \frac{[n_L - n_R]\pi}{\lambda} = \frac{\chi_{12}\pi}{n_0\lambda}. \tag{7.42}$$

Another way of viewing optical activity is the following: the incident plane wave can be represented as a superposition of R and L states. In an optically active medium these R and L states propagate with different speeds. They thus get out of phase and rotate the original polarization state. Additional, more detailed, information on optical activity can be obtained from references [4, 6]. As one example, the optical rotatory power for quartz at three different wavelengths is given in table 7.2.

Table 7.2. Specific rotatory power for quartz at three different wavelengths.

β/d	λ
$49°$ mm^{-1}	4000 Å
$31°$ mm^{-1}	5000 Å
$22°$ mm^{-1}	6000 Å

7.6 Faraday effect

It is possible to obtain results very similar to optical activity in crystals, liquids, or even gases through the application of a magnetic field. This effect is known as Faraday rotation and the amount of rotation of the plane of polarization can be controlled by the strength of the applied magnetic field. Recall the model for a bound oscillating electron:

$$m\ddot{\vec{x}} + m\Gamma\dot{\vec{x}} + m\omega_0^2\vec{x} = -e\vec{E}.$$

With no damping, $\Gamma = 0$, the displacement is

$$\vec{x} = \frac{-e\vec{E}}{m[\omega_0^2 - \omega^2]}. \tag{7.43}$$

As before, the dipole moment and polarization are $\vec{p} = -e\vec{x}$ and $\vec{P} = N_v\vec{p}$, respectively, which then gives us

$$\vec{P} = \frac{N_v e^2/m}{[\omega_0^2 - \omega^2]}\vec{E}. \tag{7.44}$$

Now consider what happens if a magnetic induction, $\vec{B}$, is applied to the sample through which the light is propagating. The equation of motion becomes:

$$m\ddot{\vec{x}} + m\Gamma\dot{\vec{x}} + m\omega_0^2\vec{x} = -e\vec{E} - e\vec{v} \times \vec{B}$$
$$= -e\vec{E} - e\dot{\vec{x}} \times \vec{B}. \tag{7.45}$$

Again for $\Gamma = 0$ and $\vec{x} \propto e^{-i\omega t}$

$$[-m\omega^2 + m\omega_0^2]\vec{x} = -e\vec{E} - e\dot{\vec{x}} \times \vec{B}$$
$$= -e\vec{E} - i\omega e\vec{x} \times \vec{B} \tag{7.46}$$

and from the fact that $\vec{P} = -N_v e\vec{x}$ we find

$$m[-\omega^2 + \omega_0^2]\vec{P} = N_v e^2\vec{E} + ie\omega\vec{P} \times \vec{B}. \tag{7.47}$$

Writing the above equation in component form and assuming the magnetic field to be applied along the z-axis, $\vec{B} = B\hat{z}$, leads to the following set of component equations:

$$\hat{x}: \quad m[\omega_0^2 - \omega^2]P_x = N_v e^2 E_x + ie\omega BP_y$$
$$\hat{y}: \quad m[\omega_0^2 - \omega^2]P_y = N_v e^2 E_y - ie\omega BP_x \tag{7.48}$$
$$\hat{z}: \quad m[\omega_0^2 - \omega^2]P_z = N_v e^2 E_z.$$

We wish to find a way to put the above equations in the form $\vec{P} = \epsilon_0 \overleftrightarrow{\chi} \vec{E}$; however, what we have thus far is more conveniently written as $\vec{E} = \overleftrightarrow{M}\vec{P}$, where the matrix $\overleftrightarrow{M}$ is given by

$$\overset{\leftrightarrow}{M} = \frac{1}{N_v e^2} \begin{pmatrix} m(\omega_0^2 - \omega^2) & -ie\omega B & 0 \\ ie\omega B & m(\omega_0^2 - \omega^2) & 0 \\ 0 & 0 & m(\omega_0^2 - \omega^2) \end{pmatrix}.$$

To arrange this in the form we are after, we must multiply through by $(\overset{\leftrightarrow}{M})^{-1}$. The inverse of the matrix, which is $\overset{\leftrightarrow}{\chi}$ can be found by standard linear algebra techniques or with the help of Maple, for example; here we simply quote the result:

$$\overset{\leftrightarrow}{\chi} = \begin{pmatrix} \chi_{11} & i\chi_{12} & 0 \\ -i\chi_{12} & \chi_{22} & 0 \\ 0 & 0 & \chi_{33} \end{pmatrix}, \tag{7.49}$$

where the matrix elements are given by

$$\chi_{11} = \left[\frac{N_v e^2}{m\varepsilon_0}\right]\left(\frac{\omega_0^2 - \omega^2}{(\omega_0^2 - \omega^2)^2 + \omega^2\omega_c^2}\right) = \chi_{22} \tag{7.50}$$

$$\chi_{33} = \left[\frac{N_v e^2}{m\varepsilon_0}\right]\left[\frac{1}{\omega_0^2 - \omega^2}\right] \tag{7.51}$$

$$\chi_{12} = \left[\frac{N_v e^2}{m\varepsilon_0}\right]\left(\frac{\omega\omega_c}{(\omega_0^2 - \omega^2)^2 + \omega^2\omega_c^2}\right), \tag{7.52}$$

where $\omega_c = eB/m$ and is called the cyclotron frequency. This is exactly the form of the optically active susceptibility matrix introduced previously, an analogy that indicates to us that the result of applying a magnetic field to this medium is the rotation of the plane of polarization of incident linearly polarized light as it traverses the medium.

We can approach this problem from a slightly different direction as well. Using the component equations we can construct the quantities $P_R \equiv P_x + iP_y$ and $P_L \equiv P_x - iP_y$. The result is

$$P_x + iP_y = \frac{N_V e^2}{m(\omega_0^2 - \omega^2)}(E_x + iE_y) + \frac{e\omega B}{m(\omega_0^2 - \omega^2)}(P_x + iP_y) \tag{7.53}$$

$$P_x - iP_y = \frac{N_V e^2}{m(\omega_0^2 - \omega^2)}(E_x - iE_y) - \frac{e\omega B}{m(\omega_0^2 - \omega^2)}(P_x - iP_y). \tag{7.54}$$

Combining terms and simplifying, we arrive at

$$P_R = \frac{N_V e^2}{m(\omega_0^2 - \omega^2) - e\omega B}E_R \tag{7.55}$$

$$P_L = \frac{N_V e^2}{m(\omega_0^2 - \omega^2) + e\omega B}E_L. \tag{7.56}$$

Thus by writing the polarization in this way we have decoupled the different equations, which is equivalent to saying that we have found the normal-mode basis for the medium. In analogy to the discussion of chapter 6, we can here expect that right-circularly polarized light incident on the medium will propagate undisturbed, as will incident left-circularly polarized light. On the other hand, in the xyz coordinate basis, we see a mixing of x- and y-components as light travels through the medium.

The simple nature of the above result allows us to write the susceptibility in diagonal (normal-mode) form immediately as

$$\begin{pmatrix} P_R \\ P_L \\ P_z \end{pmatrix} = \varepsilon_0 \begin{pmatrix} \chi_R & 0 & 0 \\ 0 & \chi_L & 0 \\ 0 & 0 & \chi_z \end{pmatrix} \begin{pmatrix} E_R \\ E_L \\ E_z \end{pmatrix}, \qquad (7.57)$$

where

$$\chi_R = \left(\frac{N_V e^2}{m\varepsilon_0} \frac{1}{\omega_0^2 - \omega^2 - \omega\omega_c} \right)$$

and

$$\chi_L = \left(\frac{N_V e^2}{m\varepsilon_0} \frac{1}{\omega_0^2 - \omega^2 + \omega\omega_c} \right),$$

where $\omega_c \equiv \frac{eB}{m}$ is the 'cyclotron frequency'.

As with an optically active medium, the amount by which the plane of polarization is rotated while traveling through the medium is determined by the difference in indices of refraction of left- and right-circularly polarized light. Since $n_{R,L} = \sqrt{1 + \chi_{R,L}}$ and because we expect only relatively small changes in the index, we can write $n_{R,L} \simeq 1 + \frac{1}{2}\chi_{R,L}$. With a bit of algebraic manipulation, and making the assumption that the incident field frequency ω is not too near resonance or that the magnetic field is not too strong, the difference in indices is found to be

$$n_L - n_R \simeq -\frac{N_v e^2}{2m\varepsilon_0\omega_0} \frac{\omega_c/2}{(\omega_0 - \omega)^2 - \omega_c^2/4}, \qquad (7.58)$$

which is roughly proportional to the applied magnetic field, B. For $\omega \ll \omega_0$,

$$n_L - n_R \simeq -\frac{N_v e^3}{4m^2\varepsilon_0\omega_0^3} \left(1 + 2\frac{\omega^2}{\omega_0^2}\right) B,$$

which is independent of the frequency to lowest order, with a correction going as the square of the frequency relative to the resonant frequency.

Suppose that a magnetic field is applied to a dielectric in the z-direction and that a linearly polarized beam of light is sent through the dielectric in the same direction as the applied magnetic field. The above analysis tells us that the magnetic field causes the medium to become optically active. The amount of rotation β of the plane of polarization will be proportional to the magnetic induction B, with the proportionality constant V known as the Verdet constant of the material and is given by $\beta = VBd$ where d is the length of the medium through which the light beam travels. In one of the homework problems you are asked to determine V for these conditions in terms of the optical parameters of this section.

To make concrete the size of the effect we are talking about here, at a wavelength of 589 nm, water at a temperature of $T = 20°$ has a Verdet constant of $V = 2.18 \times 10^{-5}\frac{\text{deg}}{\text{G}\cdot\text{mm}}$ and flint glass has a Verdet constant of $V = 5.28 \times 10^{-5}\frac{\text{deg}}{\text{G}\cdot\text{mm}}$. In contrast, a material used to make commercial Faraday rotators, $Cd_{0.55}Mn_{0.45}Te$ has a Verdet constant ($\lambda = 621.6$ nm) of $V = 1.5 \times 10^{-2}\frac{\text{deg}}{\text{G}\cdot\text{mm}}$. In the latter case, immersing a 1 mm long rod of $Cd_{0.55}Mn_{0.45}Te$ in a magnetic field of 3000 Gauss will rotate the polarization of a laser field at 621.6 nm by 45°. Since the sense of rotation does not depend on the direction of propagation through the crystal, any reflected field from an optical element placed after the rotator will be rotated by a further 45°, for a total rotation of 90°. If the initial polarization direction was set with a linear polarizer, any reflected light which returns from the 'experiment' downstream will be polarized orthogonally to the first polarizer, thanks to the Faraday rotator. Such a system is often used to isolate a laser from reflections that could, if re-entrant into the laser cavity, cause unstable operation.

In figure 7.17 another variation for measuring the Faraday effect is shown, here for a gaseous sample in a cell. Paramagnetic molecular species also show a 'resonant Faraday effect', measured by a technique known as magnetic rotation spectroscopy [7, 8]. Laser light incident on a sample cell, and near the resonance frequency for a molecular transition, will have its polarization rotated as shown in the figure. The polarizers are set such that in the absence of the sample, no light is transmitted by the second polarizer ('crossed polarizers'). Near resonance the Faraday effect causes a rotation of the polarization, thus allowing some light to pass the second polarizer.

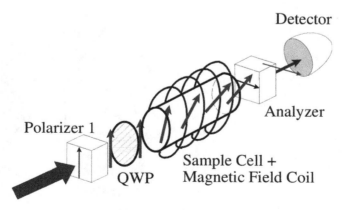

Figure 7.17. Schematic representation of a Faraday effect experiment. The polarization direction of the input laser field is rotated as it passes through the sample (in this case, in a cell) and is detected by measuring the power transmitted by a polarizer nominally crossed with the input polarizer.

7.7 The *k*-vector surface of quartz

Crystalline quartz presents an interesting example of the properties we have studied in this chapter. Crystalline quartz is both optically active and birefringent. The susceptibility tensor is given by equation (7.34) and the equivalent equation (analogous to equation (7.23)) for the *k*-vector surface is

$$
\begin{vmatrix}
\left(\dfrac{n_1\omega}{c}\right)^2 - k_y^2 - k_z^2 & k_xk_y + i\chi_{12}\left[\dfrac{\omega}{c}\right]^2 & k_xk_z \\[2ex]
k_yk_x - i\chi_{12}\left[\dfrac{\omega}{c}\right]^2 & \left(\dfrac{n_2\omega}{c}\right)^2 - k_x^2 - k_z^2 & k_yk_z \\[2ex]
k_zk_x & k_zk_y & \left(\dfrac{n_3\omega}{c}\right)^2 - k_x^2 - k_y^2
\end{vmatrix}.
$$

The *k*-vector surface for quartz is plotted in figure 7.18. The two surfaces no longer refer to orthogonal linear polarizations as they did in the case of a purely birefringent uniaxial crystal, but instead refer to orthogonal elliptical polarizations. The figure shows the types of polarization for various directions of propagation. Along the optic axis, the two surfaces do not touch as they do for an ordinary uniaxial crystal. Their separation depends upon the value of χ_{12} and thus the separation is a measure of the optical rotatory power.

7.8 Off-axis waveplates

We wish to consider the following problem. You have in the laboratory a phase retarder specified to be a quarter-waveplate for light of wavelength $\lambda = 780$ nm.

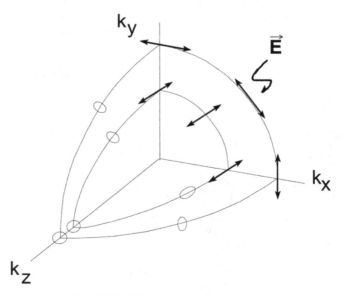

Figure 7.18. The *k*-vector surface of quartz.

The thickness of the plate is d and the indices are n_o and n_e. Assume that all light sources are monochromatic. We have learned that the $\pi/2$ relative phase shift of the plate depends on the thickness and on the difference in indices of refraction for o- and e-rays. Is it possible to rotate the waveplate about an axis lying parallel to the surface of the plate such that the wavelength at which the plate provides a quarter-wave delay will be shifted to a longer or shorter value? That is, can we make out of our waveplate a quarter-waveplate for $\lambda = 800$ nm? For $\lambda = 760$ nm? What happens to the output beam of light if this method is used?

We first consider the waveplate with normally incident light. For quarter-wave retardation we have

$$\Gamma = 2\pi\frac{(n_e - n_o)d}{\lambda} = \frac{\pi}{2}.$$

For a given material and wavelength, this sets the thickness d of the plate. Assuming some typical numbers, we might find a thickness of 20 μm or so, which is much too thin for a practical optical element. Instead, the most common option is to use a thicker plate such that the phase retardation is many multiples of 2π plus an additional phase shift of $\pi/2$ (quarter-waveplate) or π (half-waveplate).

Looking at figure 7.19(b) we see that tilting the waveplate results in a change in the pathlength for the light traversing the crystal and, more importantly, that the optical pathlength for the ordinary and extraordinary rays differ not only due to differing indices of refraction, but also due to their angles of refraction. We can construct the distance through the crystal from figure 7.19:

$$d = L_o \cos\theta_o = L_e \cos\theta_e,$$

where L_e (L_o) is the actual distance traveled by the e-wave (o-wave) through the plate. Therefore the new relative phase retardation becomes

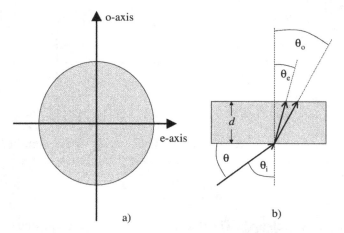

a)   b)

Figure 7.19. For the analysis of the effect of tilting a phase retarder or waveplate, we consider the principal axes to lie in the plane of the plate, as shown in (a). Looked at edge-on, the ordinary and extraordinary waves split apart when the waveplate is tilted, i.e. when the light is not normally incident.

$$\Gamma' = \frac{2\pi}{\lambda_0}(n_e L_e - n_o L_o) = \left(\frac{n_e d}{\cos\theta_e} - \frac{n_o d}{\cos\theta_o}\right)$$

$$= \frac{2\pi d}{\lambda_0}\left(\frac{n_e}{\sqrt{1 - \sin^2\theta_e}} - \frac{n_o}{\sqrt{1 - \sin^2\theta_o}}\right). \tag{7.59}$$

We will also make use of the two versions of Snell's law

$$n_i \sin\theta_i = n_o \sin\theta_o$$
$$n_i \sin\theta_i = n_e(\theta)\sin\theta_e$$

to write (assuming $n_i = 1.00$)

$$\Gamma' = \frac{2\pi d}{\lambda_0}\left(\frac{n_e}{\sqrt{1 - \sin^2\theta_i/n_e^2(\theta)}} - \frac{n_o}{\sqrt{1 - \sin^2\theta_i/n_o^2}}\right).$$

Now we can use equation (7.32) along with figure 7.19 to write θ_o and θ_e in terms of the incidence angle θ_i. We also have $\theta = \frac{\pi}{2} - \theta_i$. Putting this all together gives

$$\Gamma' = \frac{2\pi d}{\lambda_0}\left[\frac{n_e}{\sqrt{1 - \sin^2\theta_i\left(\frac{\sin^2\theta_i}{n_o^2} + \frac{\cos^2\theta_i}{n_e^2}\right)}} - \frac{n_o}{\sqrt{1 - \sin^2\theta_i/n_o^2}}\right].$$

We will assume that any tilt of the waveplate is small. For reasons to be seen below, this is a good assumption if we are going to use this technique in the lab. Thus we can use the binomial expansion to write

$$\Gamma' \simeq \frac{2\pi d}{\lambda_0}\left[n_e\left(1 + \frac{1}{2}\sin^2\theta_i\left(\frac{\sin^2\theta_i}{n_o^2} + \frac{\cos^2\theta_i}{n_e^2}\right)\right) - n_o\left(1 + \frac{1}{2}\frac{\sin^2\theta_i}{n_o^2}\right)\right]$$

$$\simeq \frac{2\pi d}{\lambda_0}(n_e - n_o) + \frac{\pi d}{\lambda_0}\left[n_e\frac{\theta_i^4}{n_o^2} + n_e\frac{\theta_i^2 - \theta_i^4/2}{n_e^2} - \frac{n_o\theta_i^2}{n_o^2}\right] \tag{7.60}$$

$$\simeq \Gamma_0' + \frac{\pi d}{\lambda_0}\left(\frac{1}{n_e} - \frac{1}{n_o}\right)\theta_i^2.$$

As we change our light source to another wavelength, the original $\pi/2$ phase shift of the quarter-waveplate changes. The term Γ_0' reflects this and is given by

$$\Gamma_0' = \frac{2\pi d(n_e - n_o)}{\lambda'},$$

where λ' is the new wavelength. The thickness of the plate d is the one constant and is determined from the initial quarter-wave retardation condition,

$$2\pi m + \frac{\pi}{2} = \frac{2\pi}{\lambda_0}(n_e - n_o)d$$

so that

$$d = \frac{\lambda_0}{n_e - n_o}\left(m + \frac{1}{4}\right).$$

Using this expression in our result for the new phase shift of the tilted waveplate

$$\Gamma_0' = \frac{2\pi(n_e - n_o)}{\lambda'}\left(\frac{\lambda_0}{n_e - n_o}\left(m + \frac{1}{4}\right)\right) = \frac{2\pi\lambda_0}{\lambda'}\left(m + \frac{1}{4}\right).$$

Thus we can conclude that for $\lambda' > \lambda_0$ the phase shift is smaller than it is for the original design wavelength, and for $\lambda' < \lambda_0$ the phase shift is larger.

The correction term in equation (7.60) depends on whether the crystal is uniaxial positive or negative. For $n_e > n_o$ the correction is negative, so only for $\lambda' < \lambda_0$ can we return to $\Gamma = \pi/2$; likewise, for $n_o > n_e$ we can achieve a quarter-wave phase shift for $\lambda' > \lambda_0$. For the sake of concreteness we can assume a waveplate made of quartz, for which $n_e = 1.553$ and $n_o = 1.544$, so we can hope to tilt the waveplate and have it work as a quarter-waveplate for $\lambda = 760$ nm. Substituting our expressions for d and Γ_0' into equation (7.60) we find

$$\Gamma' = \frac{2\pi\lambda_0}{\lambda'}\left(m + \frac{1}{4}\right) + \frac{\pi\lambda_0\left(m + \frac{1}{4}\right)}{\lambda_0(n_e - n_o)}\left(\frac{1}{n_e} - \frac{1}{n_o}\right)\theta_i^2$$

$$2\pi\left(n + \frac{1}{4}\right) = \frac{2\pi\lambda_0}{\lambda'}\left(m + \frac{1}{4}\right) - \frac{2\pi\left(m + \frac{1}{4}\right)}{2n_e n_o}\theta_i^2.$$

(7.61)

In the above, n represents the fact that one could, in principle, induce a phase shift of multiple orders by tilting the waveplate, i.e. $n \neq m$; a quick check of the size of the correction term shows that this will be impossible in practice, so we can set $n = m$. This leaves us with the result for the tilt angle $\theta_i = 0.36$ rad.

More generally, we can return to the expression for Γ_0', equation (7.60) and make a plot of the angle needed to achieve quarter-wave retardation as a function of wavelength, as shown in figure 7.20. We see that, as expected, the quarter-wave retardation can only be achieved for shorter wavelengths as long as the waveplate rotation angle is small. However, at larger rotation angles it would in principle be possible to have quarter-wave retardation for longer wavelengths as well. Note that the tilt angle is measured in radians, so the required tilt would be on the order of 60°.

Now we turn briefly to a calculation of the beam walk-off, i.e. the distance between the e- and o-rays as they emerge from the waveplate. At normal incidence, as we saw earlier in the chapter, there is no lateral displacement between the rays. For our tilted waveplate, the disadvantage to using this method to 'tune' the phase retardation for different wavelengths is that there will be a small walk-off effect. Looking at figure 7.19(b), we can define the walk-off as

$$\Delta x = x_o - x_e = d \tan \theta_o - d \tan \theta_e,$$

where $\theta_o = \arcsin\left(\frac{\sin \theta_i}{n_o}\right)$ and $\theta_e = \arcsin\left(\frac{\sin \theta_i}{n_e(\theta)}\right)$. We can plot this quantity as a function of incidence angle, with the result being shown in figure 7.21. To calculate the walk-off amount we need to assume a thickness of the waveplate; for figure 7.21 we have taken $d = 1$ mm.

In the end we can conclude from this last calculation that the amount of walk-off is minimal, even for fairly substantial tilt angles. Therefore one can change the

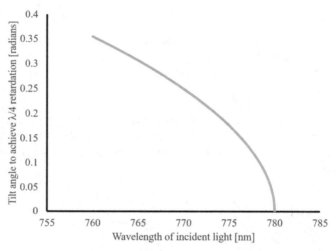

Figure 7.20. A plot of incident light wavelength versus the tilt angle needed to provide a phase retardation of $\pi/2$. We have assumed $n_e = 1.553$ and $n_o = 1.544$.

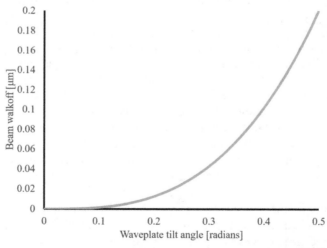

Figure 7.21. A plot of the walk-off displacement for e- and o-rays after traversing a waveplate of thickness $d = 1$ mm as a function of the tilt angle. We have assumed $n_e = 1.553$ and $n_o = 1.544$.

wavelength at which quarter-wave retardation occurs by this simple method. One must keep in mind, however, that other practical considerations will play a role in the desirability of doing so. For example, waveplates coated for normal-incidence anti-reflection will begin to create losses for oblique angles of incidence.

This extended example should serve to combine several concepts we have discussed not only in this chapter, but also in chapter 2. With the end of this chapter we also conclude what might be thought of as the background material for the understanding of the interactions between radiation and matter. The following chapters consist of applications of these ideas and can be seen as the goal toward which we have been working.

7.9 Problems

1. Initially unpolarized light passes in turn through three linear polarizers with transmission axes at $0°$, $30°$, and $60°$, respectively, relative to the horizontal. What is the irradiance of the product light, expressed as a percentage of the unpolarized light irradiance?

2. A beam of monochromatic light is incident normally upon a polarization filter. It is found that when the polarization filter is rotated in its plane, the transmitted intensity changes periodically, going through two maxima and two minima for rotation of $360°$. How can we determine whether the light is partially polarized or elliptically polarized?

3. A uniaxial crystal of indices n_0 and n_e is cut so that the optic axis is perpendicular to the surface. Show that for light from the outside at an angle of incidence Θ_i, the angle of refraction of the e-ray is

$$\tan \Theta_e = \frac{n_e}{n_0} \frac{\sin \Theta_i}{\sqrt{n_e^2 - \sin^2 \Theta_i}}.$$

4. A calcite retarder is positioned between two parallel linear polarizers. Determine the minimum thickness of the plate and its necessary orientation if no light ($\lambda = 589.3$ nm) is to emerge from the arrangement when the incident beam is unpolarized.

5. A linearly polarized electromagnetic wave at $\lambda = 6328$ Å is incident normally at $x = 0$ on the yz face of a quartz crystal so that it propagates along the x-axis (so-called 'x-cut'). Suppose the wave is polarized initially so that it has equal components along y and z.
   (a) What is the state of polarization at the plane defined by the x-coordinate where $(k_z - k_y)x = \pi/2$?
   (b) A plate which satisfies the condition in (a) is known as a quarter-waveplate because the difference in the phase shift for the two orthogonal polarization states is a quarter of 2π. Find the thickness of a quartz quarter-waveplate at $\lambda = 6328$ Å.
   (c) What is the state of polarization at the plane defined by the x-coordinate, where $\frac{2\pi}{\lambda}(n_e - n_0)x = \pi$.

6. Linearly polarized monochromatic light is incident normally upon a quarter-waveplate, the plane of polarization being at an angle ψ to the y-axis. Assume $n_y < n_x$. Determine the state of polarization of the transmitted light and the direction of rotation of the optical E-vector for the following values of ψ:
    (a) $\psi = \pi/4$.
    (b) $\psi = \pi/2$.
    (c) $\psi = 3\pi/4$.

7. A beam of white, linearly polarized light is incident normally upon a plate of quartz 0.865 mm thick cut so that the surface is parallel to the optic axis. The plane of polarization is at 45° to the axes of the plate. The principal indices of refraction of quartz for sodium light are $n_0 = 1.544$ and $n_e = 1.553$.
    (a) Which wavelengths between 6000 Å and 7000 Å emerge from the plate linearly polarized?
    (b) Which wavelengths emerge circularly polarized?
    (c) Suppose that the beam emerging from the plate passes through an analyzer whose transmission axis is perpendicular to the plane of vibration of the incident light. Which wavelengths are missing in the transmitted beam?

8. A monochromatic beam of light with wavelength 532 nm is incident on a piece of quartz at an angle of 45°. The optic axis of the quartz plate (thickness $d = 1.5$ mm) is parallel to the surface. Find the lateral walk-off between the ordinary and extraordinary rays upon exiting the crystal.

9. A Fresnel prism is made of quartz as shown in figure 7.22. The angles of the component prisms are $70° - 20° - 90°$ degrees. Determine the angle between the emerging right- and left-circularly polarized rays for sodium light. For sodium wavelengths, the indices for right-handed quartz are $n_R = 1.54420$ and $n_L = 1.54427$ and for the left-handed quartz they are just reversed, $n_R = 1.54427$ and $n_L = 1.54420$.

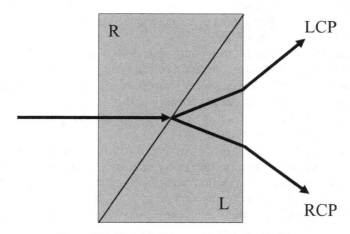

Figure 7.22. Fresnel prism considered in problem 9.

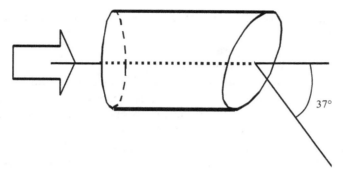

Figure 7.23. Quartz piece described in problem 10.

10. A narrow beam of linearly polarized monochromatic light is incident normally upon a piece of quartz cut as shown in figure 7.23. The optic axis of the quartz is parallel to the incident beam and the normal to the exit face is at 37° to the axis. Determine the angular separation of the two beams emerging from the quartz ($n_R = 1.54420$, $n_L = 1.54427$).

11. The indices of refraction for calcite and quartz for the sodium yellow line (589 nm) are calcite $n_0 = 1.658$ and $n_e = 1.486$; quartz: $n_0 = 1.544$ and $n_e = 1.553$. Calculate the thickness of a quarter-waveplate made of these materials.

12. A half-waveplate has a phase retardation of π. Assume that a plate is oriented so that the azimuth angle (i.e. the angle between the x-axis and the slow axis of the plate) is ψ.

    (a) Find the polarization state of the transmitted beam, assuming that the incident beam is linearly polarized in the y-direction and propagating along the z-axis.

    (b) Show that a half-waveplate will convert right-hand circularly polarized light into left-hand circularly polarized light, and vice versa, regardless of the azimuth angle of the plate.

    (c) Lithium tantalate (LiTaO$_3$) is a uniaxial crystal with indices of $n_0 = 2.1391$ and $n_e = 2.1432$ at $\lambda = 1 \,\mu$m. Find the half-waveplate thickness at this wavelength, assuming the plate is cut in such a way that the surfaces are perpendicular to the x-axis of the principal coordinate (i.e. x-cut).

13. A plane parallel plate is cut from a calcite crystal to be 3 cm thick with the optic axis parallel to the faces of the plate. Light that is unpolarized at a wavelength of 598 nm is incident on the face of the crystal at an angle of 45°. Compute the separation of the ordinary and the extraordinary rays as they emerge from the plate on the opposite side.

14. A medium consists of bound electrons with spring constants as shown in figure 7.3. Find an expression for the indices of refraction n_x, n_y, and n_z in terms of the spring constants. Assume that the damping is zero.

15. Show that $\vec{k} \cdot \vec{E}$ is not necessarily zero.

16. Find the direction of power flow, given by $\vec{S}$, of wave propagating in a uniaxial crystal.
    (a) For an ordinary ray.
    (b) For an extraordinary ray.

17. Since a sheet of Polaroid is not an ideal polarizer, not all the energy of the E-vibrations parallel to the transmission axis are transmitted, nor are all E-vibrations perpendicular to the TA absorbed. Suppose a field fraction α is transmitted in the first case, and a fraction β is transmitted in the second.
    (a) Extend Malus' law by calculating the irradiance transmitted by a pair of such polarizers with angle Θ between their transmission axes. Assume light initially polarized at $45°$ of irradiance I_0. Show that Malus' law follows in the ideal case.
    (b) Let $\alpha = 0.98$ and $\beta = 0.05$ for a sheet of Polaroid. Compare the irradiance with that of an ideal polarizer when unpolarized light is passed through two such sheets having a relative angle between TAs of $0°$, $30°$, $45°$, and $90°$.

18. Describe what happens to unpolarized light incident on birefringent material when the OA is oriented as shown in figure 7.24. You will want to comment on the following considerations: Single or double refracted rays? Any phase retardation? Any polarization of refracted rays?

19. In each of the following cases, deduce the nature of the light that is consistent with the analysis performed. Assume a 100% efficient polarizer.
    (a) When a polarizer is rotated in the path of the light, there is no intensity variation. With QWP in front of (coming first) the rotating polarizer, one finds a variation in intensity but no angular position of the polarizer that gives zero intensity.
    (b) When a polarizer is rotated in the path of the light, there is some intensity variation but no position of the polarizer giving zero

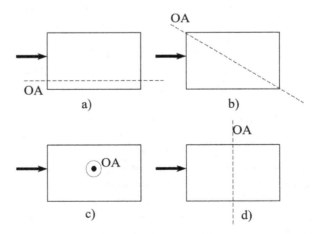

Figure 7.24. Illustration of various relative orientations of the optic axis of a crystal with respect to an incident wavevector.

intensity. The polarizer is set to give maximum intensity. A QWP is allowed to intercept the beam first with its OA parallel to the transmission axis of the polarizer. Rotation of the polarizer now can produce zero intensity.

20. What thickness of quartz is required to give an optical rotation of 10° for light of 396.8 nm? What is the specific rotation of quartz for this wavelength? The refractive indices for quartz at this wavelength, for left- and right-circularly polarized light, are $n_L = 1.55821$ and $n_R = 1.55810$, respectively.

21. Compute the angle through which linearly polarized light is rotated in traveling through a 3 mm thick quartz plate cut with faces perpendicular to the optic axis. The incident light has a wavelength of 397 nm, which corresponds to indices of 1.55821 and 1.55810 for left- and right-circular polarization, respectively.

22. Show that when a uniaxial crystal is made into a prism with its apex parallel to the optic axis, both ordinary and extraordinary rays obey the ordinary law of refraction with indices n_0 and n_e, respectively.

23. Show that in a uniaxial crystal the largest angle $\Delta\theta = \theta_e - \theta_0$ between the ray direction and the direction of the wave normal $\hat{k}$ occurs when $\hat{k}$ makes an angle with the optic axis of

$$\theta_0 = \tan^{-1}\left[\frac{n_e}{n_0}\right]$$

and that

$$(\tan \Delta\theta)_{\max} = \frac{n_0^2 - n_e^2}{2n_e n_0}.$$

24. Consider the calcite prisms in figure 7.25 used with unpolarized light at normal incidence as shown. Use $n_0 = 1.668$ and $n_e = 1.491$ and calculate the range of α for which only one polarized component is critically reflected at

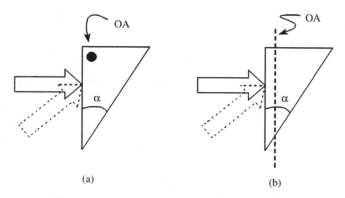

(a)   (b)

Figure 7.25. Two prisms as described in problem 24.

the second surface. Which one is it? Are prisms (a) and (b) equivalent in this regard? Are they equivalent for the dashed rays?

25. The phenomenon of double refraction in an anisotropic crystal may be used to produce polarized light. Consider a light beam incident on a plane boundary from inside of a calcite crystal ($n_0 = 1.668$, $n_e = 1.491$). Suppose that the c (symmetry) axis of the crystal is normal to the plane of incidence.

    (a) Find the range of the internal angle of incidence such that an ordinary wave is totally reflected and the transmitted wave is thus completely polarized.

    (b) Use the principle described in (a) to design a calcite Glan prism shown in figure 7.26 and find the range of the apex angle α.

26. Consider an air-spaced polarizing prism made of calcite in figure 7.26. It works by totally reflecting the ordinary ray at the air gap, while passing the extraordinary ray. There is only a finite range of the angle of incidence for which both of these effects occur. The prism is designed so that if the o-ray is incident below the normal at an angle greater than Θ_0, it ceases to be totally reflected, whereas if the e-ray is incident above the normal at an angle greater than Θ_0, it ceases to be transmitted and is also totally reflected. Use these two criteria to determine the maximum angular spread $2\Theta_0$ and prism angle α.

27. When the air gap in the prism of problem 26 is replaced by oil ($n = 1.494$), the angular aperture $2\Theta_0$ can be considerably increased provided that the angle α in figure 7.26 is made much larger. This gives what is called the Glan–Thompson polarizer. With $\alpha = 76.4°$, find the maximum value of the angle of incidence for the o-ray for which it will still be totally reflected.

28. Explain, in quantitative terms, how the Babinet compensator in figure 7.15 works.

29. Find the specific rotatory power in terms of B_0 induced by a magnetic induction $\vec{B} = B_0\hat{z}$ for a medium consisting of N_v bound electrons/vol. and for plane polarized light of wavelength λ propagating parallel to the magnetic induction.

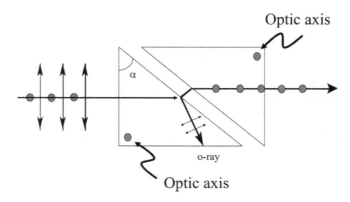

Figure 7.26. Two calcite prisms with either an air gap or an oil-filled gap.

30. Calculate the Verdet coefficient (section 7.6) in terms of the distance through the medium (d), the magnetic induction (B), and a proportionality constant, V.
31. Show that the relative phase delay between right- and left-circularly polarized light components in an optically active or Faraday rotating medium is equivalent to a rotation in the plane of polarization for linearly polarized light.

References

[1] Wikipedia: Rasmus Bartholin https://en.wikipedia.org/wiki/Rasmus_Bartholin (Accessed: 11 Jan. 2021)
[2] Wikipedia: Jean-Baptiste Biot https://en.wikipedia.org/wiki/Jean-Baptiste_Biot (Accessed: 11 Jan. 2021)
[3] Fowles G 1975 *Introduction to Modern Optics* 2nd edn (Mineola, NY: Dover)
[4] Yariv A 1989 *Quantum Electronics* 3rd edn (New York: Wiley)
[5] Born M and Wolf E 1999 *Principles of Optics* 7th edn (New York: Macmillan)
[6] Mason S F 1968 Optical activity and molecular dissymmetry *Contemp. Phys.* **9** 239–56
[7] Brecha R J, Pedrotti L M and Krause D 1997 Magnetic rotation spectroscopy of molecular oxygen with a diode laser *J. Opt. Soc. Am.* B **14** 1921–30
[8] Brecha R J and Pedrotti L M 1999 Analysis of imperfect polarizer effects in magnetic rotation spectroscopy *Opt. Exp.* **5** 101–13

IOP Publishing

Optical Radiation and Matter

Robert J Brecha and J Michael O'Hare

Chapter 8

Electro-optic effects

In the preceding chapter the dielectric medium was characterized by several alternative tensor quantities: ε_{ij}, χ_{ij}, and B_{ij}. These quantities varied with frequency, particularly in the neighborhood of a lattice or electronic resonance. The treatments also assumed that these quantities were independent of the fields in the dielectric medium; relaxing that assumption will be the goal of this chapter. Effectively, we wish to calculate changes to the indices of refraction for a material in the presence of applied, low-frequency fields. Since the deviations from field-free behavior are small, the effects can be represented by expanding the dielectric parameters in a power series of the electric field, E_i, the polarization, P_i, or the strain, S_{ij}.

8.1 Historical introduction

In section 7.6 we investigated how a magnetic field applied to a material can change the characteristics of the interaction of light with that medium. Faraday had discovered this effect in the 1840s and was convinced, without any mathematical theory but essentially due to reasons of symmetry, that there must also be similar electric-field-induced effects. It was not until 1875 that the Scottish physicist John Kerr (1824–1907) [1] found that some normally isotropic dielectrics became birefringent when an electric field was applied, the first example of an electro-optic effect. The applied electric field can be constant or at a low (compared to optical) frequency. Relating this back to chapter 7, we immediately see that this means it should be possible to create, for example, waveplates of variable retardation, with that retardation depending on an externally controllable parameter, the electric field. The effect found by Kerr was proportional to the square of the applied electric-field strength. F R Pockels (1865–1913), a German physicist, then discovered that some birefringent dielectric materials showed an even larger electro-optic effect, with the quiescent indices of refraction being changed by an amount proportional to the electric field rather than the square of the field. It is this latter effect that will be the main focus of our attention in the present chapter.

Another extremely important electro-optic effect, but one that goes beyond the scope of this book, is that of liquid crystals, in which even relatively small electric fields applied to these materials can cause a 90° polarization rotation for light propagating through the device. Coupled with polarizers and a reflecting mirror, liquid crystals offer highly controllable light intensity modulators. Liquid crystals form the heart of all flat-panel displays, ranging from pocket calculators and watch faces to notebook computer screens.

8.2 Optical indicatrix revisited

In chapter 7 we introduced the notion of the optical indicatrix. Here it will be convenient to simplify the notation used in that chapter. We begin by assuming that we have found the principal axes for the crystal; thus we can write the ellipsoid equation as

$$B_{11}x^2 + B_{22}y^2 + B_{33}z^2 = 1, \tag{8.1}$$

where we have simply defined the coefficients B as

$$B_{ii} = \frac{1}{\varepsilon_{ii}} = \frac{1}{n_i^2}. \tag{8.2}$$

If we return to equation (7.30) we can start with the constant energy density, U, and rewrite the ellipsoid equation in a more useful form. We have already seen the relation between $\vec{D}$ and $\vec{E}$, i.e. $\vec{D} = \varepsilon_0 \overleftrightarrow{\varepsilon} \vec{E}$, where $\overleftrightarrow{\varepsilon}$ is a tensor quantity. A shorthand notation for this relation, essentially written out in terms of the components of the vectors, is given by

$$D_i = \varepsilon_0 \sum_j \varepsilon_{ij} E_j. \tag{8.3}$$

A standard further shortening of this notation is obtained by noting that the sum on the right-hand side above is taken over the index that is repeated (there are two js). The summation sign itself is then eliminated, with the summation over repeated indices assumed from this point on.

Next, we rewrite the relations for $\vec{D}$ and $\vec{E}$ formally as

$$\vec{E} = \frac{1}{\varepsilon_0}\left(\overleftrightarrow{\varepsilon}\right)^{-1}\vec{D}, \tag{8.4}$$

which becomes, in our new notation

$$E_i = \frac{1}{\varepsilon_0}\left(\overleftrightarrow{\varepsilon}\right)^{-1}_{ij} D_j = \frac{1}{\varepsilon_0}B_{ij}D_j. \tag{8.5}$$

Here we have defined $\overleftrightarrow{B} = (\overleftrightarrow{\varepsilon})^{-1}$; we will simply assume here that these tensors are well-behaved mathematically such that the inverse exists.

In a general coordinate frame the ellipsoid equation can now be expressed as

$$U = \frac{1}{2}\vec{D} \cdot \vec{E} \tag{8.6}$$

or

$$\equiv B_{ij}q_iq_j = 1, \tag{8.7}$$

where, again, $q_i \equiv \frac{D_i}{\sqrt{2U\varepsilon_0}}$.

They can be written out explicitly as

$$B_{11}x^2 + B_{22}y^2 + B_{33}z^2 + 2B_{12}xy + 2B_{13}xz + 2B_{23}yz = 1. \tag{8.8}$$

Since B_{ij} is the relative dielectric impermeability tensor which is the inverse of the dielectric tensor, we can also write

$$B_{ij}\varepsilon_{jk} = \delta_{ik}. \tag{8.9}$$

Note that only in the principal axis system are the tensor elements themselves reciprocals, i.e.

$$B_{ii} = \frac{1}{\varepsilon_{ii}}.$$

For a lossless, optically inactive medium, the B_{ij} tensor is symmetric, just as ε_{ij} is, so that we may write the ellipsoid equation in the following form:

$$B_1x^2 + B_2y^2 + B_3z^2 + 2B_4yz + 2B_5xz + 2B_6xy = 1, \tag{8.10}$$

where we have use the so-called *contracted indices* notation

$$(11) \rightarrow 1 \quad ; \quad (22) \rightarrow 2 \quad ; \quad (33) \rightarrow 3$$

$$(23) \rightarrow 4 \quad ; \quad (13) \rightarrow 5 \quad ; \quad (12) \rightarrow 6.$$

In this chapter we will make use of this form for the index ellipsoid. More specifically, our focus will be on the changes induced in the ellipsoid as a consequence of a static or low-frequency electric field being applied to a birefringent material.

8.3 Electro-optic effects

The field-dependent optical effect that we will investigate in this chapter is the electro-optic (EO) effect. The EO effect refers to the change in optical dielectric properties induced by an electric field whose frequency is well below that of the lattice resonances. The EO effect is usually defined in terms of the B_{ij} matrix which describes the optical indicatrix. This is done by making a power series expansion of B_{ij} in terms of the electric field or the polarization:

$$\begin{aligned} B_{ij}(\vec{E}) - B_{ij}(0) \equiv \Delta B_{ij} &= r_{ijk}E_k + s_{ijkl}E_kE_l + \dots \\ &= f_{ijk}P_k + g_{ijkl}P_kP_l. \end{aligned} \tag{8.11}$$

In equation (8.11) we have employed the convention of repeated indices meaning summation, for example, the term $r_{ijk}E_k$ actually stands for

$$r_{ijk}E_k \rightarrow \sum_k r_{ijk}E_k.$$

Equation (8.11) also shows the B_{ij} written in terms of the polarization vector. The EO coefficients defined in terms of polarization and E-field are simply related in the principal axis system by [2]

$$r_{ijk} = \varepsilon_0(\varepsilon_{kk} - 1)f_{ijk} \qquad (8.12)$$

$$s_{ijkl} = \varepsilon_0(\varepsilon_{kk} - 1)(\varepsilon_{ll} - 1)g_{ijkl}. \qquad (8.13)$$

In the power series expansion of equation (8.11) the coefficients r_{ijk} and f_{ijk} are associated with the Pockels, or linear electro-optic effect, while the Kerr, or quadratic electro-optic effect is described by the s_{ijkl} and g_{ijkl} coefficients.

It can be shown [2] that if (i) the medium is lossless and (ii) the frequencies of the E-fields represented by indices k and l are much less than the optical fields represented by i and j, then the i and j indices can be permuted as can the k and l indices. The permutation symmetry reduces the number of elements of r_{ijk} and f_{ijk} from 27 to 18, and s_{ijkl} and g_{ijkl} from 81 to 36. We will once again use the contracted indices notation introduced above, now with $(23) \equiv (32) \rightarrow (4)$, $(13) \equiv (31) \rightarrow (5)$, and $(12) \equiv (21) \rightarrow (6)$. These contracted matrix elements do not follow the usual tensor transformation and multiplication properties. The full third-rank tensor can be arranged in the following matrix:

$$\begin{pmatrix} r_{111} & r_{112} & r_{113} \\ r_{221} & r_{222} & r_{223} \\ r_{331} & r_{332} & r_{333} \\ r_{231} & r_{232} & r_{233} \\ r_{321} & r_{322} & r_{323} \\ r_{131} & r_{132} & r_{133} \\ r_{311} & r_{312} & r_{313} \\ r_{121} & r_{122} & r_{123} \\ r_{211} & r_{212} & r_{213} \end{pmatrix},$$

which in the contracted notation is

$$\begin{pmatrix} r_{11} & r_{12} & r_{13} \\ r_{21} & r_{22} & r_{23} \\ r_{31} & r_{32} & r_{33} \\ r_{41} & r_{42} & r_{43} \\ r_{51} & r_{52} & r_{53} \\ r_{61} & r_{62} & r_{63} \end{pmatrix}.$$

It is this form for the electro-optic coefficients that we will use in all that follows. Although we have the mathematically sophisticated concept of the manipulation of third-rank tensors, for our purposes the use of this tool will be very straightforward.

As will be illustrated presently, the mathematical manipulations involved will be simply the multiplication of a 6 × 3 matrix (electro-optic coefficients) by a 3 × 1 vector (input electric-field components) to arrive at a 6 × 1 vector, the components of which represents field-induced changes to the impermeability tensor.

8.3.1 Crystal symmetry effects on the tensor r_{ijk}

The symmetry properties of a particular crystal will determine the properties of the electro-optic tensor, r_{ijk}. Because of these symmetry properties, it turns out that not all of the elements will be independent and many will actually be zero. We will examine some of the symmetry properties just to see how they can affect the elements of the electro-optic tensor. Let us designate the effect of any symmetry operation on r_{ijk} and r'_{ijk} and let us first examine the symmetry operation of inversion.

(i) Inversion (I): The symmetry property of inversion consists of making the transformation $x \rightarrow -x$, $y \rightarrow -y$, and $z \rightarrow -z$. If the crystal does not appear different after this coordinate transformation, then the crystal is said to possess inversion symmetry. The effect of the inversion operation on a third-rank tensor is

$$r_{ijk}(I) = r'_{ijk} = -r_{ijk}.$$

Now if the crystal is invariant under inversion, then the inversion operation on the coefficient r_{ijk}, designated as $r_{ijk}(I)$, just gives back the coefficient unchanged, that is

$$r_{ijk}(I) = r_{ijk}(E),$$

where E is the identity operation. But $r_{ijk}(I) = -r_{ijk}$, therefore the only solution is $r_{ijk} = 0$. Therefore, all third-rank tensors must vanish in the 11 crystal classes that possess a *center of inversion*. Thus we arrive at the important conclusion that only crystals that lack an inversion center exhibit a linear EO (Pockels) effect.

(ii) Further reductions result from application of other symmetry elements. Consider for example the group $\bar{4}2m$ that describes KH_2PO_4 (KDP); $\bar{4}2m$ symmetry is also denoted as D_{2d}. Figure 8.1 below illustrates the symmetry properties of $\bar{4}2m$ symmetry. The large circle, representing the projection of the crystal onto a plane, lies in the plane of the page. The filled and open circles can be thought of as the end points of symmetry lines that run through the crystal, indicate elements that lie above and below the plane of the page, respectively, whereas lines of mirror symmetry are represented by the horizontal and vertical straight lines. The solid elliptical shapes lie in the plane of the page and represent axes of twofold symmetry, i.e. axes about which a rotation of 180° (360°/2) about that axis reproduces that same picture. Another symmetry property for this crystal group is that of a fourfold rotation followed by an inversion through the center point of the crystal (this is the meaning of the $\bar{4}$). An introduction to crystal symmetry and, in particular, to the implications for light propagation, is found in [3]. We mention these points here only to help illustrate that the properties of the electro-optic coefficients are fundamentally derived from the inherent symmetry properties of the crystals themselves; since there are a limited number of crystal symmetry classes that exist in nature (32) the number of possible electro-optic tensors is similarly limited in scope.

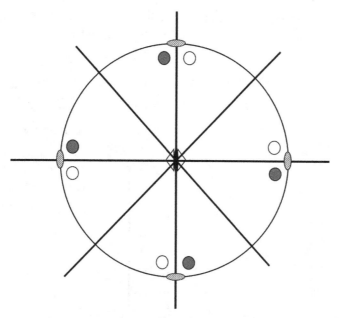

Figure 8.1. Symmetry properties for $\bar{4}2m$ symmetry class crystals. The symmetry operation symbols are explained in the text. Crystallographic axes are shown here by the pair of 45° lines.

Let us see how the full third-rank tensor transforms under some of these operations. From above, the full third-rank tensor is

$$\begin{pmatrix} r_{111} & r_{112} & r_{113} \\ r_{221} & r_{222} & r_{223} \\ r_{331} & r_{332} & r_{333} \\ r_{231} & r_{232} & r_{233} \\ r_{321} & r_{322} & r_{323} \\ r_{131} & r_{132} & r_{133} \\ r_{311} & r_{312} & r_{313} \\ r_{121} & r_{122} & r_{123} \\ r_{211} & r_{212} & r_{213} \end{pmatrix}.$$

The twofold axis about x, a rotation about which we denote as $C_2(x)$, transforms x into x', y into y', and z into z', where

$$x' \to x \quad y' \to -y \quad z' \to -z.$$

Thus $C_2(x)$ requires that all elements above with indices 2 and/or 3 an odd number of times vanish (Why?). For example,

$$r_{112}(C_2) = -r_{112} = r_{112}(E) = 0,$$

where E is the identity operation. The twofold axes $C_2(y)$, $C_2(z)$ have similar effects and together leave only those elements containing one index each (e.g. r_{123}).

8-6

The mirror reflection symmetry plane, which is perpendicular to x_1 and at 45° to x and y, transforms x into x', y into y', and z into z', where

$$x' \rightarrow -y \quad y' \rightarrow -x \quad z' \rightarrow z,$$

so that the indices 1 and 2 can be permuted; for example,

$$r_{123}(\sigma) = r_{213} = r_{123}(E),$$

where σ stands for the mirror reflection plane shown in figure 8.1. Thus the r_{ijk} matrix becomes

$$
\begin{pmatrix}
0 & 0 & 0 \\
0 & 0 & 0 \\
0 & 0 & 0 \\
r_{231} & 0 & 0 \\
r_{321} & 0 & 0 \\
0 & r_{231} & 0 \\
0 & r_{321} & 0 \\
0 & 0 & r_{123} \\
0 & 0 & r_{213}
\end{pmatrix}
$$

and in the contracted notation, taking into account the i and j permutation, the EO tensor for D_{2d} or $\bar{4}2m$ symmetry is

$$
\begin{pmatrix}
0 & 0 & 0 \\
0 & 0 & 0 \\
0 & 0 & 0 \\
r_{41} & 0 & 0 \\
0 & r_{41} & 0 \\
0 & 0 & r_{63}
\end{pmatrix}.
$$

As another example, the cubic $\bar{4}3m$ class, which describes GaAs, has a threefold axis about [111] in addition to the C_2 and σ elements of $\bar{4}2m$. Since C_3 transforms x into x', y into y', and z into z' according to

$$x' \rightarrow y \quad y' \rightarrow z \quad z' \rightarrow x,$$

all three indices can be permuted in sequence (e.g. $r_{123} = r_{231}$) and the only non-vanishing elements are $r_{231} = r_{132} = r_{123}$ or in contracted notation $r_{41} = r_{52} = r_{63}$. Thus for $\bar{4}3m$ symmetry, the r_{ijk} matrix in contracted notation is

$$
\begin{pmatrix}
0 & 0 & 0 \\
0 & 0 & 0 \\
0 & 0 & 0 \\
r_{41} & 0 & 0 \\
0 & r_{41} & 0 \\
0 & 0 & r_{41}
\end{pmatrix}.
$$

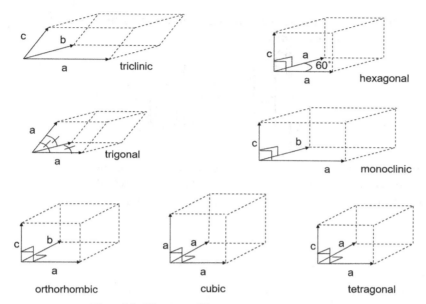

Figure 8.2. The seven different types of crystal lattice.

In figure 8.2 we show the seven families of crystal 'building blocks' in terms of the angles between unit cell vectors and angles. Again, our purpose in this section has been to provide the roughest of introductions to the study of symmetry properties of crystals as a way of justifying the more operational point of view we will be taking with respect to the electro-optic effect.

8.3.2 Deformation of the optical indicatrix

From equation (8.11) we see that if an electric field is applied to a crystal, the optical indicatrix will be changed from the zero-field case, $B_{ij}(0)$, because the B_{ij} coefficients are altered. Thus we say that the optical indicatrix is deformed. From the first section of this chapter, recall that the optical indicatrix is determined, in contracted notation, by the equation,

$$B_1 x^2 + B_2 y^2 + B_3 z^2 + 2B_4 yz + 2B_5 xz + 2B_6 xy = 1. \tag{8.14}$$

In the principle axis system with $E = 0$

$$B_4 = B_5 = B_6 = 0$$

and

$$B_1 = \frac{1}{n_1^2} \quad B_2 = \frac{1}{n_2^2} \quad B_3 = \frac{1}{n_3^2}. \tag{8.15}$$

When $E \neq 0$, the indicatrix is changed according to

$$\Delta B_\alpha = \sum_{j=1}^{3} r_{\alpha j} E_j,$$

where α ranges from 1 to 6 in the contracted notation format. The above equation can be expressed in matrix form:

$$\begin{pmatrix} \Delta B_1 \\ \Delta B_2 \\ \Delta B_3 \\ \Delta B_4 \\ \Delta B_5 \\ \Delta B_6 \end{pmatrix} = \begin{pmatrix} r_{11} & r_{12} & r_{13} \\ r_{21} & r_{22} & r_{23} \\ r_{31} & r_{32} & r_{33} \\ r_{41} & r_{42} & r_{43} \\ r_{51} & r_{52} & r_{53} \\ r_{61} & r_{62} & r_{63} \end{pmatrix} \begin{pmatrix} E_1 \\ E_2 \\ E_3 \end{pmatrix}. \tag{8.16}$$

In general, the principal axes of the new ellipsoid do not coincide with (x, y, z), the principal axes of the system in the zero applied field case. To find the new principal axes we must find the eigenvalues of the B-matrix, which is written in contracted notation:

$$\begin{pmatrix} B_1 & B_6 & B_5 \\ B_6 & B_2 & B_4 \\ B_5 & B_4 & B_3 \end{pmatrix}.$$

Triclinic—1

$$\begin{pmatrix} r_{11} & r_{12} & r_{13} \\ r_{21} & r_{22} & r_{23} \\ r_{31} & r_{32} & r_{33} \\ r_{41} & r_{42} & r_{43} \\ r_{51} & r_{52} & r_{53} \\ r_{61} & r_{62} & r_{63} \end{pmatrix}$$

Monoclinic—2

$$\begin{pmatrix} 0 & r_{21} & 0 \\ 0 & r_{22} & 0 \\ 0 & r_{23} & 0 \\ r_{41} & 0 & r_{43} \\ 0 & r_{52} & 0 \\ r_{61} & 0 & r_{63} \end{pmatrix}$$

Orthorhombic—222

$$\begin{pmatrix} 0 & 0 & 0 \\ 0 & 0 & 0 \\ 0 & 0 & 0 \\ r_{41} & 0 & 0 \\ 0 & r_{52} & 0 \\ 0 & 0 & r_{63} \end{pmatrix}$$

Orthorhombic—mm2

$$\begin{pmatrix} 0 & 0 & r_{13} \\ 0 & 0 & r_{23} \\ 0 & 0 & r_{33} \\ 0 & r_{42} & 0 \\ r_{51} & 0 & 0 \\ 0 & 0 & 0 \end{pmatrix}$$

Trigonal—3

$$\begin{pmatrix} r_{11} & -r_{22} & r_{13} \\ -r_{11} & r_{22} & r_{13} \\ 0 & 0 & r_{33} \\ r_{41} & r_{51} & 0 \\ r_{51} & -r_{41} & 0 \\ -r_{22} & -r_{11} & 0 \end{pmatrix}$$

Tr—32

$$\begin{pmatrix} r_{11} & 0 & 0 \\ -r_{11} & 0 & 0 \\ 0 & 0 & 0 \\ r_{41} & 0 & 0 \\ 0 & -r_{41} & 0 \\ 0 & -r_{11} & 0 \end{pmatrix}$$

Trigonal—3m

$$\begin{pmatrix} 0 & -r_{22} & r_{13} \\ 0 & r_{22} & r_{13} \\ 0 & 0 & r_{33} \\ 0 & r_{51} & 0 \\ r_{51} & 0 & 0 \\ -r_{22} & 0 & 0 \end{pmatrix}$$

Hexagonal—6

$$\begin{pmatrix} 0 & 0 & r_{13} \\ 0 & 0 & r_{13} \\ 0 & 0 & r_{33} \\ r_{41} & r_{51} & 0 \\ r_{51} & -r_{41} & 0 \\ 0 & 0 & 0 \end{pmatrix}$$

Hexagonal—$\bar{6}$

$$\begin{pmatrix} r_{11} & -r_{22} & 0 \\ -r_{11} & r_{22} & 0 \\ 0 & 0 & 0 \\ 0 & 0 & 0 \\ 0 & 0 & 0 \\ -r_{22} & -r_{11} & 0 \end{pmatrix}$$

Hexagonal—6mm

$$\begin{pmatrix} 0 & 0 & r_{13} \\ 0 & 0 & r_{13} \\ 0 & 0 & r_{33} \\ 0 & r_{51} & 0 \\ r_{51} & 0 & 0 \\ 0 & 0 & 0 \end{pmatrix}$$

Hexagonal—$\bar{6}$m2

$$\begin{pmatrix} 0 & -r_{22} & 0 \\ 0 & r_{22} & 0 \\ 0 & 0 & 0 \\ 0 & 0 & 0 \\ 0 & 0 & 0 \\ -r_{22} & 0 & 0 \end{pmatrix}$$

Tetragonal—$\bar{4}$2m

$$\begin{pmatrix} 0 & 0 & 0 \\ 0 & 0 & 0 \\ 0 & 0 & 0 \\ r_{41} & 0 & 0 \\ 0 & r_{41} & 0 \\ 0 & 0 & r_{63} \end{pmatrix}$$

Cubic—432

$$\begin{pmatrix} 0 & 0 & 0 \\ 0 & 0 & 0 \\ 0 & 0 & 0 \\ 0 & 0 & 0 \\ 0 & 0 & 0 \\ 0 & 0 & 0 \end{pmatrix}$$

Hexagonal—622

$$\begin{pmatrix} 0 & 0 & 0 \\ 0 & 0 & 0 \\ 0 & 0 & 0 \\ r_{41} & 0 & 0 \\ 0 & -r_{41} & 0 \\ 0 & 0 & 0 \end{pmatrix}$$

Tetragonal—4

$$\begin{pmatrix} 0 & 0 & r_{13} \\ 0 & 0 & r_{13} \\ 0 & 0 & r_{33} \\ r_{41} & r_{51} & 0 \\ r_{51} & -r_{41} & 0 \\ 0 & 0 & 0 \end{pmatrix}$$

Tetragonal—$\bar{4}$

$$\begin{pmatrix} 0 & 0 & r_{13} \\ 0 & 0 & -r_{13} \\ 0 & 0 & 0 \\ r_{41} & -r_{51} & 0 \\ r_{51} & r_{41} & 0 \\ 0 & 0 & r_{63} \end{pmatrix}$$

Tetragonal—422

$$\begin{pmatrix} 0 & 0 & 0 \\ 0 & 0 & 0 \\ 0 & 0 & 0 \\ r_{41} & 0 & 0 \\ 0 & -r_{41} & 0 \\ 0 & 0 & 0 \end{pmatrix}$$

Tetragonal—4mm

$$\begin{pmatrix} 0 & 0 & r_{13} \\ 0 & 0 & r_{13} \\ 0 & 0 & r_{33} \\ 0 & r_{51} & 0 \\ r_{51} & 0 & 0 \\ 0 & 0 & 0 \end{pmatrix}$$

Cubic—$\bar{4}$3m and 23

$$\begin{pmatrix} 0 & 0 & 0 \\ 0 & 0 & 0 \\ 0 & 0 & 0 \\ r_{41} & 0 & 0 \\ 0 & r_{41} & 0 \\ 0 & 0 & r_{41} \end{pmatrix}$$

Table 8.1. Some values of electro-optic coefficients.

Material	Room temp. EO coeff.	Index	Symmetry class
	$(10^{-12}$ m V$^{-1})$		
KDP (KH$_2$PO$_4$)	$r_{41} = 8.6$	$n_o = 1.51$	$\bar{4}2m$
	$r_{63} = 10.6$	$n_e = 1.47$	
ADP (NH$_4$H$_2$PO$_4$)	$r_{41} = 28$	$n_o = 1.52$	$\bar{4}2m$
	$r_{63} = 8.5$	$n_e = 1.48$	
Quartz (SiO$_2$)	$r_{41} = 0.2$	$n_o = 1.54$	32
	$r_{63} = 0.93$	$n_e = 1.55$	
GaAs	$r_{41} = 1.6$	$n_o = 3.34$	$\bar{4}3m$
LiTaO$_3$	$r_{33} = 33$	$n_o = 2.175$	3m
	$r_{13} = 7.5$	$n_e = 2.180$	
	$r_{51} = 20$		
	$r_{22} = 1$		
LiNbO$_3$	$r_{13} = 8.6$	$n_o = 2.272$	3
	$r_{33} = 30.8$	$n_e = 2.187$	
	$r_{51} = 28$		
	$r_{22} = 3.4$		
BaTiO$_3$	$r_{33} = 28$	$n_o = 2.480$	4m
	$r_{13} = 8$	$n_e = 2.426$	
	$r_{51} = 820$		
KNbO$_3$	$r_{33} = 25$	$n_1 = 2.279$	mm2
	$r_{13} = 8.6$	$n_2 = 2.329$	
	$r_{23} = 2$	$n_3 = 2.167$	
	$r_{42} = 270$		
	$r_{51} = 23$		
LiIO$_3$	$r_{13} = 4.1$	$n_o = 1.883$	6
	$r_{41} = 1.4$	$n_e = 1.734$	
	$r_{33} = 6.4$		
	$r_{51} = 3.3$		

In table 8.1 we give the values for the electro-optic coefficients for some commonly used materials, along with the crystal symmetry information and the indices of refraction in the absence of applied fields.

Example 1: The EO effect in KDP
From the list of electro-optic matrices, the *r*-matrix for KDP, and an example of a tetragonal, $\bar{4}2m$ crystal, is

$$\bar{r} = \begin{pmatrix} 0 & 0 & 0 \\ 0 & 0 & 0 \\ 0 & 0 & 0 \\ r_{41} & 0 & 0 \\ 0 & r_{41} & 0 \\ 0 & 0 & r_{63} \end{pmatrix}. \tag{8.17}$$

The index ellipsoid is therefore

$$\frac{x^2}{n_o^2} + \frac{y^2}{n_o^2} + \frac{z^2}{n_e^2} + 2r_{41}E_x yz + 2r_{41}E_y xz + 2r_{63}E_z xy = 1. \tag{8.18}$$

If we choose the applied field to be $\vec{E} = E_z \hat{z}$, then the index ellipsoid becomes

$$\frac{x^2}{n_o^2} + \frac{y^2}{n_o^2} + \frac{z^2}{n_e^2} + 2r_{63}E_z xy = 1. \tag{8.19}$$

The problem is thus to find a new coordinate system x', y', and z' such that

$$\frac{x'^2}{n_{x'}^2} + \frac{y'^2}{n_{y'}^2} + \frac{z'^2}{n_{z'}^2} = 1. \tag{8.20}$$

Thus x', y', and z' are the directions of the major axes of the ellipsoid in the presence of the field E_z and $n_{x'}$, $n_{y'}$, and $n_{z'}$ are the new indices along these new principal axes. That is, they are the new principal axes for the 'deformed ellipsoid'. In this particular example, it is fairly clear that in order to put the B-matrix in diagonal form, we choose x', y', and z' such that z' is parallel to z (figure 8.3). Because of symmetry in x and y, x' and y' are related to x and y by a 45° rotation, that is

$$x = x' \cos 45° - y' \sin 45°$$

$$y = x' \sin 45° + y' \cos 45°.$$

More generally, we can find these results for the principal indices by diagonalizing the B-matrix. Thus, to find x', y', and z' in the new principle axis system, find the eigenvalues of the new B_{ij} matrix according to

$$\begin{pmatrix} \dfrac{1}{n_o^2} - \lambda & r_{63}E_z & 0 \\[2mm] r_{63}E_z & \dfrac{1}{n_o^2} - \lambda & 0 \\[2mm] 0 & 0 & \dfrac{1}{n_e^2} - \lambda \end{pmatrix} = 0,$$

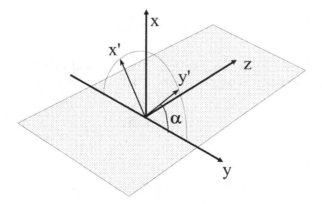

Figure 8.3. Illustration of the coordinate rotation about the z-axis as it arises in the example problem for KDP.

which we can evaluate as

$$\left[\left(\frac{1}{n_o^2} - \lambda\right)^2 - (r_{63}E_z)^2\right]\left(\frac{1}{n_e^2} - \lambda\right) = 0.$$

This has solutions

$$\left(\frac{1}{n_o^2} - \lambda\right)^2 = (r_{63}E_z)^2$$

and

$$\left(\frac{1}{n_e^2} - \lambda\right) = 0.$$

Therefore

$$\lambda_{1,2} = \frac{1}{n_o^2} \pm r_{63}E_z$$

$$\lambda_3 = \frac{1}{n_e^3}.$$

If we were to solve for the eigenvectors using these eigenvalues we would find that x', y' are related to x, y by a 45° rotation about the z-axis. Substitution into the index ellipsoid equation gives

$$\left(\frac{1}{n_o^2} - r_{63}E_z\right)x'^2 + \left(\frac{1}{n_o^2} + r_{63}E_z\right)y'^2 + \frac{z'^2}{n_e^2} = 1, \tag{8.21}$$

which gives new indices of

$$\frac{1}{n_{x'}^2} = \frac{1}{n_o^2} + r_{63}E_z \tag{8.22}$$

and

$$\frac{1}{n_{y'}^2} = \frac{1}{n_o^2} - r_{63}E_z. \tag{8.23}$$

The changes in the index of refraction from the zero-field values (n_o in this case) are very small; therefore we can write approximately

$$n_{x'} \simeq n_o - \frac{n_o^3}{2}r_{63}E_z, \tag{8.24}$$

$$n_{y'} \simeq n_o + \frac{n_o^3}{2}r_{63}E_z, \tag{8.25}$$

and

$$n_{z'} = n_e. \tag{8.26}$$

Equations (8.24)–(8.26) give the new principal indices of refraction for the physical situation of an electric field applied along the z-axis of a KDP crystal.

Example 2: Electro-optic effect in LiNbO$_3$

Now that we have a feeling for the general method used to find the new principal axes and corresponding indices of refraction, we will take a look at a somewhat more complicated case and treat it in more detail. The crystal lithium niobate (LiNbO$_3$) is commonly used in nonlinear optical applications. It belongs to the 3m crystal symmetry class. From the information given in table 8.1 we find that the ordinary and extraordinary indices of refraction are (in a zero field) $n_o = 2.272$ and $n_e = 2.187$, respectively. The non-zero electro-optic coefficients are $r_{33} = 30.8 \times 10^{-12}$, $r_{13} = 8.6 \times 10^{-12}$, $r_{22} = 3.4 \times 10^{-12}$, and $r_{51} = 28 \times 10^{-12}$ m V^{-1}. In addition we can go to the compilation of electro-optic matrices and find the form for this case to be

$$\bar{r} = \begin{pmatrix} 0 & -r_{22} & r_{13} \\ 0 & r_{22} & r_{13} \\ 0 & 0 & r_{33} \\ 0 & r_{51} & 0 \\ r_{51} & 0 & 0 \\ -r_{22} & 0 & 0 \end{pmatrix}.$$

Assuming that the initial index ellipsoid diagonal in the absence of an applied field is

$$\frac{x^2}{n_o^2} + \frac{y^2}{n_o^2} + \frac{z^2}{n_e^2} = 1,$$

we remind ourselves again that initially $B_1 = B_2 = 1/n_o^2$, $B_3 = 1/n_e^2$, and $B_4 = B_5 = B_6 = 0$. If we apply a field in the z-direction only (we will follow this up by looking at an applied field in the x- and y-directions as well), we find upon using equation (8.16) that $\Delta B_1 = r_{13}E_z$, $\Delta B_2 = r_{13}E_z$, $\Delta B_3 = r_{33}E_z$, and $\Delta B_4 = \Delta B_5 = \Delta B_6 = 0$. Thus it is very simple to write down the new index ellipsoid equation:

$$\left(\frac{1}{n_o^2} + r_{13}E_z\right)x^2 + \left(\frac{1}{n_o^2} + r_{13}E_z\right)y^2 + \left(\frac{1}{n_e^2} + r_{33}E_z\right)z^2 = 1.$$

Here we see that the new ellipse has not been rotated in the xyz space, but has simply been compressed such that the crystal is still uniaxial (the x- and y-directions still have the same index of refraction, which is different from n_z).

Now we investigate what happens if we return to the zero-field case and then apply an E-field along the x-direction. We find from equation (8.16) that $\Delta B_5 = r_{51}E_x$

and $\Delta B_6 = -r_{22}E_x$, with all other elements being zero. The new ellipsoid equation is much more complicated than we have seen up until now:

$$\frac{1}{n_o^2}x^2 + \frac{1}{n_o^2}y^2 + \frac{1}{n_e^2}z^2 + 2r_{51}E_xxz - 2r_{22}E_xxy = 1.$$

Writing out the B-matrix explicitly we have

$$B = \begin{pmatrix} \dfrac{1}{n_o^2} & -r_{22}E_x & r_{51}E_x \\ -r_{22}E_x & \dfrac{1}{n_o^2} & 0 \\ r_{51}E_x & 0 & \dfrac{1}{n_e^2} \end{pmatrix}.$$

Thus we must resort to the general method for finding the eigenvalues of the above matrix and then find the eigenvectors, which will give the directions for the coordinate axes for the new index ellipsoid principal axes. Since the results of this calculation would not be worth the effort expended, we will leave this with just the general equation which would have to be solved to find the eigenvalues, a cubic in the variable λ, the eigenvalue. This result is

$$\lambda^3 - \left(\frac{1}{n_e^2} + \frac{2}{n_o^2}\right)\lambda^2 - \left(r_{51}^2E_x^2 + \frac{2}{n_e^2n_o^2} + \frac{1}{n_o^4} - r_{22}^2E_x^2\right)\lambda \tag{8.27}$$

$$+\frac{r_{51}^2E_x^2}{n_o^2} - \frac{1}{n_e^2n_o^2} + \frac{r_{22}^2E_x^2}{n_e^2} = 0. \tag{8.28}$$

Finally, starting again from the field-free condition, we consider applying a field along the y-direction. Here we find $\Delta B_1 = -r_{22}E_y$, $\Delta B_2 = r_{22}E_y$, and $\Delta B_4 = r_{51}E_y$. The new index ellipsoid is thus

$$\left(\frac{1}{n_o^2} - r_{22}E_y\right)x^2 + \left(\frac{1}{n_o^2} + r_{22}E_y\right)y^2 + \frac{1}{n_e^2}z^2 + 2r_{51}E_yyz = 1.$$

By inspection we see that in the new principal axes system, the x'-axis is the same as the original x-axis. Therefore the new index ellipsoid is oriented along a rotated $y'z'$-axis pair, where we designate the rotation angle by α (see figure 8.4). In terms of the new coordinates, we can write

$$y = y'\cos\alpha - z'\sin\alpha$$
$$z = y'\sin\alpha + z'\cos\alpha.$$

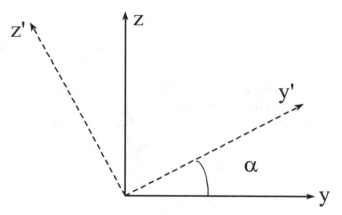

Figure 8.4. Coordinate axis transformation for the lithium niobate example described in the text. This is for the E-field applied along the crystal y-axis.

Substituting this into the equation for the ellipsoid above, we find after gathering terms that

$$\left(\frac{1}{n_o^2} - r_{22}E_y\right)x^2$$

$$+\left(\left(\frac{1}{n_o^2} + r_{22}E_y\right)\cos^2\alpha + \frac{1}{n_e^2}\sin^2\alpha + 2r_{51}E_y\cos\alpha\sin\alpha\right)y'^2$$

$$+\left(\left(\frac{1}{n_o^2} + r_{22}E_y\right)\sin^2\alpha + \frac{1}{n_e^2}\cos^2\alpha - 2r_{51}E_y\cos\alpha\sin\alpha\right)z'^2$$

$$-2\left(\left(\frac{1}{n_o^2} + r_{22}E_y - \frac{1}{n_e^2}\right)\cos\alpha\sin\alpha - r_{51}E_y(\cos^2\alpha - \sin^2\alpha)\right)y'z' = 1.$$

In the new rotated coordinate system no cross terms may appear; therefore the coefficient of the $y'z'$ term must vanish. This gives a condition on the angle of rotation α, namely

$$\tan 2\alpha = \frac{2r_{51}E_y}{\frac{1}{n_o^2} - \frac{1}{n_e^2} + r_{22}E_y}.$$

The coefficients of the remaining terms in the ellipsoid equation give the new indices of refraction. The general expressions are very messy, but if we use a numerical or symbolic manipulation package, it is possible to find an analytic results for the new ellipsoid coefficients, given by

$$\frac{1}{n_o^2} - r_{22}E_y$$

and

$$\frac{r_{22}E_y}{2} + \frac{n_o^2 + n_e^2}{2n_o^2 n_e^2} \pm \frac{\sqrt{E_y^2 n_o^4 n_e^4 (r_{21}^2 + 4r_{51}^2) + (n_o^2 + n_e^2)^2 - 2r_{22}E_y n_e^2 n_o^2 (n_o^2 - n_e^2)}}{2n_o^2 n_e^2}.$$

Furthermore, we can reduce the general result to a reasonable approximation by noticing that the terms in the square root involving the electro-optic coefficients are very small for reasonable values of the applied field, and therefore we can write

$$\frac{1}{n_{y'}^2} \simeq \frac{1}{n_o^2} + \frac{r_{22}E_y}{2}$$

$$\frac{1}{n_{z'}^2} \simeq \frac{1}{n_e^2} + \frac{r_{22}E_y}{2}.$$

To conclude this example, we repeat that the results obtained for the new coordinate axes of the index ellipsoid, and therefore the birefringence experienced by a field propagating through the crystal, will vary strongly depending on the symmetry of the crystal and the direction of the applied field.

Example 3: EO modulation using cubic crystals

As another example of the electro-optic effect, let us look at the EO effect in zincblende type cubic crystals. GaAs has $\bar{4}3m$ symmetry (zincblende) and the EO matrix is

$$\bar{\bar{r}} = \begin{pmatrix} 0 & 0 & 0 \\ 0 & 0 & 0 \\ 0 & 0 & 0 \\ r_{41} & 0 & 0 \\ 0 & r_{41} & 0 \\ 0 & 0 & r_{41} \end{pmatrix}. \tag{8.29}$$

The index ellipsoid for $\vec{E}$ along the $\langle 111 \rangle$ direction is

$$\frac{x^2 + y^2 + z^2}{n_o^2} + 2r_{41}(E_x \, yz + E_y \, xz + E_z \, xy) = 1, \tag{8.30}$$

but

$$E_x = E_y = E_z = \frac{E}{\sqrt{3}}. \tag{8.31}$$

Thus we can write

$$\frac{x^2 + y^2 + z^2}{n_o^2} + \frac{2r_{41}}{\sqrt{3}}[yz + xz + xy] = 1. \tag{8.32}$$

Note, for $E = 0$, the symmetry is cubic and therefore $n_o = n_e$. We can find the new principal axes by diagonalizing the B-matrix. Only the result for the index ellipsoid will be given here:

$$[x'^2 + y'^2]\left(\frac{1}{n_o^2} - \frac{r_{41}}{\sqrt{3}}E\right) + \left(\frac{1}{n_o^2} + \frac{2r_{41}}{\sqrt{3}}E\right)z'^2 = 1. \qquad (8.33)$$

This gives new principal indices of refraction of

$$n_{x'} = n_{y'} \simeq n_o + \frac{n_o^3 r_{41}}{2\sqrt{3}}E$$

$$n_{z'} = n_o - \frac{n_o^3 r_{41}}{\sqrt{3}}E.$$

An octant of the new index ellipsoid is shown in figure 8.5.

8.4 Electro-optic retardation

Now we wish to consider the application of the electro-optic effect to practical devices. In essence we will show how one can obtain a variable waveplate, which is a function of voltage applied to an electro-optically active crystal.

8.4.1 The longitudinal electro-optic effect

Using KDP as our example again, consider a KDP crystal with propagation along the z-axis. The index ellipsoid is

$$\frac{x^2}{n_o^2} + \frac{y^2}{n_o^2} + \frac{z^2}{n_e^2} + 2r_{63}E_z xy = 1.$$

Using the results for the index ellipsoid and the principal indices found in the previous section,

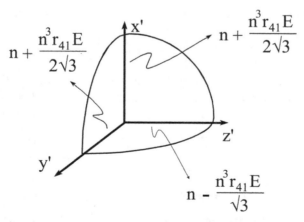

Figure 8.5. Sketch of the new principal indices of refraction for the example of the cubic electro-optic effect.

$$\left[\frac{1}{n_o^2} + r_{63}E_z\right]x'^2 + \left[\frac{1}{n_o^2} - r_{63}E_z\right]y'^2 + \frac{z'^2}{n_e^2} = 1$$

$$n_{x'} = n_o - \frac{n_o^3}{2}r_{63}E_z$$

$$n_{y'} = n_o + \frac{n_o^3}{2}r_{63}E_z.$$

According to the method described in chapter 7 we need to find the intersection ellipse formed by the plane $z = 0$ and the above indicatrix, which is

$$\frac{x'^2}{n_{x'}^2} + \frac{y'^2}{n_{y'}^2} = 1.$$

For retardation, consider light propagating along z and polarized along x':

$$\begin{aligned}
E_{x'} &= A \exp\left[i(k_z z - \omega t)\right] \\
&= A \exp\left[i\left(\frac{\omega}{c}n_{x'}z - \omega t\right)\right] \\
&= A \exp\left[i\left(\frac{\omega}{c}\left(n_o - \frac{n_o^3}{2}r_{63}E_z\right)z - \omega t\right)\right].
\end{aligned} \tag{8.34}$$

Similarly, for polarization along y',

$$\begin{aligned}
E_{y'} &= A \exp\left[i\left(\frac{\omega}{c}n_{y'}z - \omega t\right)\right] \\
&= A \exp\left[i\left(\frac{\omega}{c}\left(n_o + \frac{n_o^3}{2}r_{63}E_z\right)z - \omega t\right)\right].
\end{aligned} \tag{8.35}$$

Now at the output plane, $z = L$, the phase difference between the components is

$$\Gamma = \phi_{y'} - \phi_{x'} = \frac{\omega}{c}n_o^3 r_{63}E_z L = \frac{2\pi}{\lambda}n_o^3 r_{63}V. \tag{8.36}$$

This result allows us to find the quarter-wave voltage, $V_{\pi/2}$, for the configuration below using a KDP crystal; that is, we want to find the applied dc voltage needed to yield quarter-wave retardation for the configuration shown in figure 8.6:

$$\Gamma = \frac{\pi}{2} = \frac{\omega}{c}n_o^3 r_{63}V_{\pi/2} = \frac{2\pi}{\lambda_0}n_o^3 r_{63}V_{\pi/2}. \tag{8.37}$$

Therefore $V_{\pi/2} = \lambda_0/4n_o^3 r_{63}$.

Now for KDP we have, from table 8.1, the following parameters:

$$n_o = 1.51; \quad r_{63} = 10.6 \times 10^{-12} \text{ m V}^{-1}; \quad n_o^3 r_{63} = 3.65 \times 10^{-11} \text{ m V}^{-1}.$$

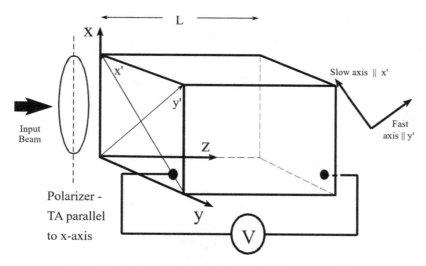

Figure 8.6. Configuration for the longitudinal electro-optic effect.

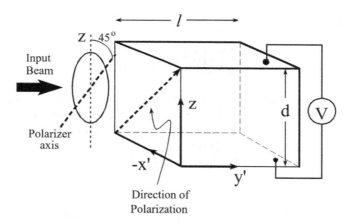

Figure 8.7. Configuration for the transverse electro-optic effect.

If we let $\lambda_0 = 1.06\ \mu$m (Nd:YAG laser wavelength), the quarter-wave voltage is found to be $V_{\pi/2} = \dfrac{1.06 \times 10^{-6}}{4(3.65 \times 10^{-11}\text{m V}^{-1})} = 7.26 \times 10^3$ volts.

8.4.2 The transverse electro-optic effect

If the dc voltage is applied as shown in figure 8.7 and the light wave propagates in the y'-direction, the configuration is called the transverse electro-optic effect. Again, the indicatrix is

$$\frac{x'^2}{n_{x'}^2} + \frac{y'^2}{n_{y'}^2} + \frac{z'^2}{n_{z'}^2} = 1.$$

The intersection ellipse is obtained from the plane at $y' = 0$ and the indicatrix. Recalling that $n_{z'} = n_e$, we have for the intersection ellipse

$$\frac{x'^2}{n_{x'}^2} + \frac{z^2}{n_e^2} = 1.$$

(8.38)

The phase retardation is then

$$\Gamma = \phi_{x'} - \phi_{z'} = \left(\left(n_o - \frac{n_o^3}{2}r_{63}E_z\right)l - n_e l\right)\frac{\omega}{c}$$

$$= \left((n_o - n_e) - \frac{n_o^3}{2}r_{63}\frac{V}{d}\right)\frac{\omega}{c}l.$$

(8.39)

The quarter-wave voltage would be obtained from

$$\Gamma = \frac{\pi}{2} = \frac{2\pi}{\lambda_0}l\left((n_o - n_e) - \frac{n_o^3}{2}r_{63}\frac{V_{\pi/2}}{d}\right)$$

(8.40)

giving a quarter-wave voltage of

$$V_{\pi/2} = \frac{2d}{n_o^3 r_{63}}\left((n_o - n_e) - \frac{\lambda_0}{4l}\right).$$

(8.41)

We would like to compare the relative quarter-wave voltages for the longitudinal and transverse configurations. To be consistent we will consider again a wavelength of 1.06 μm, but now we must also specify the crystal dimensions. Numbers which are not unreasonable would be (for KDP): a length $l = 5.0$ cm and a cross-section of 0.3 cm×0.3 cm (i.e. $d = 0.3$ cm). First we evaluate the voltage-independent term in equation (8.40):

$$\frac{2\pi}{\lambda}l\,(n_o - n_e) = \frac{2\pi}{1.06 \times 10^{-6}\text{ m}} \cdot 5.0\text{ cm} \cdot (1.51 - 1.47) = 3773.58\pi.$$

Thus the static contribution already leads to many multiples of 2π in relative phase lag. Once the modulus of the phase shift is extracted, i.e. 1.58π, we can compute the voltage necessary to achieve the phase retardation desired. Here,

$$\frac{\pi}{2} = 1.58\pi - \frac{2\pi}{\lambda_0}l\left(\frac{n_o^3}{2}r_{63}\frac{V_{\pi/2}}{d}\right)$$

$$V_{\pi/2} = \frac{1.08\lambda_0 d}{l\,n_0^3 r_{63}} = 1882\text{ volts},$$

which is significantly less than the quarter-wave voltage found for the longitudinal configuration. In fact, we should be a bit more clever and realize that the polarity of the applied voltage is irrelevant, as well as that the phase shifts 0 and 2π are equivalent. Therefore we could just as well look for the voltage such that the phase shift goes in the other direction, i.e. shifts by -0.42π. If we plug in the numbers here, it turns out not to make much difference, yielding a voltage of $V_{\pi/2} = 1650$ V. In general, however, it is possible to switch voltages on an electro-optic device to change the sense of the phase shift.

8.5 Electro-optic amplitude modulation

Using the results of the preceding sections, we can design an amplitude modulator. Consider the case for the longitudinal EO effect shown previously, except that in this case we place a polarizer in the output plane with an orientation of 90° to the input polarizer (see figure 8.8).

At the input plane ($z = 0$) we have $E_{x'} = E_0 \cos \omega t$ and $E_{y'} = E_0 \cos \omega t$, where $E_0 = E_{x'}(0) = E_{y'}(0)$. The incident intensity is

$$I_i \propto |\vec{E}_i|^2 = |E_{x'}(0)|^2 + |E_{y'}(0)|^2 = 2E_0^2. \tag{8.42}$$

At $z = L$, the x'- and y'-components have acquired a relative phase shift of Γ radians:

$$E_{x'}(L) = E_0 \qquad E_{y'}(L) = E_0 e^{-i\Gamma}.$$

The total field, E_y, at the output polarizer is

$$E_y = - E_0 \sin 45° + E_0 e^{-i\Gamma} \cos 45°$$
$$= - \frac{E_0}{\sqrt{2}}[1 - e^{-i\Gamma}], \tag{8.43}$$

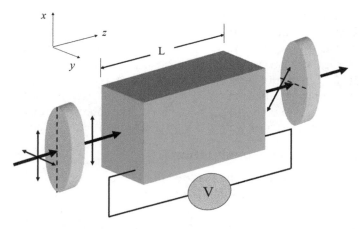

Figure 8.8. Basic configuration for electro-optic amplitude modulation.

with an output intensity of

$$I_0 \propto |E_y|^2 = \frac{E_0^2}{2}[e^{-i\Gamma} - 1][e^{i\Gamma} - 1]$$

$$= \frac{E_0^2}{2}[1 - e^{-i\Gamma} - e^{i\Gamma} + 1]$$

$$= \frac{E_0^2}{2}[2 - (e^{i\Gamma} + e^{-i\Gamma})] \qquad (8.44)$$

$$= E_0^2[1 - \cos\Gamma]$$

$$= 2E_0^2 \sin^2\frac{\Gamma}{2}.$$

Again, taking KDP as our example, we have

$$\Gamma = \frac{2\pi}{\lambda_0}n_0^3 r_{63}V.$$

Now, if $\Gamma = \pi$, from equation (8.36) we have

$$\pi = \frac{2\pi}{\lambda_0}n_o^3 V_\pi$$

and

$$V_\pi = \frac{\lambda_0}{2n_o^3 r_{63}}.$$

Finally, the output intensity can thus be written as

$$I_{\text{out}} = I_{\text{in}} \sin^2\frac{\Gamma}{2} = I_{\text{in}} \sin^2\left[\frac{\pi}{2}\frac{V}{V_\pi}\right], \qquad (8.45)$$

where I_{in} is the input intensity.

We see that for an applied voltage of $V = V_\pi$, the device passes all of the incident irradiance. For an applied voltage of zero, the device passes none of the incident radiation. If $0 < V < V_\pi$ the device will pass incident irradiance according to equation (8.45). Alternatively, if we add a dc voltage, called the bias voltage, such that about 50% of the input light is transmitted, it is possible to operate in a fairly linear regime of input modulation voltage versus output light intensity. As shown in figure 8.9, applying a voltage of $V_{\text{bias}} + V_m \sin \omega t$, the intensity transmitted through the analyzing polarizer (figure 8.8) will vary sinusoidally about the value $I_0/2$.

8.6 Electro-optic phase modulation

The electro-optic effect can also be used for phase modulation of the optical signal.

In figure 8.10 we have, for a KDP type crystal,

$$\phi_{x'} = \frac{\omega}{c}n_{x'}L = \frac{\omega}{c}\left(n_0 - \frac{n_o^3 r_{63}}{2}E_z\right)L. \qquad (8.46)$$

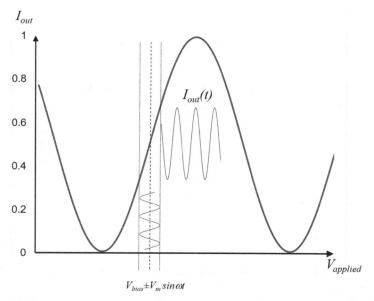

Figure 8.9. Applying a sinusoidal voltage to an electro-optic modulator in amplitude modulation mode, the output intensity of light can also be made to vary sinusoidally.

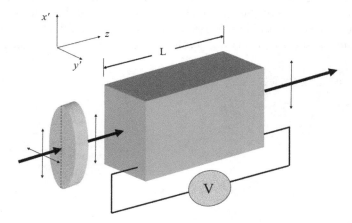

Figure 8.10. Basic configuration using the longitudinal electro-optic effect for phase modulation.

Let the applied ac field be given by $E_z = E_m \sin \omega_m t$; at the input plane ($z = 0$), the optical field is $E_{\text{in}} = E_0 \cos \omega t$ and at the output plane ($z = L$) we have

$$E_{\text{out}} = E_0 \cos \left(\frac{\omega}{c} \left[n_o - \frac{n_o^3}{2} r_{63} E_m \sin \omega_m t \right] L - \omega t \right). \tag{8.47}$$

Dropping the constant phase factor, the output optical field, E_{out}, is

$$\begin{aligned} E_{\text{out}} &= E_0 \cos \left[\omega t + \delta \sin \omega_m t \right] \\ &= E_0 \cos \omega t \cos[\delta \sin(\omega_m t)] - E_0 \sin \omega t \sin[\delta \sin(\omega_m t)], \end{aligned} \tag{8.48}$$

where

$$\delta = \frac{\omega n_o^3 r_{63} E_m L}{2c} = \frac{\pi n_o^3 r_{63} E_m L}{\lambda_0} \tag{8.49}$$

and δ is called the phase modulated index. Next, if we use a Bessel function expansion of $\cos(\delta \sin \omega_m t)$ and $\sin(\delta \sin \omega_m t)$ [4]

$$\cos(\delta \sin \omega_m t) = J_0(\delta) + 2J_2(\delta)\cos 2\omega_m t + 2J_4(\delta)\cos 4\omega_m t + \ldots$$

$$\sin(\delta \sin \omega_m t) = 2J_1(\delta)\sin \omega_m t + 2J_3(\delta)\sin 3\omega_m t + \ldots$$

so that equation (8.48) becomes

$$E_{\text{out}} = E_0 J_0(\delta)\cos \omega t$$
$$+ E_0 \sum_{k=1} [J_k(\delta)\cos(\omega - k\omega_m t)t + (-1)^k J_k(\delta)\cos(\omega + k\omega_m)t]. \tag{8.50}$$

Equation (8.50) is the expression for the phase modulated optical E-field at the output plane of the crystal in terms of a series of Bessel functions. The Bessel functions $J_k(\delta)$ give the amplitudes of each of the sidebands. A few of these functions are plotted in figure 8.11. The modulation index δ can be calculated using equation (8.49) and the numerical values for the given configuration. Once the modulation index is known the value of the various Bessel functions can be determined and thus one knows how large the sidebands are. Figure 8.12 is a rough sketch of the electric field at the output end of the crystal. Note that the first sidebands on either side of

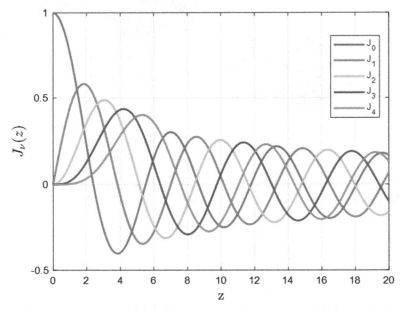

Figure 8.11. Plot of the first five Bessel functions, $J_0(\delta) - J_4(\delta)$, as a function of the modulation index δ.

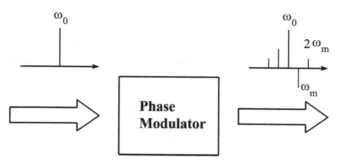

Figure 8.12. Illustration of the creation of sidebands by electro-optic modulation. It is important to note that the first sidebands at $\omega_0 \pm \omega_m$ are out of phase.

the central (carrier) frequency are out of phase. For relatively small modulation amplitudes only these sidebands will make a noticeable contribution (see the example below). Thus when this field is incident on a detector (which detects the magnitude-squared of the field) there will be no evidence of the modulation frequency. One way of looking at this is that beats between the $+1$ sideband and the carrier frequency will destructively interfere with beats from the -1 sideband and the carrier.

Example. We will now take a look at an example using the preceding ideas. Returning to the case of a KDP crystal in the longitudinal configuration we find $n_0 = 1.51$, $r_{63} = 10.6 \times 10^{-12}$ m V^{-1}. Assuming we are using a He–Ne laser at $\lambda_0 = 633$ nm, we wish to find out how large the modulation index is for a given applied voltage, say 500 V. Plugging in the values we have

$$\delta = \frac{\pi (1.51)^3 (10.6 \times 10^{-12} \text{ m V}^{-1})(500 \text{ V})}{633 \times 10^{-9} \text{ m}} = 0.09.$$

We can use a program such as MatLab to calculate $J_0(\delta = 0.09) - J_4(\delta = 0.09)$. These are

$$J_0(0.09) = 0.998$$
$$J_1(0.09) = 0.045$$
$$J_2(0.09) = 0.001$$
$$J_3(0.09) = 1.52 \times 10^{-5}$$
$$J_4(0.09) = 1.71 \times 10^{-7}.$$

Thus we see that the statement made above holds true, namely that only the first sideband contributes significantly and it is even very small compared to the central carrier frequency.

8.7 The quadratic electro-optic effect

We now proceed to consider a more general form of the electro-optic effect, one that will come into play even for highly symmetric crystals. From equation (8.11) we have

$$\Delta B_{ij}(E) = B_{ij}(E) - B_{ij}(0)$$
$$= \underbrace{r_{ijk}E_k}_{\text{linear EO}} + \underbrace{s_{ijkl}E_kE_l}_{\text{quadratic EO}} + \ldots.$$

We will designate the linear electro-optic effect by $\Delta B_{ij}^{(1)}$ and the quadratic effect by $\Delta B_{ij}^{(2)}$.

If we assume that the linear electro-optic effect is zero, the index ellipsoid for the quadratic EO effect in contracted notation becomes

$$x^2\left(\frac{1}{n_x^2} + s_{11}E_x^2 + s_{12}E_y^2 + s_{13}E_z^2 + 2s_{14}E_yE_z + 2s_{15}E_xE_z + 2s_{16}E_xE_y\right)$$

$$+ \; y^2\left(\frac{1}{n_y^2} + s_{21}E_x^2 + s_{22}E_y^2 + s_{23}E_z^2 + 2s_{24}E_yE_z + 2s_{25}E_xE_z + 2s_{26}E_xE_y\right)$$

$$+ \; z^2\left(\frac{1}{n_z^2} + s_{31}E_x^2 + s_{32}E_y^2 + s_{33}E_z^2 + 2s_{34}E_yE_z + 2s_{35}E_xE_z + 2s_{36}E_xE_y\right) \quad (8.51)$$

$$+ \; 2yz\left(s_{41}E_x^2 + s_{42}E_y^2 + s_{43}E_z^2 + 2s_{44}E_yE_z + 2s_{45}E_xE_z + 2s_{46}E_xE_y\right)$$

$$+ \; 2zx\left(s_{51}E_x^2 + s_{52}E_y^2 + s_{53}E_z^2 + 2s_{54}E_yE_z + 2s_{55}E_xE_z + 2s_{56}E_xE_y\right)$$

$$+ \; 2xy\left(s_{61}E_x^2 + s_{62}E_y^2 + s_{63}E_z^2 + 2s_{64}E_yE_z + 2s_{65}E_xE_z + 2s_{66}E_xE_y\right)$$

$$= \; 1.$$

Equation (8.51) can be generalized by the following matrix:

$$\begin{pmatrix} \Delta B_1 \\ \Delta B_2 \\ \Delta B_3 \\ \Delta B_4 \\ \Delta B_5 \\ \Delta B_6 \end{pmatrix} = \begin{pmatrix} s_{11} & s_{12} & s_{13} & s_{14} & s_{15} & s_{16} \\ s_{21} & s_{22} & s_{23} & s_{24} & s_{25} & s_{26} \\ s_{31} & s_{32} & s_{33} & s_{34} & s_{35} & s_{36} \\ s_{41} & s_{42} & s_{43} & s_{44} & s_{45} & s_{46} \\ s_{51} & s_{52} & s_{53} & s_{54} & s_{55} & s_{56} \\ s_{61} & s_{62} & s_{63} & s_{64} & s_{65} & s_{66} \end{pmatrix} \begin{pmatrix} E_x^2 \\ E_y^2 \\ E_z^2 \\ 2E_yE_z \\ 2E_xE_z \\ 2E_xE_y \end{pmatrix}.$$

The quadratic electro-optic effect s-matrices for the different crystal symmetry classes are given below.

Isotropic:

$$\begin{pmatrix}
s_{11} & s_{12} & s_{12} & 0 & 0 & 0 \\
s_{12} & s_{11} & s_{12} & 0 & 0 & 0 \\
s_{12} & s_{12} & s_{11} & 0 & 0 & 0 \\
0 & 0 & 0 & \frac{s_{11}-s_{12}}{2} & 0 & 0 \\
0 & 0 & 0 & 0 & \frac{s_{11}-s_{12}}{2} & 0 \\
0 & 0 & 0 & 0 & 0 & \frac{s_{11}-s_{12}}{2}
\end{pmatrix}$$

Triclinic: 1, $\bar{1}$

$$\begin{pmatrix}
s_{11} & s_{12} & s_{13} & s_{14} & s_{15} & s_{16} \\
s_{21} & s_{22} & s_{23} & s_{24} & s_{25} & s_{26} \\
s_{31} & s_{32} & s_{33} & s_{34} & s_{35} & s_{36} \\
s_{41} & s_{42} & s_{43} & s_{44} & s_{45} & s_{46} \\
s_{51} & s_{52} & s_{53} & s_{54} & s_{55} & s_{56} \\
s_{61} & s_{62} & s_{63} & s_{64} & s_{65} & s_{66}
\end{pmatrix}$$

Monoclinic: 2, m, 2/m

$$\begin{pmatrix}
s_{11} & s_{12} & s_{13} & 0 & s_{15} & 0 \\
s_{21} & s_{22} & s_{23} & 0 & s_{25} & 0 \\
s_{31} & s_{32} & s_{33} & 0 & s_{35} & 0 \\
0 & 0 & 0 & s_{44} & 0 & s_{46} \\
s_{51} & s_{52} & s_{53} & 0 & s_{55} & 0 \\
0 & 0 & 0 & s_{64} & 0 & s_{66}
\end{pmatrix}$$

Orthorhombic: 2mm, 222, mmm

$$\begin{pmatrix}
s_{11} & s_{12} & s_{13} & 0 & 0 & 0 \\
s_{21} & s_{22} & s_{23} & 0 & 0 & 0 \\
s_{31} & s_{32} & s_{33} & 0 & 0 & 0 \\
0 & 0 & 0 & s_{44} & 0 & 0 \\
0 & 0 & 0 & 0 & s_{55} & 0 \\
0 & 0 & 0 & 0 & 0 & s_{66}
\end{pmatrix}$$

Trigonal: 3, $\bar{3}$

$$\begin{pmatrix}
s_{11} & s_{12} & s_{13} & s_{14} & s_{15} & -s_{61} \\
s_{12} & s_{11} & s_{13} & -s_{14} & -s_{15} & s_{61} \\
s_{31} & s_{31} & s_{33} & 0 & 0 & 0 \\
s_{41} & -s_{41} & 0 & s_{44} & s_{45} & -s_{51} \\
s_{51} & -s_{51} & 0 & -s_{45} & s_{44} & s_{41} \\
s_{61} & -s_{61} & 0 & -s_{15} & s_{14} & \frac{s_{11}-s_{12}}{2}
\end{pmatrix}$$

Tetragonal: 422, 4m, $\bar{4}$2m, 4/mm

$$\begin{pmatrix}
s_{11} & s_{12} & s_{13} & 0 & 0 & 0 \\
s_{12} & s_{11} & s_{13} & 0 & 0 & 0 \\
s_{31} & s_{31} & s_{33} & 0 & 0 & 0 \\
0 & 0 & 0 & s_{44} & 0 & 0 \\
0 & 0 & 0 & 0 & s_{55} & 0 \\
0 & 0 & 0 & 0 & 0 & s_{66}
\end{pmatrix}$$

Trigonal: 32, 3m, $\bar{3}$m

$$\begin{pmatrix}
s_{11} & s_{12} & s_{13} & s_{14} & 0 & 0 \\
s_{12} & s_{11} & s_{13} & -s_{14} & 0 & 0 \\
s_{13} & s_{13} & s_{33} & 0 & 0 & 0 \\
s_{41} & -s_{41} & 0 & s_{44} & 0 & 0 \\
0 & 0 & 0 & 0 & s_{44} & s_{41} \\
0 & 0 & 0 & 0 & s_{14} & \frac{s_{11}-s_{12}}{2}
\end{pmatrix}$$

Tetragonal: 4, $\bar{4}$ 4/m

$$\begin{pmatrix}
s_{11} & s_{12} & s_{13} & 0 & 0 & s_{16} \\
s_{21} & s_{11} & s_{13} & 0 & 0 & -s_{16} \\
s_{31} & s_{31} & s_{33} & 0 & 0 & 0 \\
0 & 0 & 0 & s_{44} & s_{45} & 0 \\
0 & 0 & 0 & -s_{45} & s_{44} & 0 \\
s_{61} & -s_{61} & 0 & 0 & 0 & s_{66}
\end{pmatrix}$$

Hexagonal: 6, $\bar{6}$ 6/m

$$\begin{pmatrix}
s_{11} & s_{12} & s_{13} & 0 & 0 & -s_{61} \\
s_{12} & s_{11} & s_{13} & 0 & 0 & s_{61} \\
s_{31} & s_{31} & s_{33} & 0 & 0 & 0 \\
0 & 0 & 0 & s_{44} & s_{45} & 0 \\
0 & 0 & 0 & -s_{45} & s_{44} & 0 \\
s_{61} & -s_{61} & 0 & 0 & 0 & \frac{1}{2}(s_{11}-s_{12})
\end{pmatrix}$$

Cubic: 23, m3

$$\begin{pmatrix}
s_{11} & s_{12} & s_{13} & 0 & 0 & 0 \\
s_{13} & s_{11} & s_{12} & 0 & 0 & 0 \\
s_{12} & s_{13} & s_{11} & 0 & 0 & 0 \\
0 & 0 & 0 & s_{44} & 0 & 0 \\
0 & 0 & 0 & 0 & s_{44} & 0 \\
0 & 0 & 0 & 0 & 0 & s_{44}
\end{pmatrix}$$

Hexagonal: 622, 6m, $\bar{6}$m2, 6/mmm

$$\begin{pmatrix} s_{11} & s_{12} & s_{13} & 0 & 0 & 0 \\ s_{12} & s_{11} & s_{13} & 0 & 0 & 0 \\ s_{31} & s_{31} & s_{33} & 0 & 0 & 0 \\ 0 & 0 & 0 & s_{44} & 0 & 0 \\ 0 & 0 & 0 & 0 & s_{44} & 0 \\ 0 & 0 & 0 & 0 & 0 & \frac{1}{2}(s_{11} - s_{12}) \end{pmatrix}$$

Cubic: 432, m3m, $\bar{4}$3m

$$\begin{pmatrix} s_{11} & s_{12} & s_{12} & 0 & 0 & 0 \\ s_{12} & s_{11} & s_{12} & 0 & 0 & 0 \\ s_{12} & s_{12} & s_{11} & 0 & 0 & 0 \\ 0 & 0 & 0 & s_{44} & 0 & 0 \\ 0 & 0 & 0 & 0 & s_{44} & 0 \\ 0 & 0 & 0 & 0 & 0 & s_{44} \end{pmatrix}$$

As an example, we consider a field applied in the z-direction in an isotropic medium, $\vec{E} = E_z \hat{z}$. The index ellipsoid becomes

$$x^2\left(\frac{1}{n^2} + s_{12}E^2\right) + y^2\left(\frac{1}{n^2} + s_{12}E^2\right) + z^2\left(\frac{1}{n^2} + s_{11}E^2\right) = 1. \qquad (8.52)$$

Since the x^2 and y^2 coefficients are the same, the index ellipsoid can be written

$$\frac{x^2 + y^2}{n_o^2} + \frac{z^2}{n_e^2} = 1 \qquad (8.53)$$

with principal indices of refraction given by

$$n_o = n - \frac{1}{2}n^3 s_{12}E^2 \qquad (8.54)$$

$$n_e = n - \frac{1}{2}n^3 s_{11}E^2. \qquad (8.55)$$

From this we see that the birefringence, i.e. the difference between n_e and n_o, is given by

$$n_e - n_o = \frac{1}{2}n^3(s_{12} - s_{11})E^2. \qquad (8.56)$$

Often one writes $n_e - n_o = K\lambda E^2$ where $K \equiv$ Kerr constant, with some values of K for different materials given in table 8.2 below.

8.8 A microscopic model for electro-optic effects

In previous chapters we have used a simple model of a charged mass on a spring to make a connection between macroscopic parameters and a microscopic-level picture of the behavior of a material. In this section we pursue this technique one step further and consider how one can describe electro-optic effects with a microscopic model. We will consider an anharmonic oscillator and use methods of perturbation theory to find an expression for a susceptibility and changes in the complex dielectric constant which will correspond to the linear electro-optic (Pockels) effect. In principle, another type of nonlinearity could be used to model a crystal which

Table 8.2. Some Kerr coefficients.

Substance	λ (μm)	n	$K(10^{-14}$ m V$^{-2})$
Benzene	0.546	1.503	0.49
	0.633	1.496	0.414
CS$_2$	0.546	1.633	3.88
	0.633	1.619	3.18
	1.000	1.596	1.84
	1.600	1.582	1.11
CCl$_4$	0.633	1.456	0.074
	0.546	1.460	0.086
Water	0.589	1.33	5.1

shows no linear electro-optic effect but does have a contribution due to the Kerr or quadratic electro-optic effect. Further details on this approach can be found in [5].

8.8.1 Pockels effect

We consider a modification to our basic driven, damped harmonic oscillator model to include the simplest nonlinear force term as a function of the coordinate x. (We will consider only a one-dimensional case.) The potential corresponding to the nonlinearity is given by $V = -ax^3/3$ and thus the force is $F_{NL} = max^2$. The equation of motion for the electron is then

$$m\ddot{x}^2 + m\Gamma\dot{x} + m\omega_0^2 x + ma\,x^2 = -e\,E = -(E_\omega e^{-i\omega t} + E_0), \qquad (8.57)$$

where we have assumed a harmonically varying incident electric field as usual, but now with the change that a dc field is added as well. To solve for the behavior of the displacement, and thus for the polarization and susceptibility, we take a perturbative approach in which we first ignore the nonlinearity and solve for the first order term $x^{(1)}$. This is then substituted back into the equation including the nonlinear term and used to solve for the second-order correction, $x^{(2)}$. Consistent to this order, the displacement is given by

$$x = x^{(1)} + x^{(2)}.$$

Ignoring the term in x^2 in equation (8.57) we define a new variable $x' = x + \dfrac{e}{m\omega_0^2}E_0$ and assume that $x' \sim x_0' e^{-i\omega t}$. Eliminating the common term $e^{-i\omega t}$ we have

$$-\omega^2 x_0' - i\omega\Gamma x_0' + \omega_0^2 x_0' = -\frac{e}{m}E_\omega.$$

We can solve this for x_0' to find

$$x_0' = \frac{-(e/m)E_\omega}{\omega_0^2 - \omega^2 - i\omega\Gamma}$$

or, reverting back to the original variable $x^{(1)}$, we have

$$x^{(1)} = -\frac{E}{M}\left[\frac{E_\omega\, e^{-i\omega t}}{\omega_0^2 - \omega^2 - i\omega\Gamma} + \frac{E_0}{\omega_0^2}\right].$$

The response of the system in the linear approximation is simply the sum of the responses to the oscillating and dc fields, as one would expect.

To proceed we substitute this back into the nonlinear term ax^2:

$$ax^2 = a\frac{e^2}{m^2}\left[\frac{E_\omega^2\, e^{-2i\omega t}}{(\omega_0^2 - \omega^2 - i\omega\Gamma)^2} + \frac{E_0^2}{\omega_0^4} + \frac{2E_0 E_\omega\, e^{-i\omega t}}{\omega_0^2(\omega_0^2 - \omega^2 - i\omega\Gamma)}\right].$$

To remain consistent with all of the rest of the terms in the full expression for $x(t)$, we can only retain the term above which varies as $e^{-i\omega t}$. The other two terms represent fields oscillating at frequencies 2ω and $\omega = 0$ (dc). Rewriting the full expression we arrive at

$$(\omega_0^2 - \omega^2 - i\omega\Gamma)x\, e^{-i\omega t} + a\frac{e^2}{m^2}\left(\frac{2E_0 E_\omega\, e^{-i\omega t}}{\omega_0^2(\omega_0^2 - \omega^2 - i\omega\Gamma)}\right)e^{-i\omega t}$$

$$= -\frac{e}{m}E_\omega\, e^{-i\omega t} - \frac{e}{m}E_0.$$

Solving for x leads us to our final result for the oscillator amplitude,

$$x = -\underbrace{\frac{e}{m}\frac{E_\omega\, e^{-i\omega t} + E_0}{\omega_0^2 - \omega^2 - i\omega\Gamma}}_{x^{(1)}} - \underbrace{\left(\frac{e}{m}\right)^2\frac{2aE_0 E_\omega\, e^{-i\omega t}}{\omega_0^2(\omega_0^2 - \omega^2 - i\omega\Gamma)^2}}_{x^{(2)}}.$$

Now we wish to go back and think about the polarization of the medium and relate that to the susceptibility and to indices of refraction. For the linear term $x^{(1)}$ we can write the polarization for both the constant and the frequency dependent parts as

$$P_\omega^{(1)} = -N_V\, ex_\omega^{(1)} = \frac{N_V\, e^2}{m}\left[\frac{E_\omega\, e^{-i\omega t}}{\omega_0^2 - \omega^2 - i\omega\Gamma}\right] \tag{8.58}$$

$$P_0^{(1)} = \frac{N_V\, e^2}{m}\left(\frac{E_0}{\omega_0^2}\right). \tag{8.59}$$

Likewise, for the second-order term we can write

$$P_\omega^{(2)} = -N_V\, ex_\omega^{(2)} = \frac{N_V\, e^3}{m^2}\frac{2aE_0 E_\omega\, e^{-i\omega t}}{\omega_0^2(\omega_0^2 - \omega^2 - i\omega\Gamma)^2}. \tag{8.60}$$

Further, we can write the susceptibilities corresponding to each term by using $P = \varepsilon_0 \chi E$:

$$\chi_\omega^{(1)} = \frac{N_V\, e^2}{m\varepsilon_0}\left[\frac{1}{\omega_0^2 - \omega^2 - i\omega\Gamma}\right] \tag{8.61}$$

$$\chi_0^{(1)} = \frac{N_V \, e^2}{m \varepsilon_0 \omega_0^2} \tag{8.62}$$

$$\chi_\omega^{(2)} = \frac{N_V \, e^3}{m^2 \varepsilon_0} \frac{2 a E_0 \left(\dfrac{e}{m}\right)^2}{\omega_0^2 (\omega_0^2 - \omega^2 - i\omega\Gamma)^2}. \tag{8.63}$$

To gain information about the indices of refraction and, most importantly, the changes in index due to the application of the field E_0, we concentrate on the last term above. We know that $\varepsilon_\omega = 1 + \chi_\omega$ and therefore that changes in ε_ω are given by $\Delta\varepsilon_\omega = \Delta\chi_\omega$ (we omit here the 'tilde' used to denote complex quantities). In addition we are interested in the real part only, since we are considering the index of refraction changes. Thus the change in the permittivity caused by the second-order term (i.e. by the interaction of the optical field and the static field) is given by

$$\Re\Delta\varepsilon_\omega = \frac{N_V \, e^3}{m^2 \varepsilon_0} \frac{2 a E_0 \left\{ \left(\omega_0^2 - \omega^2\right)^2 - \omega^2\Gamma^2 \right\}}{\omega_0^2 \left[\left(\omega_0^2 - \omega^2\right)^2 + \omega^2\Gamma^2 \right]^2}. \tag{8.64}$$

The changes in the impermeability tensor $B \simeq 1/n^2$ are found from the above by taking

$$\Delta\left(\frac{1}{n^2}\right) = \Delta\left(\frac{1}{\varepsilon}\right) = -\frac{\Delta\varepsilon}{\varepsilon^2} = -\frac{N_V \, e^3}{m^2 \varepsilon^2 \varepsilon_0} \frac{2 a E_0 \left\{ \left(\omega_0^2 - \omega^2\right)^2 - \omega^2\Gamma^2 \right\}}{\omega_0^2 \left[\left(\omega_0^2 - \omega^2\right)^2 + \omega^2\Gamma^2 \right]^2}. \tag{8.65}$$

Finally we recall that the electro-optic coefficient is the proportionality constant relating the changes in the impermeability tensor to the strength of the applied field,

$$\Delta B = rE, \tag{8.66}$$

since we are working only in one dimension here. Thus the electro-optic coefficient can be deduced from this simple harmonic oscillator model to be

$$r = -\frac{N_V \, e^3}{m^2 \varepsilon^2 \varepsilon_0} \frac{2 a \left\{ \left(\omega_0^2 - \omega^2\right)^2 - \omega^2\Gamma^2 \right\}}{\omega_0^2 \left[\left(\omega_0^2 - \omega^2\right)^2 + \omega^2\Gamma^2 \right]^2}. \tag{8.67}$$

This form brings out a point not made earlier in our discussion of the electro-optic effect, namely that there is resonant behavior to the coefficients, just as there was for the linear susceptibility. If we go far from resonance, effectively setting $\Gamma = 0$, we can write the above as

$$r(\Gamma = 0) \simeq \frac{2 a N_V \, e^3}{m^2 \varepsilon^2 \varepsilon_0 \omega_0^2 \left(\omega_0^2 - \omega^2\right)^2} \tag{8.68}$$

$$= \frac{2ae\chi_0^{(1)}}{me^2\left(\omega_0^2 - \omega^2\right)^2} \tag{8.69}$$

$$= \frac{2ame_0^2}{e^3N_V^2} \frac{\chi_0^{(1)}\left\{\chi_\omega^{(1)}\right\}^2}{\varepsilon^2}, \tag{8.70}$$

which demonstrates that the electro-optic effect can indeed be thought of as a nonlinear 'wave-mixing' effect, a topic on which one could spend a whole semester.

8.9 High-frequency modulation

Now we consider in somewhat more detail the example of electro-optic modulation of the index of refraction of a crystal. In section 8.6 we ignored one important point, which we will now take up: What limits the frequency with which we can modulate radiation? To answer this question, consider what the result would be if an applied modulation field were to change significantly while the optical field is propagating through the crystal. For example, in a crystal such as $LiNbO_3$ with index of refraction $n = 2.2$ and length $l = 3\,\text{cm}$, the propagation time would be $\Delta t = 2.2 \times 10^{-10}$ s. A modulating field with frequency $\nu_m = 2.2$ GHz would oscillate through a complete cycle during the transit time of the optical field, which in turn would result in a net zero phase shift, since the modulating field has opposite sign during the two halves of a cycle. Therefore, we must conclude that the efficiency of phase modulation will decrease as the frequency increases.

To see this result explicitly it will be necessary to modify the result given in equation (8.48) so that we may take into account the changing magnitude of the modulation signal during propagation through the crystal, i.e. we must integrate such that

$$\phi_{x'} = -\frac{\pi n_o^3 r_{63}E_m}{\lambda} \int_0^L \sin \omega_m t\, dz \tag{8.71}$$

$$= -\frac{c\pi n_o^3 r_{63}E_m}{n_o\lambda} \int_0^{nL/c} \sin \omega_m t\, dt, \tag{8.72}$$

where we have changed the integral over distance to one over the propagation time. Carrying out the integration and doing some algebraic manipulation leads to

$$\phi_{x'} = -\frac{\pi n_o^3 r_{63}E_m\omega_m L}{\lambda} \frac{\sin^2\left(\frac{n_o\omega_m L}{2c}\right)}{\frac{n_o\omega_m L}{2c}}. \tag{8.73}$$

In figure 8.13 we show a plot of the electro-optically induced phase shift as a function of the parameter $x = \frac{n_o\omega_m L}{2c}$. We see that the maximum phase modulation occurs for approximately $x = 1$. For higher modulation frequencies (or longer crystals) the modulation decreases, as we argued above.

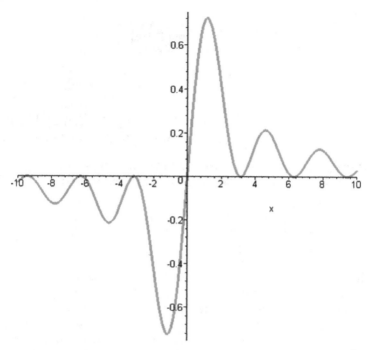

Figure 8.13. A plot of the phase modulation $\phi_{x'}$ as a function of the parameter $x = \frac{n_o \omega_m L}{2c}$. The maximum phase modulation occurs for approximately $x = 1$. For either a higher modulation frequency or a longer crystal than that giving $x = 1$ the phase modulation begins to decrease.

Next we consider a method commonly used to circumvent the modulation-frequency limit, then investigate in turn the modulation-frequency limits on the new technique and, finally, two methods to further extend the bandwidth for electro-optic modulation. As a general heuristic argument, if we could launch the modulation signal down the electro-optic crystal such that it effectively accompanies the optical signal as the latter traverses the medium, in the reference frame of the optical signal the limit discussed previously could be avoided. The electrodes on the crystal are configured such that the modulation signal (perhaps at frequencies of many GHz) propagates as if in a 'waveguide', a topic we have not addressed [6]. In fact, such a system is known as a traveling-wave modulator and can significantly extend the range of possible modulation frequencies available for a given crystal. A limit does arise, however, due to the fact that the index of refraction for the modulation signal in the waveguide and for the optical signal in the crystal may differ and thus propagate at different speeds, eventually getting out of phase.

For an optical signal $E = E_0 \cos(kz - \omega t)$, where $k = n_o \omega / c$, with n_o the index of refraction for the optical field, a modulation signal $u(z, t) = u_m \cos(k_m z - \omega_m t)$ is 'seen'. At a certain time Δt later, at a distance into the crystal $\Delta z = v \Delta t$ the two are slightly separated and the optical field sees the modulation signal as

$$u = u_m \cos(k_m \Delta z - \omega_m (t_0 + \Delta t)) \tag{8.74}$$

$$u_m \cos\left(k_m(v\Delta t) - \omega_m t_0 - \omega_m\left(\frac{\Delta z}{v}\right)\right) \tag{8.75}$$

$$u_m \cos\left(k_m\left(1 - \frac{v_m}{v}\right)\Delta z - \omega_m t_0\right). \tag{8.76}$$

Thus the optical wave sees a signal that changes by one full cycle with a spatial period of $\Lambda = \frac{2\pi}{k_m\left(1 - \frac{v_m}{v}\right)} = \frac{\lambda_m}{n_m - n}$. The maximum phase modulation will occur for one-half this spatial period, and one can find the exact modulation amount again by integrating over the propagation distance in the crystal to find

$$\phi_{x'} = \beta\frac{\sin\left(\frac{k_m \alpha L}{2}\right)}{\frac{k_m \alpha L}{2}} \sin\left(\frac{k_m \alpha L}{2} - \omega_m t_0\right), \tag{8.77}$$

where we have defined $\alpha = 1 - v_m/v$. It should be clear that this function is maximum for the argument equal to zero, since that argument is proportional to the difference in indices for the modulation signal and the optical signal; if those two are equal, the waves travel with the same speed through the crystal. Otherwise, there is a frequency bandwidth–length limitation for effective modulation. For example, a common electro-optic material that might be used as a traveling-wave modulator is lithium niobate ($LiNbO_3$) which has a bandwidth of less than 10 GHz-cm [7], still significantly better than that achieved in the non-traveling-wave configuration.

To close out this example, we qualitatively describe two further proposals for increasing the modulation bandwidth of electro-optics modulators. In the first [7] one can take advantage of a technique known as 'periodic poling' in which a crystal is fabricated so that properties such as nonlinear optical coefficients and electro-optic coefficients vary (in fact, switch sign but not magnitude) in a periodic way through the crystal. To take advantage of such a periodically poled crystal as an electro-optic modulator, the index mismatch must be known at a given desired modulation frequency and optical frequency, then the crystal can be fabricated such that every time the modulation function is ready to 'turn over' the fields enter a new domain in which the electro-optic coefficient changes sign and the modulation phase shift therefore continues to increase. Estimates of the efficacy of this technique are that it could lead to a modulation of frequency of several hundred GHz for optical signals with wavelength $\lambda = 1.55$ μm [7], a common wavelength for telecommunications applications.

A second proposal that has been shown to work experimentally involves a somewhat different configuration [8], essentially taking advantage of interference effects for light traveling through a waveguide along two different paths in a polymeric material. The advantage to this technique is that the possibility exists of engineering the waveguide such that $n_m = n_o$, thereby achieving optimal coupling between the optical and modulation fields.

The last two examples take us relatively far from our starting point of learning about the basics of electro-optic modulation, but do show how one can cleverly take advantage of some of these principles to push this technology to new limits.

8.10 Frequency modulation (FM) spectroscopy

In section 4.7 we presented an example of a fairly straightforward experiment for measuring the absorption of a medium as a function of concentration or of frequency. Here we use the concept of phase modulation (or equivalently, frequency modulation) and re-visit that same experimental set-up. We will refer in what follows to the schematic shown in figure 8.14. The point of this technique is twofold. First, since the detected signal, as we shall see, is zero when the laser is far from an atomic resonance and non-zero near resonance, we gain in sensitivity. Experimentally it is almost always advantageous to look for small deviations from zero signal as opposed to small deviations from a large constant signal. Second, the fact that our actual detected signal is the result of a narrow-bandwidth filtering of a relatively high (modulation) frequency, we further gain in sensitivity because there is little in the way of ambient noise at these higher frequencies.

We begin with the expression we found for the phase modulated electric field written in terms of Bessel functions, equation (8.50) where we now retain only the carrier and the first sidebands:

$$E_{\text{in}} = E_0 J_0(\delta)\cos \omega t + E_0[J_1(\delta)\cos(\omega - \omega_m t)t - J_1(\delta)\cos(\omega + \omega_m)t]. \qquad (8.78)$$

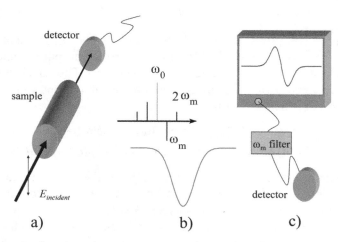

a)   b)   c)

Figure 8.14. Schematic of an absorption spectroscopy experiment using a frequency-modulated laser. (a) The incident laser, with carrier frequency ω_0 and frequency sidebands at $\omega_0 \pm \omega_m$ and at higher multiples of ω_m is partially transmitted by the absorbing sample. (b) Qualitative view of the frequency spectrum of the laser (top) and of the absorption feature with lineshape $g(\omega)$. The laser frequency is scanned across the frequency range over which there is absorption. (c) The optical signal on the detector is filtered to remove any frequencies other than those close to the difference frequency ω_m between the carrier and the sidebands. The filtered signal is displayed on an oscilloscope.

The field is written here as E_{in} because it is now the input field to the 'experiment'; in equation (8.50) we were looking at the field after the modulator and called it E_{out} for that reason. It must be pointed out that our approximation of keeping only the first sidebands means that we are interested in a limit of small modulation amplitude [9].

The transmission through the sample may be represented as a function of frequency by

$$t(\omega) \equiv e^{-\delta(\omega)-i\phi(\omega)}, \tag{8.79}$$

with $\delta(\omega)$ being the (field) attenuation coefficient and $\phi(\omega)$ a dispersive phase shift. If we write the field using the complex field representation, we find for the field transmitted through the medium

$$E_T = E_0 e^{i\omega t} \tag{8.80}$$

$$\times [e^{-\delta(\omega)-i\phi(\omega)}J_0(\delta) + e^{-\delta(\omega-\omega_m)-i\phi(\omega-\omega_m)}J_1(\delta)e^{-i\omega_m t} \\ - e^{-\delta(\omega+\omega_m)-i\phi(\omega+\omega_m)}J_1(\delta)e^{i\omega_m t}]. \tag{8.81}$$

As usual, we are interested in looking at the transmitted intensity, proportional to the magnitude-squared of the above and given by

$$T(\omega) \equiv \frac{|E_T|^2}{|E_{in}|^2} = J_0^2(\delta)e^{-2\delta(\omega)} + J_1^2(\delta)e^{-2\delta(\omega-\omega_m)} + J_1^2(\delta)e^{-2\delta(\omega+\omega_m)}$$
$$+ J_0(\delta)J_1(\delta)e^{-i\omega_m t}e^{-(\delta(\omega-\omega_m)+\delta(\omega))}e^{-i(\phi(\omega)-\phi(\omega-\omega_m))} + \text{c. c.}$$
$$- J_0(\delta)J_1(\delta)e^{-i\omega_m t}e^{-(\delta(\omega+\omega_m)+\delta(\omega))}e^{-i(\phi(\omega)-\phi(\omega+\omega_m))} - \text{c. c.}$$
$$+ J_1^2(\delta)e^{-\delta(\omega-\omega_m)}e^{-\delta(\omega+\omega_m)}(e^{i(\phi(\omega-\omega_m)-\phi(\omega+\omega_m))}e^{2i\omega_m t} + \text{c. c.}).$$

To start analyzing this result we recall the premise of our experiment, as shown in figure 8.14, namely, that we will be filtering the detector output so that we are only sensitive to fields oscillating at an angular frequency ω_m. Thus the last terms above, proportional to $e^{2i\omega_m t}$, can be dropped, as can the first terms that have no $e^{i\omega_m t}$ dependence. We are then left with

$$T(\omega_m) = 2J_0(\delta)J_1(\delta)e^{-\delta(\omega)}[e^{-\delta(\omega-\omega_m)}\cos(\omega_m t + (\phi(\omega) - \phi(\omega - \omega_m)))] \\ -2J_0(\delta)J_1(\delta)e^{-\delta(\omega)}[e^{-\delta(\omega+\omega_m)}\cos(\omega_m t + (\phi(\omega) - \phi(\omega + \omega_m)))].$$

We can expand the cosine terms using the trigonometric identity $\cos(a + b) = \cos a \cos b - \sin a \sin b$ and then make the approximations that $e^{-\delta} \simeq 1 - \delta$ and that the differences between ϕ values is small over the frequency ranges of interest. Keeping only terms less than second order in small quantities, we finally arrive at

$$T(\omega_m) = 2J_0(\delta)J_1(\delta) \tag{8.82}$$

$$\times [(\delta(\omega + \omega_m) - \delta(\omega - \omega_m))\cos \omega t \\ - (\phi(\omega + \omega_m) - \phi(\omega - \omega_m))\sin \omega_m t]. \tag{8.83}$$

Equation (8.83) is the main result of this example. It illustrates the points we set out to make, since we now have a signal that oscillates at the modulation frequency, which is high compared to typical laboratory noise sources. Frequencies on the order of many MHz to GHz are typical in this application. We are also able to choose the phase of our detection, although the details of how we might do so are beyond the scope of this example. In any case, we can pick out either the cosine or the sine term; if we are interested in measuring the absorption, we would here set our detection system to look at the cosine term. Finally we note that the signal we detect is proportional to the difference between the attenuation of the two sidebands. Thus if we are far from resonance no signal is recorded, as neither sideband is attenuated (and of course, neither is the carrier frequency, which lies between the sidebands). If we are exactly on resonance, and if the absorption lineshape is symmetric about the resonance frequency, we will also see no signal on our oscilloscope. As the carrier frequency is scanned through resonance, the output signal will change sign, as first one sideband then the other is more strongly attenuated. To a first approximation, the shape of the signal in an FM spectroscopy experiment is thus the derivative of the absorption lineshape itself and one sometimes refers to derivative spectroscopy.

One final comment can be made before closing out this section and chapter. We began this example by assuming a single frequency laser field was being frequency modulated by an electro-optic crystal, thus giving rise to the sidebands. For many lasers, however, it is possible to modulate the laser frequency directly. For example, semiconductor diode lasers operate by injecting current from an external power supply; one can modulate the injection current directly to create the sidebands needed for FM spectroscopy. In other laser systems it is possible to modulate a mirror, thereby changing the laser frequency slightly and in a controllable manner.

8.11 Problems

1. Show that crystals of the point group $\bar{4}$ become biaxial when an electric field is applied along the optic axis (z-axis).
2. Find the equation for the index ellipsoid for crystals of the point group $\bar{4}$ when an electric field is applied along the optic axis (z-axis).
3. Consider a crystal of point group symmetry $\bar{4}$ with an electric field applied along the optic axis.
    (a) Show that the new principal axes are obtained by rotating the old principal axes by an angle θ about the z-axis, where

    $$\tan 2\theta = \frac{r_{63}}{r_{13}}.$$

    (b) What is the birefringence for a light beam propagating along the z-axis?
4. Find the phase retardation for light normally incident on a $\bar{4}3m$ crystal with an electric field E applied along a cubic axis. Let ℓ be the thickness of the crystal. Assume the field is applied in a direction orthogonal to the propagation direction.

5. A voltage V is applied across an electro-optic crystal of thickness d. The phase retardation will be proportional to V and electrically tunable. Find the phase retardation for the following crystals in the longitudinal EO configuration.
   (a) GaAs (cubic symmetry, $\bar{4}$3m).
   (b) BaTiO$_3$ (tetragonal symmetry, 4mm)
   (c) AD*P (tetragonal symmetry, $\bar{4}$2m).
   (d) KNbO$_3$ (orthorhombic symmetry, mm2).

6. Design an electro-optic device which is capable of beam deflection.

7. Which uniaxial crystal classes allow a linear electro-optic effect when the dc field is orthogonal to the optic axis and the optical field is propagating along the optic axis?

8. Consider a LiNbO$_3$ crystal such that a radio frequency (RF) field with E-vector parallel to the z-axis is applied to the crystal. Both the optical beam and the RF field are propagating in the y-direction. Obtain an expression for the phase modulation index δ.

9. Consider a crystal of barium titanate (BaTiO$_3$). Below its critical temperature, barium titanate is a ferroelectric and belongs to the point group 4mm. If a static electric field E_0 is applied along the $\langle 110 \rangle$ direction when this crystal is below its transition temperature, obtain the new principal indices of refraction.

10. When the barium titanate crystal in the previous problem is above its transition temperature, its crystal symmetry is $m3m$ (cubic). If a static electric field is applied along the $\langle 110 \rangle$ direction when the crystal is above its transition temperature, obtain the new principal indices of refraction. Take $n = 2.42$, $s_{11} - s_{12} = 2290 \times 10^{-18}$ m^2 V^{-2} and $s_{44} = 3100 \times 10^{-18}$ m^2 V^{-2}.

11. Suppose you are given a sample of GaAs and a dc electric field is applied along the z-axis.
    (a) Determine the new principal indices of refraction.
    (b) What would be the physical arrangement needed to use this as a phase modulator and determine the phase change induced by the applied voltage V? Assume a wavelength of $\lambda = 10.6$ μm.
    (c) Determine the half-wave voltage for phase modulation.

12. For the previous problem, assume that the modulating electric field is now sinusoidal. Determine the phase modulation index.

13. Suppose that we wish to use the crystal in problem 11 as an amplitude modulator. A front polarizer aligned along the x-axis is placed in front of the crystal so that equal amplitudes of x'- and y'- components are created.
    (a) Determine the phase difference between the fields.
    (b) Determine the half-wave voltage.

14. Suppose the modulation voltage in problem 13 is sinusoidal and is given by $V_m \sin \omega_m t$. Obtain the phase retardation.

15. Consider the GaAs crystal of problem 11. By placing a fixed quarter-waveplate in series and an analyzer orientated perpendicular to the front polarizer, we can convert phase modulation to amplitude modulation. Determine the appropriate arrangement of the quarter-waveplate and determine the transmission of this device.

16. The nth order Bessel function is given by

$$J_n(x) = \sum_{k=0}^{\infty} \frac{(-1)^k (x/2)^{n+2k}}{k!\, \Gamma(n+k+1)}.$$

Consider the phase modulation of KDP expressed by equations (8.49) and (8.50). Determine the output intensity:

    (a) By retaining the $n = 0$ term only in the Bessel function expansion.

    (b) By retaining the $n = 0$ and $n = 1$ terms in the expansion.

17. Consider the crystal class $\bar{4}2m$ and suppose that a dc electric field is applied along the $<010>$ direction. In addition, suppose that the optical field propagates along the optic axis. Show that this arrangement gives rise to an electro-optic effect that is quadratic in the dc electric field.

18. Use the Bessel function expansion of equation (8.50) to plot the ratio of the second harmonic $(2\omega_m)$ of the output intensity to the fundamental in terms of δ.

19. Show that for an isotropic material

$$s_{44} = \frac{s_{11} - s_{22}}{2}.$$

20. Suppose that a dc electric field is applied in the $<110>$ direction in a isotropic material so that the E-field is given by

$$\vec{E} = \frac{1}{\sqrt{2}} E_0 (\hat{x} + \hat{y}).$$

Diagonalize the index ellipsoid to find the new principal indices of refraction.

21. Discuss how you could use the transverse EO effect to make a frequency modulation device. Give a quantitative explanation of how the device would work.

22. Plot the output versus applied voltage of the amplitude modulation device of section 8.5.

23. It is desired to make an amplitude modulation device whose output is sinusoidal about a bias point of 50% transmission. How would you make such a device?

References

[1] Wikipedia: John Kerr (physicist) https://en.wikipedia.org/wiki/John_Kerr_(physicist) (Accessed: 11 Jan. 2021)

[2] Kaminow I 1974 *An Introduction to Electro-optic Devices* (New York: Academic)

[3] Wood E 1977 *Crystals and Light: An Introduction to Optical Crystallography* 2nd rev edn (New York: Dover)

[4] Spiegel M 1968 *Mathematical Handbook of Formulas and Tables: Schaum's Outline Series* (New York: McGraw-Hill)

[5] Agulló-López F, Cabrera J M and Agulló-Rueda F 1994 *Electro-optics—Phenomena, Materials and Applications* (San Diego, CA: Academic)

[6] Yariv A and Yeh P 1984 *Optical Waves in Crystals* (New York: Wiley)

[7] Lu Y, Xiao M and Salamo G J 2001 Wide-bandwidth high-frequency electro-optic modulator based on periodically poled LiNbO$_3$ *Appl. Phys. Lett.* **78** 1035–7

[8] Lee M, Katz H E, Erben C, Gill D M, Gopalan P, Heber J D and McGee D J 2002 Broadband modulation of light by using an electro-optic polymer *Science* **298** 1401–3

[9] Supplee J M, Whittaker E A and Lenth W 1994 Theoretical description of frequency modulation and wavelength modulation spectroscopy *Appl. Opt.* **33** 6294–302

IOP Publishing

Optical Radiation and Matter

Robert J Brecha and J Michael O'Hare

Chapter 9

Acousto-optic effects

In chapter 8 we saw how an applied electric field can alter the optical properties of a crystal and lead to various device applications. In this chapter, we will be concerned with the alteration of the optical properties of a medium by an acoustical wave. Thus we say that the subject of acousto-optics deals with the interaction of an optical wave and an acoustic wave in a material medium. When an acoustic wave propagates in a medium, it introduces a strain in the medium and this strain alters the index of refraction of the medium. In the language of chapter 7 we may say that the acoustic-wave-induced strain deforms the optical indicatrix or index ellipsoid. As with the electro-optic effect, acousto-optic effects lead to many device applications. We will approach this subject in a somewhat backwards fashion, starting, after a brief historical introduction, with some general results and a qualitative description of acousto-optic effects. In the course of the chapter we will delve more deeply into the physics needed to arrive at the results used at the beginning.

9.1 Historical introduction

The first relatively complete treatment of the effect of an acoustic wave on the propagation characteristics of an optical wave traveling through a medium was made by Leon Brillouin (1889–1969) [1] in 1921. Experimental confirmation of the acousto-optic diffraction of light waves by sound came a decade later, and was nearly simultaneously observed by Debye and Sears [2] and by Lucas and Biquard [3]. Essentially, the observed effect is the Bragg diffraction of a light wave from a periodic grating structure created in a material by the propagating sound wave.

The system Brillouin considered was that of visible light traversing a liquid in which an ultrasonic wave had been launched. However, as we have often seen to be the case, the introduction of the laser provided a coherent source of light, which, along with advances in materials science, gave the impetus in the 1960s to the use of acousto-optic effects in a variety of practical applications. Acousto-optic modulators may be used to deflect beams or to shift the frequency of a laser beam by an amount

doi:10.1088/978-0-7503-2624-7ch9

equal to the ultrasound frequency, often in the range of tens to hundreds of MHz; one of the main goals of this chapter will be to investigate these two phenomena. It is interesting to note that the acoustic waves in an acousto-optic device are often created by applying a time-varying voltage to a separate crystal, called a transducer. The transducer reacts to the applied voltage by expanding and contracting (the piezoelectric effect), thus giving rise to the acoustic signal that propagates in the material through which the optical wave is propagating. The piezoelectric effect is a mechanical version of the electro-optic effect, and the coefficients that determine the amount of length change of a piezoelectric crystal obey exactly the same symmetry properties as those describing the electro-optic effect. However, here we will not be concerned with the exact characteristics of the piezoelectric effect.

9.2 Interaction of light with acoustic waves

We begin with a schematic picture of a possible acousto-optic device. Consider a crystal, which is transparent optically, with dimensions of, say, 5 mm height × 10 mm length × 10 mm width. On one end of the crystal a transducer is mounted, usually a piezoelectric crystal, as illustrated in figure 9.1. In piezoelectricity the physical dimensions of a material change under the action of an applied electric field. The transducer is driven at a frequency on the order of tens to hundreds of MHz. The time-varying length changes of the transducer lead to compression and rarefaction waves in the crystal itself. These changes in the crystal lead to a variation in the index of refraction in a periodic way, since the index depends on the density of the medium. The spatial dependence of the index of refraction can be imagined roughly as the creation of a diffraction grating in the crystal. Incident light is thus scattered from this 'grating' in such a way that two (or more) output beams are seen, corresponding to different orders of constructive interference.

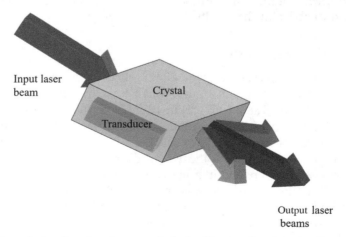

Figure 9.1. A schematic drawing of an acousto-optic device. The transducer creates a time-varying, spatially varying index of refraction in the crystal. This leads to some scattering of the incident laser radiation into the diffracted beam.

As an approach to understanding the physics behind the scattering of light from the sound wave set up in the acousto-optic crystal we can imagine a stack of glass plates in air. The plates represent regions of increased density in the crystal and the air gaps are rarefied regions. Of course this is just an approximation, since the density in an actual crystal would vary sinusoidally as a function of position. Figure 9.2 illustrates this simplification. If we want to consider the scattering in a given direction from a whole stack of glass plates, we require that there be constructive interference for all the scattered rays. Thus the difference in path length between neighboring rays must be given by the distance $ABC \equiv \Delta d = m\lambda$. Using the fact that $\theta_i = \theta_r$, along with the geometry shown in figure 9.2, we arrive at $2\Lambda \sin \theta = m\lambda$, where Λ is the wavelength of sound (glass plate separation).

Let us try to get a sense of the geometry or the relative sizes of the optical beam diameter compared to the wavelength of sound in a typical medium. For lead molybdate, $PbMoO_4$, $v_s = 3750$ m s^{-1}. If the acoustic frequency is on the order of 40 MHz, we find that $\Lambda = v_s/f = 9.4 \times 10^{-5}$ m or roughly 0.1 mm. This compares to a laser beam diameter which might be on the order of a millimeter or less. Thus we have a situation as shown in figure 9.2 where the laser beam can be thought of as being incident on several glass plates at once.

To continue the glass plate analogy, we realize that some of the light will be transmitted through the plates. There will be some sort of efficiency for transfer of light from the incident beam direction to the reflected beam direction. We state without proof for now the efficiency for the acousto-optic crystal case:

$$\eta = \frac{I_{\text{diff}}}{I_{\text{inc}}} = \sin^2 \left[\frac{\pi \ell}{\sqrt{2}\lambda} \left[\frac{n^6 p^2}{\rho v_s^3} \right]^{1/2} \sqrt{I_{\text{acoustic}}} \right],$$

where ℓ is the length of the interaction of the optical beam with the crystal, λ is the wavelength of the incident light, ρ is the density of the material, v_s is the velocity of

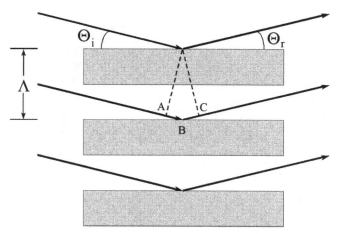

Figure 9.2. Stack of glass plates with index of refraction n used as a simplified representation of scattering of light off of an acoustic wave.

sound in the material, n is the index of refraction, and p is the photoelastic coefficient, which plays an analogous role in acousto-optics that the electro-optic coefficients played in chapter 8. These latter two quantities will be discussed in more detail in the course of this chapter. The applied acoustic intensity (W m^{-2}) is given by I_{acoustic}.

9.3 Elastic strain

To study the effects of acoustically induced strain on the optical properties of a medium, we ignore the discrete nature of the material and consider the medium as a homogeneous elastic continuum. Under external forces, the medium becomes deformed. A measure of the deformation is given by the strain. Internal forces, called stresses, will develop opposing the deformation. A material will be elastic up to a certain value of strain and within this limit, called the elastic limit, the stresses and strains are directly related through Hooke's law. In this section we present an overview of the principles involved in thinking about the deformations caused in a crystal by an applied acoustic wave. The main result of this treatment will be a sort of algorithm: given an acoustic field of a certain amplitude in one direction and propagating along a certain axis, we will be able to figure out the relevant strain (tensor) components. The approach we use follows that of reference [4].

Consider a small element of volume about a point P as shown in figure 9.3. Let x_i ($i = 1, 2, 3$) be the coordinates of a particle in the undeformed lattice, and let $x_i'(x_1, x_2, x_3)$ be the coordinates of the same material point in the deformed state. We use this new notation because it will allow us to compactly write down various useful relations. In the end x_1 is no different than our usual Cartesian 'x', x_2 is y, and x_3 is z. Let ds be the length of PQ and ds' be the length of $P'Q'$. The deformation of the lattice can then be characterized by the difference in the length of these increments, i.e.

$$
\begin{aligned}
ds'^2 - ds^2 &= (dx_i')^2 - (dx_i)^2 \\
&= dx_i' dx_i' - dx_i dx_i,
\end{aligned}
\tag{9.1}
$$

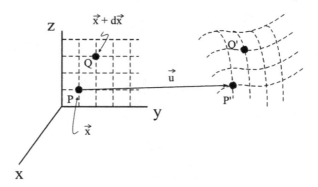

Figure 9.3. Strain deformation of an elastic medium. Coordinates of a point in an undeformed and deformed state.

where the repeated indices indicate summation, i.e.

$$dx_i dx_i = dx_1^2 + dx_2^2 + dx_3^2 \tag{9.2}$$

and similarly for $dx_i' dx_i'$. Now, since the increments designated by dx_i' are functions of x_1, x_2, and x_3 we have

$$dx_i' = \frac{\partial x_i'}{\partial x_j} dx_j. \tag{9.3}$$

Again, this is a shorthand notation; to be clear, we write this expression out more fully one last time:

$$\frac{\partial x_i'}{\partial x_j} dx_j \equiv \frac{\partial x'}{\partial x} dx + \frac{\partial x'}{\partial y} dy + \frac{\partial x'}{\partial z} dz. \tag{9.4}$$

The economy of the new notation should be clear. With the above we can now write

$$ds'^2 - ds^2 = \frac{\partial x_i'}{\partial x_j} dx_j \frac{\partial x_i'}{\partial x_k} dx_k - dx_i^2$$
$$= \left[\frac{\partial x_i'}{\partial x_j} \frac{\partial x_i'}{\partial x_k} - \delta_{jk} \right] dx_j dx_k, \tag{9.5}$$

where, again, the repeated indices indicated summation. Now we define the displacement component $u_i = x_i' - x_i$ of the vector $\vec{u}$, as shown in figure 9.3, which is just the displacement of the material point P. Using this expression for u_i, we have

$$\frac{\partial x_i'}{\partial x_j} = \frac{\partial u_i}{\partial x_j} + \frac{\partial x_i}{\partial x_j} = \frac{\partial u_i}{\partial x_j} + \delta_{ij} \tag{9.6}$$

$$ds'^2 - ds^2 = \left[\left(\frac{\partial u_i'}{\partial x_j} + \delta_{ij} \right) \left(\frac{\partial u_i'}{\partial x_k} + \delta_{ik} \right) - \delta_{jk} \right] dx_j dx_k$$
$$= \left[\frac{\partial u_j'}{\partial x_k} + \frac{\partial u_k'}{\partial x_j} + \frac{\partial u_i'}{\partial x_j} \frac{\partial u_i'}{\partial x_k} \right] dx_j dx_k. \tag{9.7}$$

For most applications, we will only be interested in small deformation, i.e. small displacements, so that the term in equation (9.7) that is the product of two partial derivatives may be dropped. Therefore, the measure of the deformation becomes

$$ds'^2 - ds^2 = \left[\frac{\partial u_j'}{\partial x_k} + \frac{\partial u_k'}{\partial x_j} \right] dx_j dx_k. \tag{9.8}$$

Next we define the symmetric strain tensor and the antisymmetric rotation tensor by

$$S_{ij} = \frac{1}{2} \left[\frac{\partial u_i'}{\partial x_j} + \frac{\partial u_j'}{\partial x_i} \right] \tag{9.9}$$

$$R_{ij} = \frac{1}{2}\left[\frac{\partial u_i'}{\partial x_j} - \frac{\partial u_j'}{\partial x_i}\right]. \tag{9.10}$$

Thus the elongations of a line, $\partial u_i'/\partial x_j$, can be written as the sum of the symmetric and antisymmetric strain tensors

$$\begin{aligned}\frac{\partial u_i'}{\partial x_j} &= \frac{1}{2}\left[\frac{\partial u_i'}{\partial x_j} + \frac{\partial u_j'}{\partial x_i}\right] + \frac{1}{2}\left[\frac{\partial u_i'}{\partial x_j} - \frac{\partial u_j'}{\partial x_i}\right] \\ &= S_{ij} + R_{ij}.\end{aligned} \tag{9.11}$$

From figure 9.3, after the material has been deformed, the point P (taken to be at the origin) goes into P' and will have the coordinates (u_1, u_2, u_3); and the coordinate Q at (x_1, x_2, x_3) which will go into Q' with the coordinates (for a homogeneous deformation) $(x_1 + u_1 + \vec{x}_1 \cdot \nabla u_1, x_2 + u_2 + \vec{x}_2 \cdot \nabla u_2, x_3 + u_3 + \vec{x}_3 \cdot \nabla u_3)$ as shown in more detail in figure 9.4.

The components of displacement along the three axes are

$$u_i' - u_i \equiv \Delta u_i = \frac{\partial u_i'}{\partial x_j} x_j. \tag{9.12}$$

The tensors S_{ij} and R_{ij} are material parameters and thus their values will be dependent on the particular material under consideration.

Example. Figure 9.4 shows a line, initially along the x_1 axis (x-axis) that undergoes a displacement, $\vec{x}$, in the x–y plane as well as a subsequent deformation. We wish to determine the displacement of the line. From equation (9.12) above, the displacement components are

$$\Delta u_1 = S_{11}x_1 + (S_{12} + R_{12})x_2 + (S_{13} + R_{13})x_3 \tag{9.13}$$

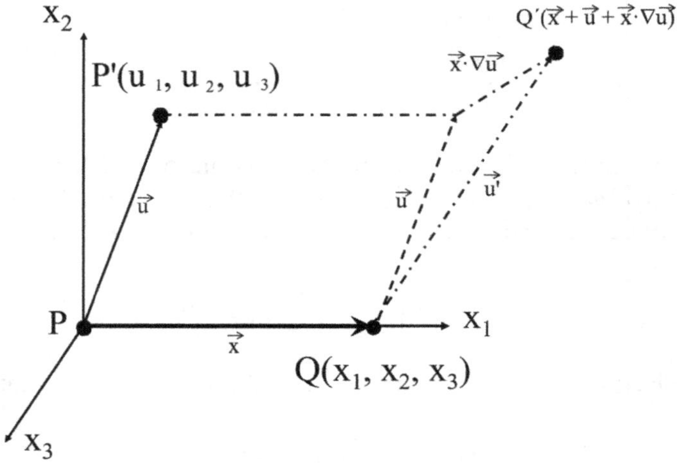

Figure 9.4. Displacement of neighboring points undergoing a deformation.

$$= S_{11}x_1 \tag{9.14}$$

$$\Delta u_2 = (S_{21} + R_{21})x_1 + S_{22}x_2 + (S_{23} + R_{23})x_3 \tag{9.15}$$

$$= (S_{21} + R_{21})x_1 \tag{9.16}$$

$$\Delta u_3 = (S_{31} + R_{31})x_1 + (S_{22} + R_{32})x_2 + S_{33}x_3 \tag{9.17}$$

$$= 0. \tag{9.18}$$

We arrive at the above since $x_2 = x_3 = 0$, since the point Q lies on the x_1-axis. Since the displacement is only in the x–y plane (x_1–x_2 plane), the derivatives involving x_3 are zero, making the third component Δu_3 vanish. Thus we conclude that the displacement along the x-axis is $S_{11}x_1$ and the rotation about the z-axis is $(S_{21} + R_{21})$ as shown in figure 9.4. In addition, suppose that we also wish to know the deformation of a line along the y-axis, for example the line PR in figure 9.4. The displacement along the y-axis is $S_{22}x_2$ and the rotation would be $(S_{21} - R_{21})$ about the z-axis toward the x-axis. This follows from the fact that $S_{21} = S_{12}$ and $R_{12} = -R_{21}$ so that, from equation (9.12), the appropriate tensor equations are

$$\Delta u_2 = S_{22}x_2$$
$$\Delta u_1 = (S_{12} + R_{12})x_2 = (S_{21} - R_{21})x_2.$$

The above example is one key piece for what follows. As we shall see shortly, for a given acoustic wave we can use the above results to find the strain tensor components.

9.4 The photoelastic effect

The linear photoelastic effect involves changes in the optical properties of insulators due to acoustic strain in a manner analogous to the linear EO effect. We can account for the strain induced changes in the optical indicatrix by using the same type of relationship as was done in treating the linear EO effect. Therefore, let us write the changes in the impermeability tensor as

$$\Delta B_{ij} = p_{ijkl} S_{kl} + p'_{ijkl} R_{kl}, \tag{9.19}$$

where S_{ij} and R_{ij} are as defined in the previous section and the coefficients p_{ijkl} are called the elements of the strain-optic tensor. For algebraic simplicity, the rotation term is often ignored. However, it has been pointed out that this is not always a good approximation [5]. For our purposes we will make this approximation as a reasonable first attempt at some problems. In any case, it will keep the algebraic complexity to a minimum. Therefore

$$\Delta B_{ij} \simeq p_{ijkl} S_{kl}. \tag{9.20}$$

We can write this out in full as

$$
\begin{pmatrix} \Delta B_1 \\ \Delta B_2 \\ \Delta B_3 \\ \Delta B_4 \\ \Delta B_5 \\ \Delta B_6 \end{pmatrix} = \begin{pmatrix} p_{11} & p_{12} & p_{13} & p_{14} & p_{15} & p_{16} \\ p_{21} & p_{22} & p_{23} & p_{24} & p_{25} & p_{26} \\ p_{31} & p_{32} & p_{33} & p_{34} & p_{35} & p_{36} \\ p_{41} & p_{42} & p_{43} & p_{44} & p_{45} & p_{46} \\ p_{51} & p_{52} & p_{53} & p_{54} & p_{55} & p_{56} \\ p_{61} & p_{62} & p_{63} & p_{64} & p_{65} & p_{66} \end{pmatrix} \begin{pmatrix} S_1 \\ S_2 \\ S_3 \\ S_4 \\ S_5 \\ S_6 \end{pmatrix},
$$

where the contracted notation has been used ($S_{11} \equiv S_1$, $S_{12} \equiv S_6$, etc).

Example. We would like to gain some feeling for the photoelastic tensor components based on symmetry arguments, much as we did for the electro-optic coefficients. Consider the p_{ijkl} for an isotropic solid. In an isotropic nondeformed solid, the index ellipsoid is

$$
\frac{1}{n^2}(x^2 + y^2 + z^2) = 1. \tag{9.21}
$$

If we apply a longitudinal strain along the x-axis, meaning that the displacement is along that axis and therefore that all derivatives for the Δu_i with respect to y and z (x_1 and x_2) are zero; the result is that S_{11} (S_1 in contracted notation) will be the only surviving term. Consequently, the index ellipsoid keeps the same principal axes in the deformed state. Therefore we conclude that

$$
p_{41} = p_{51} = p_{61} = 0 \quad \text{(contracted notation)}.
$$

Furthermore, the equivalence of y and z due to symmetry gives $p_{21} = p_{31}$. Therefore, the index ellipsoid becomes

$$
\left[\frac{1}{n^2} + p_{11} S_1 \right] x^2 + \left[\frac{1}{n^2} + p_{21} S_1 \right] (y^2 + z^2) = 1. \tag{9.22}
$$

A similar argument applied to a strain S_2 (S_{22}) and S_3 (S_{33}) gives

$$
p_{42} = p_{43} = p_{52} = p_{53} = p_{62} = p_{63} = 0.
$$

Due to the equivalence of x, y, and z it is easy to see that

$$
p_{12} = p_{21} = p_{31} = p_{13} = p_{23} = p_{32}
$$

and that

$$
p_{11} = p_{22} = p_{33}.
$$

Additional symmetry arguments lead to the following form for the p_{ijkl} matrix of an isotropic solid (in contracted form):

$$
\begin{bmatrix}
p_{11} & p_{12} & p_{12} & 0 & 0 & 0 \\
p_{12} & p_{11} & p_{12} & 0 & 0 & 0 \\
p_{12} & p_{12} & p_{11} & 0 & 0 & 0 \\
0 & 0 & 0 & \frac{1}{2}(p_{11}-p_{12}) & 0 & 0 \\
0 & 0 & 0 & 0 & \frac{1}{2}(p_{11}-p_{12}) & 0 \\
0 & 0 & 0 & 0 & 0 & \frac{1}{2}(p_{11}-p_{12})
\end{bmatrix}.
\tag{9.23}
$$

Example. As another example, consider the index ellipsoid when all S_{ij} are zero except $S_{12} = S_{21} = S_6$

$$
\Delta B_{ij} = p_{ijkl} \quad \text{or} \quad \Delta B_\alpha = p_{\alpha\beta} S_\beta \ (\alpha, \beta = 1, 2, \dots, 6).
$$

The index ellipsoid becomes

$$
\frac{x^2}{n^2} + \frac{y^2}{n^2} + \frac{z^2}{n^2} + 2p_{66} S_6 xy = 1.
\tag{9.24}
$$

This has the same form as an example of the EO effect given in chapter 8, and based on what was done there it is easy to see that the index ellipsoid in the new principal axis system is

$$
\left(\frac{1}{n^2} + p_{66} S_6\right) x'^2 + \left(\frac{1}{n^2} - p_{66} S_2\right) y'^2 + \frac{z^2}{n^2} = 1.
\tag{9.25}
$$

Following the same procedures as in chapter 6, we find that

$$
n_{x'} = n - \frac{n^3}{2} p_{66} S_6
$$
$$
n_{y'} = n + \frac{n^3}{2} p_{66} S_6
$$
$$
n_{z'} = n_z,
$$

where $p_{66} = \frac{1}{2}(p_{11} - p_{12})$.

Thus we see that the form of the strain-optic tensors, but not the magnitude, can be obtained from considerations of the crystal symmetry. The procedure for doing this is the same as was demonstrated in chapter 8. The strain-optic tensors for all classes of crystal symmetry are given in table 9.1. It turns out that the form of these tensors is exactly the same as the tensors for the quadratic electro-optic effect. Listed below are values for some of the strain-optic coefficients for various crystals while table 9.2 lists some of the other properties of various acousto-optic materials [6].

Table 9.1. Photoelastic tensors for various crystal symmetry classes.

Isotropic: (water, fused silica)

$$
\begin{pmatrix}
p_{11} & p_{12} & p_{12} & 0 & 0 & 0 \\
p_{12} & p_{11} & p_{12} & 0 & 0 & 0 \\
p_{12} & p_{12} & p_{11} & 0 & 0 & 0 \\
0 & 0 & 0 & \frac{1}{2}(p_{11}-p_{12}) & 0 & 0 \\
0 & 0 & 0 & 0 & \frac{1}{2}(p_{11}-p_{12}) & 0 \\
0 & 0 & 0 & 0 & 0 & \frac{1}{2}(p_{11}-p_{12})
\end{pmatrix}
$$

Cubic: ($\bar{4}$3m, 432, m3m) GaAs, Ge, NaCl

$$
\begin{pmatrix}
p_{11} & p_{12} & p_{12} & 0 & 0 & 0 \\
p_{12} & p_{11} & p_{12} & 0 & 0 & 0 \\
p_{12} & p_{12} & p_{11} & 0 & 0 & 0 \\
0 & 0 & 0 & p_{44} & 0 & 0 \\
0 & 0 & 0 & 0 & p_{44} & 0 \\
0 & 0 & 0 & 0 & 0 & p_{44}
\end{pmatrix}
$$

Hexagonal: ($\bar{6}$m2, 6mm, 622, 6/mmm) CdS

$$
\begin{pmatrix}
p_{11} & p_{12} & p_{13} & 0 & 0 & 0 \\
p_{12} & p_{11} & p_{13} & 0 & 0 & 0 \\
p_{31} & p_{31} & p_{33} & 0 & 0 & 0 \\
0 & 0 & 0 & p_{44} & 0 & 0 \\
0 & 0 & 0 & 0 & p_{44} & 0 \\
0 & 0 & 0 & 0 & 0 & \frac{1}{2}(p_{11}-p_{12})
\end{pmatrix}
$$

Tetragonal: (4mm, 422, 4/mmm, $\bar{4}$2m)
KDP, ADP, TeO$_2$

$$
\begin{pmatrix}
p_{11} & p_{12} & p_{13} & 0 & 0 & 0 \\
p_{12} & p_{11} & p_{13} & 0 & 0 & 0 \\
p_{31} & p_{31} & p_{33} & 0 & 0 & 0 \\
0 & 0 & 0 & p_{44} & 0 & 0 \\
0 & 0 & 0 & 0 & p_{44} & 0 \\
0 & 0 & 0 & 0 & 0 & p_{66}
\end{pmatrix}
$$

Tetragonal: (4, $\bar{4}$ 4/m)
PbMoO$_4$

$$
\begin{pmatrix}
p_{11} & p_{12} & p_{13} & 0 & 0 & p_{16} \\
p_{12} & p_{11} & p_{13} & 0 & 0 & -p_{16} \\
p_{31} & p_{31} & p_{33} & 0 & 0 & 0 \\
0 & 0 & 0 & p_{44} & p_{45} & 0 \\
0 & 0 & 0 & -p_{45} & p_{44} & 0 \\
p_{61} & -p_{61} & 0 & 0 & 0 & p_{66}
\end{pmatrix}
$$

Trigonal: (3m, 32, $\bar{3}$m)
Al$_2$O$_3$, LiTaO$_3$, LiNbO$_3$

$$
\begin{pmatrix}
p_{11} & p_{12} & p_{13} & p_{14} & 0 & 0 \\
p_{12} & p_{11} & p_{13} & -p_{14} & 0 & 0 \\
p_{13} & p_{13} & p_{33} & 0 & 0 & 0 \\
p_{41} & -p_{41} & 0 & p_{44} & 0 & 0 \\
0 & 0 & 0 & 0 & p_{44} & p_{41} \\
0 & 0 & 0 & 0 & p_{14} & \frac{1}{2}(p_{11}-p_{12})
\end{pmatrix}
$$

Table 9.2. Table of properties of some acousto-optic materials.

Material	ρ	v_s	n	p	M ($\times 10^{-15}$)
Water (H$_2$O)	1000 kg m^{-3}	1500 m s^{-1}	1.33	0.31	158
Fused quartz (SiO$_2$)	2200	5970	1.46	0.20	0.83
Lithium niobate (LiNbO$_3$)	4700	7400	2.25	0.15	1.5
Sapphire (Al$_2$O$_3$)	4000	11,000	1.76	0.17	0.16
Lead molybdate (PbMoO$_4$)	6950	3750	2.30	0.28	32
KDP (KH$_2$PO$_4$)	2340	5500	1.51	0.25	1.91
Lithium tantalate (LiTaO$_3$)	7450	6190	2.18	0.15	1.37

Example. We wish now to take a look at how to determine the impermeability change in the presence of an acoustic disturbance. Suppose an acoustic wave is propagating in an isotropic medium. Let the sound wave be a longitudinal wave propagating in the z-direction. The displacement produced by a longitudinal sound wave can be written as

$$\vec{u}(z, t) = -\Re \left[\vec{u}_0 \, e^{i(Kz - \Omega t)} \right] = -\vec{u}_0 \cos (Kz - \Omega t), \tag{9.26}$$

where K is the wave number of the sound wave, Ω is the frequency, and u_0 is its amplitude. The polarization of the wave is determined from the condition of its being either 'longitudinal' or 'shear'. In the former case the direction of relative particle displacement is along the direction of propagation. For a shear wave the displacement occurs in a direction transverse to the propagation direction. For the above example we thus have $\vec{u}_0 = u_0 \, \hat{z}$, since we are dealing with a longitudinal wave. The only component of the displacement is u_3 and it depends only on z (x_3). Therefore, from the definition of S_{ij}, equation (9.9), the only non-zero component of the strain tensor is

$$S_{33} \equiv S_3 = K u_0 \sin (Kz - \Omega t)$$
$$= S_0 \sin (Kz - \Omega t).$$

For an *isotropic* medium, the strain-optic tensor components give the following results for the change in the impermeability tensor:

$$\Delta B_{11} = p_{12} S_3 = p_{12} S_0 \sin (Kz - \Omega t)$$
$$\Delta B_{22} = p_{12} S_3 = p_{12} S_0 \sin (Kz - \Omega t)$$
$$\Delta B_{33} = p_{11} S_3 = p_{11} S_0 \sin (Kz - \Omega t)$$
$$\Delta B_{ij} = 0 \quad \text{for} \ i \neq j.$$

Therefore the new index ellipsoid becomes

$$\left[\frac{1}{n^2} + p_{12} S_0 \sin (Kz - \Omega t) \right] x^2 + \left[\frac{1}{n^2} + p_{12} S_0 \sin (Kz - \Omega t) \right] y^2$$
$$+ \left[\frac{1}{n^2} + p_{11} S_0 \sin (Kz - \Omega t) \right] z^2 = 1. \tag{9.27}$$

Since the principal axes of the index ellipsoid remain unchanged in this case, the new principal indices of refraction are easily found to be

$$n_x = n - \frac{1}{2}n^3 p_{12} S_0 \sin(Kz - \Omega t)$$

$$n_y = n - \frac{1}{2}n^3 p_{12} S_0 \sin(Kz - \Omega t) \qquad (9.28)$$

$$n_z = n - \frac{1}{2}n^3 p_{11} S_0 \sin(Kz - \Omega t).$$

We see that the index of refraction seen for x- and y-polarized optical fields are equal, whereas the index n_z is different. The amount of change in the index depends on both the material, through p_{12} and p_{11}, and on the amplitude of the acoustic wave. Our result also shows what we postulated at the beginning of this chapter—the index of refraction varies sinusoidally through the crystal.

9.5 Diffraction of light by acoustic waves

We have seen how an acoustic wave can cause a periodic modulation in the refractive index of a medium with the period of the modulation being the period of the acoustic wave. This interaction of sound waves with a light beam forms the heart of the study of acoustic-optics. Next we wish to address the question of how an electromagnetic wave will propagate in such a periodically modulated medium. At the beginning of this chapter we described qualitatively how light is scattered by the acoustic wave and that this scattering occurs because the material displacement associated with the acoustic wave causes changes in the index of refraction. The displacements may be longitudinal, as we have seen in the previous example, or they may be transverse. The longitudinal sound waves are compressional waves, whereas transverse waves are shear waves. In the discussions below, we will only consider scattering of a light wave by a periodic variation in the index of refraction without regard to whether the variation is caused by a longitudinal or transverse acoustic wave. The index variation behaves like an optical grating, of spacing $\Lambda = 2\pi v_s/\omega$ with v_s being the acoustic wave velocity, and thus it is capable of diffracting the light. This grating travels in the direction of $\vec{K}$ with the velocity v_s. This velocity, however, is much smaller than the light velocity (~ 5 orders of magnitude) so that it can be considered fixed relative to the incident light beam. The actual effect of the displacements is only noticed by a Doppler shift in the optical frequency. We will consider two types of scattering by this acoustical grating:

- Raman–Nath diffraction which is scattering from a thin medium, i.e. where the transverse dimensions of the sound wave are small and the interaction length is short; here more than one scattered wave is possible.
- Bragg diffraction which is scattering from a thick medium, i.e. the transverse dimensions of the sound wave are large and the interaction length is long; here only one scattered wave is possible.

These two types of scattering are depicted in figure 9.5.

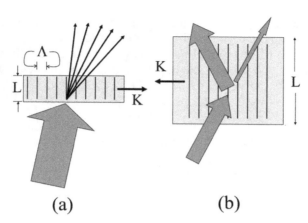

$$(a) \qquad\qquad\qquad (b)$$

Figure 9.5. Scattering of optical beams by acoustic waves: (a) Raman–Nath diffraction and (b) Bragg diffraction.

9.5.1 Raman–Nath diffraction

Let us consider a plane optical wave incident on an acoustic medium with some large rectangular cross section. Assume that the optical wave is incident on the acoustical medium at an angle of θ and that it passes through the medium for some finite distance L before exiting. Consider a thin-sheet region in a medium in which an acoustic wave induces a traveling-wave modulation in the index of refraction, given by

$$\Delta n(x,\, y,\, z,\, t) = \Delta n_o \sin(\vec{K} \cdot \vec{r} - \Omega t), \quad 0 < z < L$$
$$= 0, \qquad\qquad\quad \text{otherwise,}$$

where $\Delta n_o \propto pS$, i.e. just the elasto-optic coefficient times the strain without particularizing the polarization components, for ease in notation. Thus, for the sake of simplicity in illustrating the basic ideas, we assume that the medium is isotropic and Δn is a scalar. An incident optical wave is represented by

$$\vec{E}_i = \vec{\mathcal{E}}_0 e^{i(\vec{k} \cdot \vec{r} - \omega t)}.$$

This wave is incident onto the thin 'sheet' of sound wave at $z = 0$. In Raman–Nath scattering, the interaction length L is small enough so that this periodically perturbed sheet ($0 < z < L$) acts as a phase grating. In other words, the transmission of light through the perturbed region $0 < z < L$ only affects the phase of the plane wave and the transmitted wave can be written

$$\vec{E}_t = \vec{\mathcal{E}}_0 e^{i(\vec{k} \cdot \vec{r} - \omega t + \phi)}, \tag{9.29}$$

where ϕ is the phase shift result from passing through the perturbed region and can be written

$$\phi = \int \frac{\omega}{c} \Delta n \, d\ell. \tag{9.30}$$

The integral is over the ray path within the region $0 < z < L$. Since θ is the angle of incidence (i.e. the angle between the k vector and the direction of the sound wavevector, assumed to be parallel to the z-axis), the path length in the perturbed region is $L/\cos\theta$. If L is small enough, the integral can simply be written

$$\phi = \frac{\omega}{c}\,\Delta n_0 \frac{L}{\cos\theta}\,\sin(\vec{K}\cdot\vec{r} - \Omega t). \tag{9.31}$$

Substitution of equation (9.31) for ϕ into equation (9.29) leads to

$$\vec{E}_t = \vec{\mathcal{E}}_0 e^{i(\vec{k}\cdot\vec{r} - \omega t + \delta\,\sin(\vec{K}\cdot\vec{r} - \Omega t))}, \tag{9.32}$$

where δ is given by

$$\delta = \frac{\omega \Delta n_0 L}{c\cos\theta} = \frac{2\pi \Delta n_0 L}{\lambda \cos\theta}$$

and is referred to as the modulation index. Thus the transmitted field is phase-modulated in a manner quite similar to electro-optic phase modulation. By using the following identities for Bessel functions as was done in chapter 8,

$$\begin{aligned}
\cos(x\sin\theta) &= J_0(x) + 2J_2(x)\cos 2\theta + 2J_4(x)\cos 4\theta + \cdots \\
\sin(x\sin\theta) &= 2J_1(x)\sin\theta + 2J_3(x)\sin 3\theta + 2J_5(x)\sin 5\theta + \cdots.
\end{aligned} \tag{9.33}$$

The phase factor in equation (9.32) can be written

$$e^{i\delta\,\sin\theta} = \sum_{q=-\infty}^{\infty} J_q(\delta)e^{iq\theta}$$

and the transmitted field, equation (9.32), can be written

$$\vec{E}_t = \vec{\mathcal{E}}_0 \sum_{q=-\infty}^{\infty} J_q(\delta)e^{i[(\vec{k}+q\vec{K})\cdot\vec{r} - (\omega + q\Omega)t]}. \tag{9.34}$$

From the above expression we see that the transmitted field is just a linear superposition of plane waves with frequencies of $\omega + q\Omega$ and wavevectors of $\vec{k} + q\vec{K}$ which correspond to the qth-order diffracted beam with amplitude of $J_q(\delta)$.

The diffraction efficiency for the qth-order Raman–Nath diffraction is given by [6]

$$\eta(q) = J_q^2(\delta) = J_q^2\left[\frac{2\pi L \Delta n_0}{\lambda \cos\theta}\right]. \tag{9.35}$$

If there is no acoustic wave, i.e. $\Delta n_0 = 0$, all the light energy is found in the $q = 0$ order, that is, $\eta_0 = 1$ and $\eta_q = 0$ for $q \neq 0$. Using the identity [7]

$$\sum_{q=-\infty}^{\infty} J_q^2(\delta) = J_0(0) = 1, \tag{9.36}$$

it follows that

$$\sum_{q=-\infty}^{\infty} \eta(q) = 1.$$

In other words, the sum of the intensities of all the diffracted orders must be equal to the intensity of the incident beam so that the energy is conserved.

The intensity of the qth order is proportional to $J_q^2(\delta)$. In figure 8.1, in chapter 8, we show the five lowest order Bessel functions. From this figure it is clear that in most applications we will be interested in either the directly transmitted beam ($q = 0$) or one of the first-order scattered beams ($q = \pm 1$). This is because for small values of the peak phase shift, δ, most of the transmitted energy is contained in these orders.

It is important to observe that the peak value of $J_1^2(\delta)$ is 0.335, occurring for $\delta = 1.85$ radians. This limits the efficiency of thin acoustic cells to 33.5% when used as beam deflectors. Also, one may have to be concerned about interference from other scattered orders if the use of a wide range of scanning angles is contemplated. For example, when $\delta = 1.85$ radians we have $J_2^2(\delta = 1.85) = 0.102$ so that the first-order beam is only three times as strong as the second order beam.

9.5.2 Bragg scattering

When the transverse dimensions of the acoustic wave are large and the interaction length long, there will only be one scattered wave, as illustrated in figure 9.5(b), which results in Bragg diffraction. Reference [8] gives a coupled mode treatment of Bragg diffraction, the main elements of which will be given here. In order to proceed with such an analysis, we review the connection between the impermeability tensor and the relative dielectric tensor.

In chapter 7 we saw that in the principal axis system

$$B_{ii} = 1/\varepsilon_{ii} = 1/n_i^2$$

and also that

$$\varepsilon_{ij}B_{jk} = \delta_{ik}. \tag{9.37}$$

Pockels originally defined the electro-optic and elasto-optic effects in terms of the impermeability tensor B_{ij}. It is often more convenient, however, to use the changes in the polarization vector, in particular for nonlinear effects, and thus we need to know the connection between $\Delta\varepsilon_{ij}$ and ΔB_{ij}. Using equation (9.37) and assuming that $\Delta\varepsilon_{ij} \ll \varepsilon_{ii}, \varepsilon_{jj}$, and the fact that in the principal axis system

$$B_{ij} = \varepsilon_{ij} = 0, \quad i \neq j,$$

we have

$$\Delta\chi_{ij} = \Delta\varepsilon_{ij} = -\Delta B_{ij}\varepsilon_{ii}\varepsilon_{jj}.$$

Therefore, using the above result along with equation (9.19), the induced polarization due to elasto-optic effects is given by

$$\Delta P_i = \varepsilon_0 \Delta \chi_{ij} E_j = -\varepsilon_0 \varepsilon_{ii} \varepsilon_{jj} p_{ijkl} S_{kl} E_j.$$

We will have to sort through this tangle of subscripts; again, all repeated subscripts represent a summation.

The basic scenario for Bragg diffraction is the following:
1. An acoustical wave is input into an optical medium.
2. An optical beam is input into the same medium.
3. Under certain conditions, part of the optical beam is diffracted in a new direction and shifted in frequency by an amount equal to the acoustic frequency.

Let us start with the wave equation for an optical field propagating in a medium simultaneously supporting an acoustical wave:

$$\nabla^2 \vec{E} = \mu_0 \varepsilon_0 \varepsilon \frac{\partial^2 \vec{E}}{\partial t^2} + \mu_0 \frac{\partial^2 \Delta \vec{P}}{\partial t^2} \quad [\sigma = 0; \mu = \mu_0], \tag{9.38}$$

where $\Delta \vec{P}$ is the change in polarization caused by the sound wave. Let the optical field consist of two plane waves: one incident wave, $\vec{E}_i$, and the scattered beam, $\vec{E}_s$:

$$\vec{E}_0(\vec{r}, t) = \vec{\mathcal{E}}_0 e^{i(\vec{k}_0 \cdot \vec{r} - \omega_i t)} \tag{9.39}$$

$$\vec{E}_s(\vec{r}, t) = \vec{\mathcal{E}}_s e^{i(\vec{k}_s \cdot \vec{r} - \omega_s t)}, \tag{9.40}$$

where $\vec{k}_0$ and $\vec{k}_s$ are propagation vectors for the incident and scattered waves. The total E-field is the sum of the incident and scattered waves.

Next, we need to apply the wave equation to both the incident and scattered fields. As a temporary measure, to evaluate the ∇^2 operator in the wave equation for these two fields we define r_0 and r_s to be distances measured along $\vec{k}_0$ and $\vec{k}_s$, respectively. For example, if we let $\vec{k}$ be along the z-axis, then the ∇^2 operator can be expressed as

$$\nabla^2 = \frac{\partial^2}{\partial x^2} + \frac{\partial^2}{\partial y^2} + \frac{\partial^2}{\partial z^2} = \frac{\partial^2}{\partial x^2} + \frac{\partial^2}{\partial y^2} + \frac{\partial^2}{\partial r_i^2},$$

which leads to

$$\begin{aligned} \nabla^2 \vec{\mathcal{E}}(\vec{r}, t) &= \nabla^2 \vec{\mathcal{E}}(r_i) \exp[i(k r_i - \omega_i t)] \\ &= \frac{\partial^2}{\partial r_i^2} \vec{E}(r_i) \exp[i(k r_i - \omega_i t)] \\ &= \frac{\partial}{\partial r_i}\left[\frac{\partial \vec{\mathcal{E}}}{\partial r_i} + ik\mathcal{E}\right] \exp[i(k r_i - \omega_i t)] \\ &= \left[\frac{d^2 \mathcal{E}}{dr_i^2} + 2ik\frac{d\mathcal{E}}{dr_i} + (ik)^2 \mathcal{E}\right] \exp[i(k r_i - \omega_i t)]. \end{aligned} \tag{9.41}$$

Now we make what is called the 'slowly varying amplitude approximation' which says that the amplitude variation of the waves is small enough that we may take the following approximation:

$$\frac{d^2\mathcal{E}}{dr_i^2} \ll k\frac{d\mathcal{E}}{dr_i}.$$

This approximation is valid when the polarization source term ΔP in the wave equation is small. With this approximation the ∇^2 term becomes

$$\nabla^2 E_i = \left[2ik\frac{d\mathcal{E}}{dr_i} - k^2\mathcal{E}\right]\exp\left[i(kr_i - \omega_i t)\right]. \tag{9.42}$$

This then becomes the left-hand side (LHS) of equation (9.38). On the right-hand side (RHS) we have

$$\begin{aligned}
\Delta P_i(\vec{r}, t) &= -\varepsilon_0\varepsilon_{ii}\varepsilon_{jj}p_{ijkl}S_{kl}E_j \\
&= -\varepsilon_0\varepsilon_i\varepsilon_j p_{ijkl}S_{kl}(\vec{r}, t)\mathcal{E}_j\exp[i(\vec{k}_j \cdot \vec{r}_j - \omega_j t)],
\end{aligned} \tag{9.43}$$

where $\varepsilon_{ii} = \varepsilon_i$ and $\varepsilon_{jj} = \varepsilon_j$. The input strain field is that of an acoustic wave propagating in some arbitrary direction $\vec{K}$ at frequency Ω and can be taken as a sinusoidal function as we have already seen,

$$S_{kl}(\vec{r}, t) = S_{kl}\sin(\vec{K} \cdot \vec{r} - \Omega t), \tag{9.44}$$

and the induced polarization change is given by

$$\begin{aligned}
\Delta P_i = -\epsilon_0\epsilon_i\epsilon_j p_{ijkl}S_{kl}&\left(\frac{e^{i(\vec{K}z-\Omega t)} - e^{-i(\vec{K}z-\Omega t)}}{2i}\right) \\
&\times \mathcal{E}_j\exp\left[i\left(\vec{k}_j \cdot \vec{r}_j - \omega_j t\right)\right],
\end{aligned}$$

where we assume an acoustic wave traveling in the z-direction.

To simplify this problem somewhat, we will make several reasonable assumptions. First, we will take both the incident and scattered wavevectors to lie in the x–z plane, i.e. $\vec{k}_0 = k_{0x}\hat{x} + k_{0z}\hat{z}$ and likewise for $\vec{k}_s$. Second, we will take the optical waves to be traveling nearly in the x-direction, so that derivatives with respect to z of the field amplitudes will be neglected. Using these points and returning to Cartesian coordinates, we can write the simplified version of the wave equation for both the incident and scattered fields as

$$\begin{aligned}
&2i\left[k_{0x}\frac{\partial(\mathcal{E}_0)_i}{\partial x} + k_{0z}\frac{\partial(\mathcal{E}_0)_i}{\partial z}\right]e^{i(k_{0x}x+k_{0z}z-\omega_0 t)} \\
&+ 2i\left[k_{sx}\frac{\partial(\mathcal{E}_s)_i}{\partial x} + k_{sz}\frac{\partial(\mathcal{E}_s)_i}{\partial z}\right]e^{i(k_{sx}x+k_{sz}z-\omega_s t)} \\
&= \omega^2\mu_0\varepsilon_0\varepsilon_i\varepsilon_j p_{ijkl}S_{kl}\left(\frac{e^{i(\vec{K}z-\Omega t)} - e^{-i(\vec{K}z-\Omega t)}}{2i}\right) \\
&\times \left[(\mathcal{E}_0)_j e^{i(k_{0x}x+k_{0z}z-\omega_0 t)} + (\mathcal{E}_s)_j e^{i(k_{sx}x+k_{sz}z-\omega_s t)}\right].
\end{aligned} \tag{9.45}$$

Now we will multiply everything above by $(\mathcal{E}_0)_i^* e^{-i(k_{0x}x+k_{0y}y-\omega_0 t)}$. We wish to see how the fields change in traversing the crystal essentially along the x-direction. It is sufficient for us to look at these equations by averaging over the variable z and over time. Skipping some tedious algebra, we find first that we must require $\omega_0 = \omega_s \pm \Omega$ for the fields to not be averaged to zero. In addition, we must have $\vec{k}_0 = \vec{k}_s \pm \vec{K}$ for the same reason. The latter condition is referred to as 'phase-matching' and has the physical interpretation as representing conservation of momentum in the acousto-optic interaction. The frequency matching condition is a statement of conservation of energy. These conditions are expressed graphically in figure 9.6. With all of this, we arrive at a final result

$$\frac{d\mathcal{E}_0}{dx} = i\alpha_{ij}(\mathcal{E}_s)_j e^{i\gamma} \qquad (9.46)$$

$$\frac{d(\mathcal{E}_s)_j}{dx} = i\alpha_{ji}\mathcal{E}_i e^{-i\gamma}, \qquad (9.47)$$

where we have defined

$$\alpha_{ij} = \frac{\omega_i^2}{4}(\mu_0\epsilon_0\epsilon_i)\epsilon_j p_{ijkl} S_{kl} \qquad (9.48)$$

and $\gamma = (k_s - k_0)x$.

Equations (9.47) are the coupled mode equations for the two optical fields present in the problem, the incident field and the scattered or diffracted field. Finally, when $k_0 = k_s$, γ is zero, which occurs for exactly the Bragg reflection condition. What happens if γ is not zero? It turns out that no significant contribution of $\mathcal{E}_i$ to $\mathcal{E}_s$ will occur and vice versa for interaction lengths greater than $\pi/|k_s - k_0|$ when γ is non-zero, since the energy carried by the two fields is essentially traded back and forth. Going back to topics covered much earlier in this text, the two fields act like coupled harmonic oscillators trading energy.

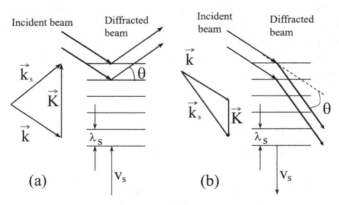

Figure 9.6. Bragg diffraction conditions: (a) counterpropagating sound waves and (b) sound waves adding energy to the optical field.

Next consider the case when $\alpha_{ij} = \alpha_{ji}$; since $\omega_s \ll \omega_i$ and since $k \simeq k_s$, the vector triangle in figure 9.6 above gives

$$k_s = 2k \sin \theta. \tag{9.49}$$

Equation (9.49) is called the Bragg diffraction condition. Going back to the coupled equations, i.e. to equation (9.47), and using $\gamma = 0$ and $\alpha_{ij} = \alpha_{ji} = \alpha$, we have

$$\frac{d(\mathcal{E}_0)_i}{dx} = i\alpha(\mathcal{E}_s)_j \tag{9.50}$$

$$\frac{d(\mathcal{E}_s)_j}{dx} = i\alpha(\mathcal{E}_0)_i. \tag{9.51}$$

The solutions to these coupled equations are fairly easily shown [8] to be (dropping the indices for simplicity),

$$\mathcal{E}_i(x) = \mathcal{E}_i(0) \cos(\alpha x) + i\mathcal{E}_s(0) \sin(\alpha x) \tag{9.52}$$

$$\mathcal{E}_s(x) = \mathcal{E}_s(0) \cos(\alpha x) + i\mathcal{E}_i(0) \sin(\alpha x). \tag{9.53}$$

Now consider the special case when there is a single frequency input, ω_i; then $\mathcal{E}_s(0) = 0$ and

$$\mathcal{E}_i(x) = \mathcal{E}_i(0) \cos(\alpha x) \tag{9.54}$$

$$\mathcal{E}_s(x) = i\mathcal{E}_i(0) \sin(\alpha x). \tag{9.55}$$

From these,

$$|\mathcal{E}_i(x)|^2 + |\mathcal{E}_s(x)|^2 = |\mathcal{E}_i(0)|^2. \tag{9.56}$$

If the interaction distance between the two beams is such that $\alpha x = \alpha x = \pi/2$, the total power in the incident beam is transferred to the diffracted beam. The fraction of the incident beam power transferred in a distance ℓ is, from equations (9.55) and (9.56) above,

$$\frac{I_{diff}}{I_{incident}} = \frac{|\mathcal{E}_s|^2}{|\mathcal{E}_i(0)|^2} = \sin^2(\alpha\ell), \tag{9.57}$$

where, from equation (9.48), α is

$$\alpha = \frac{\omega}{2}[\varepsilon_0\mu_0]^{1/2} n^3 p_{ijkl} S_{kl} = \frac{\pi n^3}{\lambda} p_{ijkl} S_{kl}$$

$$= \frac{\pi n^3}{2\lambda} pS,$$

where $pS \equiv p_{ijkl} S_{kl}$. Therefore

$$\frac{I_{diff}}{I_{incident}} = \sin^2\left[\frac{\pi n^3}{2\lambda} pS\ell\right].$$

Example. It will be instructive to present one example in some detail to illustrate how all of this comes together. We consider a LiNbO$_3$ crystal into which is launched an acoustic signal. The result is a longitudinal wave, with the displacement given by

$$\vec{u}\,(x,\,t) = \hat{x}A\cos(Kx - \Omega t).$$

From our results in section 7.2 we can calculate the components of the strain tensor given the above, i.e. using equation (9.9). This leads to $S_{11} \equiv S_1 = -KA\sin(Kx - \Omega t)$ and all other $S_{ij} = 0$. The changes in the impermeability tensor components are given by $\Delta B_{ij} = p_{ijkl}S_{kl}$ or in the contracted notation $\Delta B_\alpha = p_{\alpha\beta}\,S_\beta$. Looking at the photoelastic tensor for LiNbO$_3$ we thus find that the non-zero components are

$$\Delta B_1 = p_{11}S_1$$
$$\Delta B_2 = p_{12}S_1$$
$$\Delta B_3 = p_{13}S_1$$
$$\Delta B_4 = p_{41}S_1.$$

From this we can write down the new impermeability tensor (recalling that LiNbO$_3$) is a uniaxial crystal):

$$\bar{B} = \begin{pmatrix} \dfrac{1}{n_0^2} + p_{11}S_1 & 0 & 0 \\[2mm] 0 & \dfrac{1}{n_0^2} + p_{12}S_1 & p_{41}S_1 \\[2mm] 0 & p_{41}S_1 & \dfrac{1}{n_e^2} + p_{13}S_1 \end{pmatrix}.$$

Yariv [8] has shown how the above expressions can be written in a somewhat more practical form. First, define the acoustic intensity according to

$$I_{\text{acoustic}} = \frac{1}{2}\rho v_s^3 S^2, \tag{9.58}$$

where ρ is the mass density and v_s is the velocity of sound. Then

$$S = \left[\frac{2I_{\text{acoustic}}}{\rho v_s^3}\right]^{1/2}, \tag{9.59}$$

which then leads to the following expression for the diffraction efficiency

$$\frac{I_{\text{diff}}}{I_{\text{incident}}} = \sin^2\left[\frac{\pi\ell}{\sqrt{2}\,\lambda}\left[\frac{n^6 p^2 I_{\text{acoustic}}}{\rho v_s^3}\right]^{1/2}\right]. \tag{9.60}$$

It is also customary to define a constant called the 'figure of merit' for diffraction. The figure of merit is defined as

$$M = \frac{n^6 p^2}{\rho v_s^3} \tag{9.61}$$

so that the diffraction efficiency can be written as

$$\frac{I_{\text{diff}}}{I_{\text{incident}}} = \sin^2\left[\frac{\pi \ell}{\sqrt{2}\,\lambda}[M I_{\text{acoustic}}]^{1/2}\right]. \tag{9.62}$$

Example. As an illustration, let us look at acousto-diffraction of a He–Ne laser beam in water. Since water is an isotropic material, from the tables of acoustic properties the pertinent parameters for the problem are

$$n = 1.33$$
$$p = 0.31$$
$$v_s = 1.5 \times 10^3 \text{ m s}^{-1}$$
$$\rho = 1000 \text{ kg m}^{-3}$$
$$\lambda = 0.6328 \ \mu\text{m}.$$

With these numbers, the equation for the diffraction efficiency becomes

$$\frac{I_{\text{diff}}}{I_{\text{incident}}} = \sin^2\left[1.4\ell\sqrt{I_{\text{acoustic}}}\right].$$

Using this expression for water we can write down a working equation for other materials at other wavelengths given by the formula below:

$$\frac{I_{\text{diff}}}{I_{\text{incident}}} = \sin^2\left[1.4\frac{0.632\,8}{\lambda\,(\mu\text{m})}\ell\sqrt{M_W I_{\text{acoustic}}}\right], \tag{9.63}$$

where $M_W = M/M_{\text{H}_2\text{O}}$ = diffraction figure of merit relative to H_2O.

At small diffraction efficiencies, the diffracted light intensity is proportional to the acoustic intensity (Why?).

Example. As a second illustration using the above working equation, let us look at acoustic scattering of He–Ne laser light in a different material, PbMoO_4 for example. Suppose we are given an acoustic power of 1 W and an acoustic beam cross section of 1 mm^2. Let the optical path of the acoustic beam be $\ell = 1$ mm. Finally from the table of acousto-optic properties we have $M_W = 0.22$. The values of the parameters in equation (9.63) lead to a diffraction efficiency of

$$\frac{I_{\text{diff}}}{I_{\text{incident}}} = 37\%.$$

9.6 Problems

1. Consider a sound wave polarized along the x-axis and propagating along the z-axis in germanium. If the displacement associated with the sound wave is

$$\vec{u}\,(z,\,t) = -\hat{x}A\,\cos(Kz - \Omega t),$$

determine the new principal indices of refraction.

2. Expand on the statement that for a longitudinal strain along the x-axis of an isotropic material,

$$p_{41} = p_{51} = p_{61} = 0.$$

3. Verify equation (9.25).

4. Find the non-zero components of the strain tensor for a longitudinal sound wave propagating along the y-axis.

5. Determine the diffraction efficiency for the first three orders in Raman–Nath diffraction if $\delta = 1.85$.

6. Since, in an isotropic medium, the refractive index seen by a particular light beam depends on the beam's direction and polarization, the index seen by the incident and diffracted beams in Bragg diffraction will in general be different and the momentum conservation triangle will no longer be isosceles. Find the Bragg diffraction condition for an anisotropic material.

7. Find the diffraction efficiency of He–Ne laser beam in KDP with an acoustic beam power of 2 W and cross section of 1 mm². Assume that the optical path of the acoustic beam is 10 mm.

8. Obtain an expression for the acoustic intensity needed in Bragg diffraction to convert 100% into diffracted power.

9. Consider a shear acoustic wave polarized along the x-axis and propagating along the y-axis in the crystal LiTaO₃.
   (a) Determine the components of the strain tensor.
   (b) Determine the impermeability tensor.

10. Suppose that in the arrangement described in the previous problem an optical wave at the He–Ne wavelength propagates in the y-direction in a 1 cm long crystal of LiTaO₃. For LiTaO₃, $p_{11} = -0.081$ and $p_{12} = 0.081$.
    (a) If the optical beam is polarized along the z-axis, find the acoustic power needed for 100% conversion of the incident power to diffracted power.
    (b) The same as part (a), except that the optical wave is polarized along the x-axis.

11. Consider a shear acoustic wave in LiNbO₃ which is polarized along the x-axis and propagating in the y-direction. The displacement associated with this wave is

$$\vec{u}\,(y,\,t) = -\hat{x}A\,\cos(Ky - \Omega t).$$

(a) Find the strain tensor associated with this wave.

(b) What is the change in the impermeability tensor, i.e. find ΔB_{ij}.

12. Obtain an expression for the Bragg diffraction efficiency similar to equation (9.62) when $k \neq k_s$.

13. Consider a shear acoustic wave polarized along the z-axis and propagating along the x-axis in PbMoO$_4$. The photoelastic constants are $p_{44} = 0.067$ and $p_{45} = -0.01$. If an optical wave of wavelength 0.633 μm propagates along the x-axis and is polarized along the z-axis:

(a) Find an expression for the Bragg diffraction efficiency.

(b) Evaluate the diffraction in part (a) if the length of the crystal is 1 cm and the acoustic power density is 2 W cm^{-2}.

14. Design an acoustic-optic beam deflector.

15. Consider the interaction of a longitudinal acoustic wave and an optical wave which are both propagating along the x-axis in LiNbO$_3$.

(a) Find the strain tensor.

(b) Find the impermeability tensor.

(c) Find an expression for the required acoustic frequency, Ω.

References

[1] Wikipedia: Léon Brillouin https://en.wikipedia.org/wiki/L%C3%A9on_Brillouin (Accessed: 11 Jan. 2021)

[2] Debye P and Sears F W 1932 On the scattering of light by supersonic waves *Proc. Natl Acad. Sci. USA* **18** 409–14

[3] Lucas R and Biquard P 1932 Propriétés optiques des milieux solides et liquides soumis aux vibrations élastiques ultra sonores (Translation: Optical properties of solids and liquids under ultrasonic vibrations) *J. Phys. Radium* **3** 464–77

[4] Sapriel J 1979 *Acousto-optics* (New York: Wiley)

[5] Nelson D F and Lax M 1970 New symmetry for acousto-optic scattering *Phys. Rev. Lett.* **24** 379–80

[6] Yariv A and Yeh P 1984 *Optical Waves in Crystals* (New York: Wiley)

[7] Spiegel M 1968 *Mathematical Handbook of Formulas and Tables: Schaum's Outline Series* (New York: McGraw-Hill)

[8] Yariv A 1989 *Quantum Electronics* 3rd edn (New York: Wiley)

IOP Publishing

Optical Radiation and Matter

Robert J Brecha and J Michael O'Hare

Appendix A

Vector theorems and identities

A.1 Theorems

- Divergence theorem:

$$\int_S \vec{A} \cdot d\vec{a} = \int_V \nabla \cdot \vec{A} \, d^3r,$$

where S is the surface which bounds the volume V.
- Stokes' theorem:

$$\oint_C \vec{A} \cdot d\vec{l} = \int_S (\nabla \times \vec{A}) \cdot d\vec{a},$$

where C is the closed curve which bounds the surface S.

A.2 Identities

$$\nabla(\vec{A} \cdot \vec{B}) = (\vec{B} \cdot \nabla)\vec{A} + (\vec{A} \cdot \nabla)\vec{B} + \vec{B} \times (\nabla \times \vec{A}) + \vec{A} \times (\nabla \times \vec{B}) \qquad (A.1)$$

$$\nabla \cdot (f \, \vec{A}) = (\nabla f) \cdot \vec{A} + f \, (\nabla \cdot \vec{A}) \qquad (A.2)$$

$$\nabla \cdot (\vec{A} \times \vec{B}) = \vec{B} \cdot (\nabla \times \vec{A}) - \vec{A} \cdot (\nabla \times \vec{B}) \qquad (A.3)$$

$$\nabla \times (f \, \vec{A}) = (\nabla f) \times \vec{A} + f(\nabla \times \vec{A}) \qquad (A.4)$$

$$\nabla \times (\vec{A} \times \vec{B}) = (\vec{B} \cdot \nabla)\vec{A} - (\vec{A} \cdot \nabla)\vec{B} + (\nabla \cdot \vec{B}) \, \vec{A} - (\nabla \cdot \vec{A})\vec{B} \qquad (A.5)$$

$$\nabla \times \nabla \times \vec{A} = \nabla(\nabla \cdot \vec{A}) - \nabla^2 \vec{A}. \qquad (A.6)$$

IOP Publishing

Optical Radiation and Matter

Robert J Brecha and J Michael O'Hare

Appendix B

Operators in different coordinate systems

B.1 Rectangular coordinates

$$\nabla f = \frac{\partial f}{\partial x}\hat{i} + \frac{\partial f}{\partial y}\hat{j} + \frac{\partial f}{\partial z}\hat{k} \tag{B.1}$$

$$\nabla \cdot \vec{A} = \frac{\partial A_x}{\partial x} + \frac{\partial A_y}{\partial y} + \frac{\partial A_z}{\partial z} \tag{B.2}$$

$$\nabla \times \vec{A} = \left(\frac{\partial A_z}{\partial y} - \frac{\partial A_y}{\partial z}\right)\hat{i} + \left(\frac{\partial A_x}{\partial z} - \frac{\partial A_z}{\partial x}\right)\hat{j}$$
$$+ \left(\frac{\partial A_y}{\partial x} - \frac{\partial A_x}{\partial y}\right)\hat{k} \tag{B.3}$$

$$\nabla^2 f = \frac{\partial^2 f}{\partial x^2} + \frac{\partial^2 f}{\partial y^2} + \frac{\partial^2 f}{\partial z^2} \tag{B.4}$$

$$\nabla^2 \vec{A} = \nabla^2 A_x \hat{i} + \nabla^2 A_y \hat{j} + \nabla^2 A_z \hat{k}. \tag{B.5}$$

B.2 Cylindrical coordinates

$$\nabla f = \frac{\partial f}{\partial \rho}\hat{\rho} + \frac{1}{\rho}\frac{\partial f}{\partial \varphi}\hat{\varphi} + \frac{\partial f}{\partial z}\hat{z} \tag{B.6}$$

$$\nabla \cdot \vec{A} = \frac{1}{\rho}\frac{\partial}{\partial \rho}\left(\rho A_\rho\right) + \frac{1}{\rho}\frac{\partial A_\varphi}{\partial \varphi} + \frac{\partial A_z}{\partial z} \tag{B.7}$$

doi:10.1088/978-0-7503-2624-7ch11

$$\nabla \times \vec{A} = \left(\frac{1}{\rho} \frac{\partial A_z}{\partial \varphi} - \frac{\partial A_\varphi}{\partial z} \right) \hat{\rho} + \left(\frac{\partial A_\rho}{\partial z} - \frac{\partial A_z}{\partial \rho} \right) \hat{\varphi}$$
$$+ \frac{1}{\rho} \left(\frac{\partial}{\partial \rho} (\rho A_\varphi) - \frac{\partial A_\rho}{\partial \varphi} \right) \hat{z}$$

(B.8)

$$\nabla^2 f = \frac{1}{\rho} \frac{\partial}{\partial \rho} \left(\rho \frac{\partial f}{\partial \rho} \right) + \frac{1}{\rho^2} \frac{\partial^2 f}{\partial \varphi^2} + \frac{\partial^2 f}{\partial z^2}.$$

(B.9)

B.3 Spherical coordinates

$$\nabla f = \frac{\partial f}{\partial r} \hat{r} + \frac{1}{r} \frac{\partial f}{\partial \theta} \hat{\theta} + \frac{1}{r \sin \theta} \frac{\partial f}{\partial \varphi} \hat{\varphi}$$

(B.10)

$$\nabla \cdot \vec{A} = \frac{1}{r^2} \frac{\partial}{\partial r} (r^2 A_r) + \frac{1}{r \sin \theta} \frac{\partial}{\partial \theta} (A_\theta \sin \theta) + \frac{1}{r \sin \theta} \frac{\partial A_\varphi}{\partial \varphi}$$

(B.11)

$$\nabla \times \vec{A} = \frac{1}{r \sin \theta} \left[\frac{\partial}{\partial \theta} (A_\varphi \sin \theta) - \frac{\partial A_\theta}{\partial \varphi} \right] \hat{r}$$
$$+ \frac{1}{r} \left[\frac{1}{\sin \theta} \frac{\partial A_r}{\partial \varphi} - \frac{\partial (r A_\varphi)}{\partial r} \right] \hat{\theta} + \frac{1}{r} \left[\frac{\partial (r A_\theta)}{\partial r} - \frac{\partial A_r}{\partial \theta} \right] \hat{\varphi}$$

(B.12)

$$\nabla^2 f = \frac{1}{r^2} \frac{\partial}{\partial r} \left(r^2 \frac{\partial f}{\partial r} \right) + \frac{1}{r^2 \sin \theta} \frac{\partial}{\partial \theta} \left(\sin \theta \frac{\partial f}{\partial \theta} \right)$$
$$+ \frac{1}{r^2 \sin^2 \theta} \frac{\partial^2 f}{\partial \varphi^2}.$$

(B.13)

IOP Publishing

Optical Radiation and Matter

Robert J Brecha and J Michael O'Hare

Appendix C

Coordinate transformations

C.1 Rectangular (x, y, z)—spherical (r, θ, φ)

$$x = r \sin \theta \cos \varphi \tag{C.1}$$

$$y = r \sin \theta \sin \varphi \tag{C.2}$$

$$z = r \cos \theta \tag{C.3}$$

$$r = \sqrt{x^2 + y^2 + z^2} \tag{C.4}$$

$$\theta = \cos^{-1}\left[\frac{z}{\sqrt{x^2 + y^2 + z^2}}\right] \tag{C.5}$$

$$\varphi = \tan^{-1}\left(\frac{y}{x}\right). \tag{C.6}$$

C.2 Rectangular (x, y, z)—cylindrical (ρ, φ, z)

$$x = r \cos \varphi \tag{C.7}$$

$$y = r \sin \varphi \tag{C.8}$$

$$z = z \tag{C.9}$$

$$r = \sqrt{x^2 + y^2} \tag{C.10}$$

$$\varphi = \tan^{-1}\left(\frac{y}{x}\right). \tag{C.11}$$

CPSIA information can be obtained
at www.ICGtesting.com
Printed in the USA
BVHW011105030721
611104BV00003B/20